Environmental Chemistry:
Chemistry of Major Environmental Cycles

Teh Fu Yen

University of Southern California, USA

ICP

Imperial College Press

Published by

Imperial College Press
57 Shelton Street
Covent Garden
London WC2H 9HE

Distributed by

World Scientific Publishing Co. Pte. Ltd.
5 Toh Tuck Link, Singapore 596224
USA office: 27 Warren Street, Suite 401-402, Hackensack, NJ 07601
UK office: 57 Shelton Street, Covent Garden, London WC2H 9HE

Library of Congress Cataloging-in-Publication Data
Yen, Teh Fu, 1927—
   Environmental chemistry : chemistry of major environmental cycles / Teh Fu Yen.
      p. cm.
   Includes index.
   ISBN 1-86094-474-4
   1. Environmental chemistry. I. Title.

   TD193.Y46 2005
   628-dc22

                                                    2004062539

British Library Cataloguing-in-Publication Data
A catalogue record for this book is available from the British Library.

Printed in Singapore by World Scientific Printers (S) Pte Ltd

This book is dedicated to my wife Shiao Ping Siao Yen, Ph.D.,
of the Jet Propulsion Laboratory,
California Institute of Technology,
for her support, understanding, and sacrifice.

FOREWORD

It is necessary to define what chemistry is. It is usually defined as the study of the composition and structure of matter, and especially the study of the changes that occur in matter. Chemistry thus touches and encompasses all other sciences. The field of chemistry in fact touches the entire universe, from far-reaching galaxies to human cells. Behind our daily human needs, such as food, water, and shelter, are the chemical processes that call for such necessities. Beyond our own needs are the chemical processes that create the major environmental changes that we as humans are so accustomed to, use for our benefit, and strive to understand and control.

The ancient Greeks' universe comprised four elements: fire, air, water, and earth. These chemical elements were used in describing the complex interactions among what today we call the five global environmental cycles.

The ancient Chinese defined five courses that align with the five major environmental cycles discussed in this book: metal (lithosphere), wood (biosphere), water (hydrosphere), fire (atmosphere), and soil (pedosphere). The Chinese courses interact with each other, and when one course is not in harmony with another, the balance between the courses is negatively affected. Similarly, the major cycles can be harmonious actors on a global stage, or imbalances can occur. This may happen between two cycles, such as the pedosphere and hydrosphere, or between any other two or among any three or four cycles.

A human being is simply an observer of these five global environmental cycles — we are a chemical soup that has fallen into the cycles of nature. However, we humans also have the ability to disrupt the natural harmony that exists among the cycles. Controlling, repairing, or leaving the movement of the cycles to themselves requires a deep understanding of the cycles themselves. This book is an introduction to the major environmental cycles, and is a gateway to understanding the technology that humans use to benefit from, repair, and maintain the environment. This is the mission of environmental chemistry.

PREFACE

It was around Earth Day of this year that this text was completed. The environment is a resource with a value that is difficult to comprehend — its value for humans is, in fact, infinite. Humans are products of the environment in contact with it daily, and understanding its chemistry is necessary to maintain sustainability so that future generations can continue to co-exist with the environment.

Degradation to the environment can lower the quality of life for humans through contact with the major environmental cycles. Environmental engineering and science seeks to remedy the present degradation caused by humans, and to create technology that will prevent further damage.

There is no denying that chemistry is essential in the understanding of environmental engineering and science. It is the basic foundation on which science and engineering can be built. This book will be useful for people who want to specialize in the field of environmental science and engineering, and it will be a valuable reference tool.

This book includes the basics of chemistry, not only aquatic but also atmospheric, as well as soil chemistry. It is a comprehensive book, covering the chemistry of energy resources and aspects of biochemistry, geochemistry, and toxicological chemistry in addition to the three important substances: air, water, and soil.

Presented in this book are the issues that face environmental engineers and scientists. The material is presented analytically so that the student, after gaining a clear insight into the chemical nature of the issues, will be able to formulate an opinion if so inclined. This book, however, does not seek to impress upon students any opinions, but only to prepare the students analytically.

Some portions of this book have been used in Environmental Chemistry by Prentice Hall, in two volumes. Permission has been given to use that material for the present version. The work was double-checked, many typos were corrected from the original version, and new material has been added. The framework from this book has been used for teaching a one-semester course in environmental chemistry for close to twenty years.

The author would like to take this opportunity to thank all of the colleagues and previous students who have taken the course and who have contributed to shaping this book. Lastly, the author would like to name Shivani Yardi, Divya Devaguptapu, Akash Tayal, Aaron Cornell, Rachel Tucker, Vivek Gupta and Sangeeta Patel for their assistance.

Teh Fu Yen
October 2004

CONTENTS

THE LITHOSPHERE

The **lithosphere**, including the internal structure below the earth's surface, is the part of the earth where all mineral resources originate. A number of illustrations will be used to introduce the lithosphere. From the continental crust down, the structure of the mantle always provides the necessary energy to form compounds from iron and the elements associated with iron above the upper mantle level (refer to Figure A-1). Different metamorphic facies are formed due to both the pressure and temperature gradients. Above the transition zone, especially in the continental margins, organic-rich sediments are formed, as shown in Figure A-2. All mineral deposits in veins and domes accumulate at the tectonic plate boundaries, as can be seen in Figure A-3. These can take the shape of belts or bands, as illustrated in Figure A-4.

All energy on earth comes from the sun and is called **stellar energy** (see Figure A-5). When stellar energy is absorbed by the atmosphere and hydrosphere, the net result is the energy in wind, waves, ocean currents, tidal currents, hydroelectric power, etc. When energy is absorbed by the biosphere and used in lifecycles by photosynthesis, the biomass produces decay that becomes buried in sediments, thereby producing fossil energy. Energy is absorbed by the lithosphere and is used in geochemical cycles and the conduction and convection from the earth's interior. These are called **geothermal** and **nuclear energy**; the latter of these is known as **fissile fuels**.

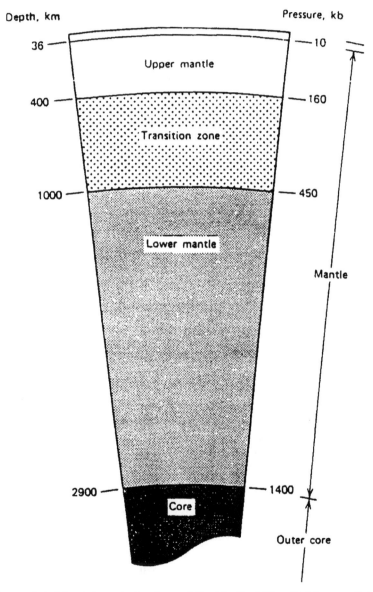

Figure A-1. The internal structure of the Earth. (After B. Mason, *Principle of Geochemistry*, 1966)

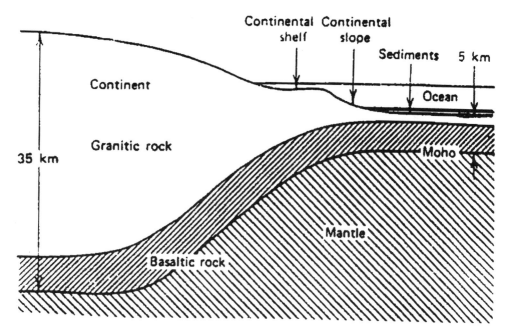

Figure A-2. The Earth's crust under the continents and oceans. The moho is Mohorovicic discontinuity which separates the highly heterogenous crust from the more homogenous upper mantle.

    In the following chapters of this section, we will briefly discuss fossil fuels, followed by solar, wind, tidal, and other non-fossil fuels. In the last group, we will concentrate on nuclear power.

    As the world has become increasingly industrialized and modernized, energy consumption has increased accordingly. After decades of careless exploitation of energy resources, the world has finally come to realize that the energy supplies that were once considered inexhaustible will shortly be nonexistent. Moreover, this increased energy consumption has resulted in an extremely adverse impact on the environment in the form of air, water, and other kinds of pollution. There are increasing conflicts between environmental conservationists, who are concerned with leaving a clean and healthy world for the future generations, and industrialists and businesspersons, who are pushing for a higher level of technological civilization and material consumption. In the following chapter, we will explore the background of energy production and consumption with an emphasis on fossil energy.

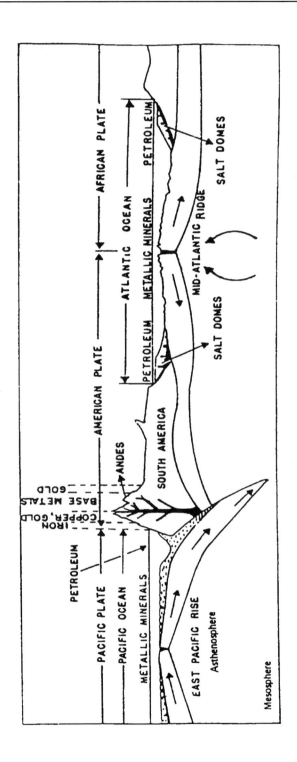

Figure A-3.   The role of the plate boundaries in the accumulation of mineral deposits.

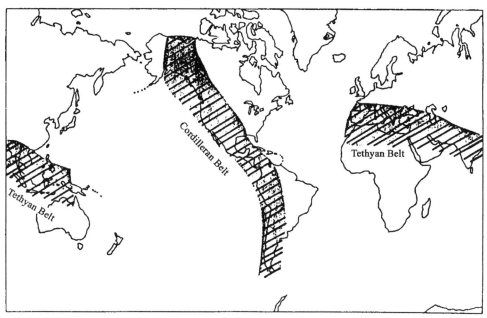

Figure A-4. Major world petroleum belts. [Source: T.F. Yen, Genesis and Degradation of Petroleum Hydrocarbons in Marine Environments in (T. M. Church ed.) Marine Chemistry in the Coastal Environments, ACS Symp. Series 18, Washington, DC, 1975 pp. 231-266.]

We will see that protecting the environment is not necessarily incompatible with meeting energy demands. We will begin with energy and power, and explain energy consumption. The introduction is followed by a brief classification of fossil fuels. Lastly, we will introduce some of the resource models used in energy technology. Energy conservation and the green process will be discussed briefly, as will alternative energy sources, mostly arising from indirect, inanimate sources as indicated by Figure A-5. The origin of fissile fuels is also discussed in succeeding chapters.

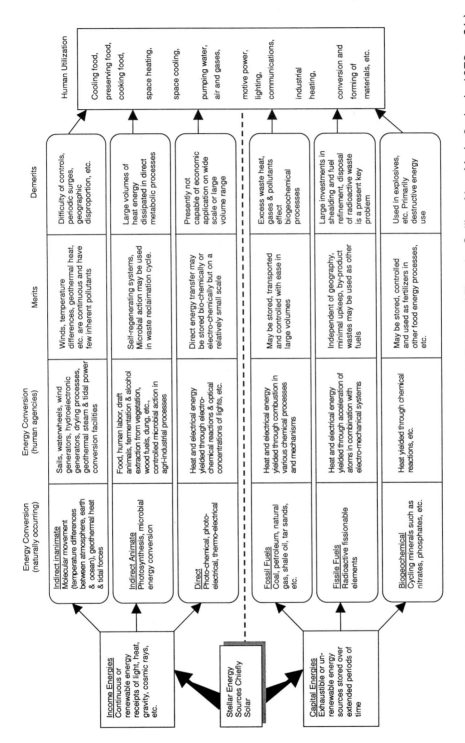

Figure A-5. Energy Systems. (Source: J. McHale, World Facts and Trends, Collier, New York, 2nd ed., 1972, p.61.)

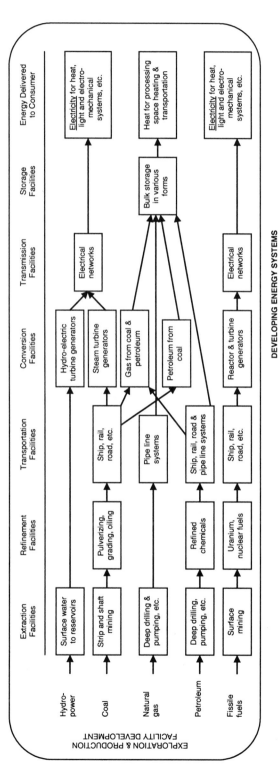

Figure A-5. Energy Systems. (Continued)

FOSSIL FUELS

1.1 ENERGY AND POWER

Solar energy is the major source of the earth's energy system, in conjunction with two other small contributors: **tidal energy** and **geothermal heat**. The hydroelectric power we utilize results from rainfall, through a process in which sunlight provides energy to raise water up to the atmosphere. The food we eat is the end-product of food chains in which sunlight provides energy for photosynthesis. The fossil fuel we burn originates from these organic compounds created through the preceding process, to which solar power makes a great contribution. As shown in Figure 1-1, a quantity of $174,000 \times 10^{12}$ W of solar energy is intercepted by the earth. Of this quantity, about 30% is reflected or scattered back into space. This scattered fraction is called the **albedo**, and it determines the total energy balance of the earth. Another 47% of this solar energy is converted directly into heat. About 23% of the total energy intercepted is used to drive the **hydrologic cycle**, the massive evaporation and precipitation of water upon which we depend for our fresh water supplies. Only a very small fraction, 0.1%, is used by green plants and algae in photosynthesis, which are vital for our food supply.

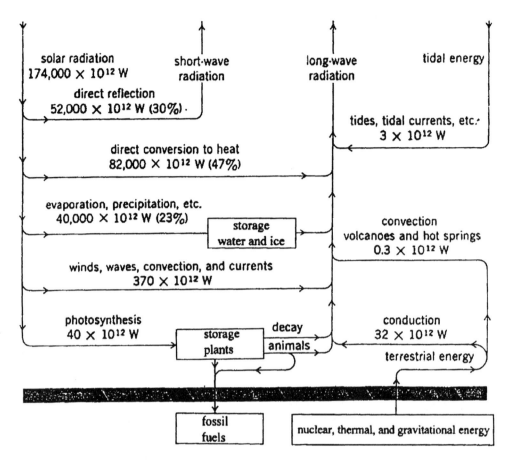

Figure 1-1. Energy flow-sheet for the Earth. [From M.K. Hubbert, U.S. energy resources: A review as of 1972, pt. 1, in A National Fuels and Energy Policy Study, U.S. 93rd Congress, 2nd Session, Senate Committee on the Interior and Insular Affairs, ser. no. 93-40 (92-75), 1974.]

Various energy unit systems have been adapted for use by different groups of people. It is troublesome, however, for people to convert a value from one unit to another. For convenience, Table 1-1 and Figure 1-2 provide some common energy equivalents and conversion factors.

For energy, it is useful to know some of the abbreviations that are commonly adopted. In addition to the units discussed in the Appendix 1, both **Quad** (10^{15}) and **Quin** (10^{18}) are used in large numbers. In the study of energy demands, various units (tons of coal, barrels of oil, cubic feet of gas, gigawatts, and so on) are often encountered. To be consistent with other studies, Btu is sometimes used as the common unit of energy. On that basis, one Quad is equal to 1×10^{15} Btu and one Quin is equal to 1×10^{18} Btu.

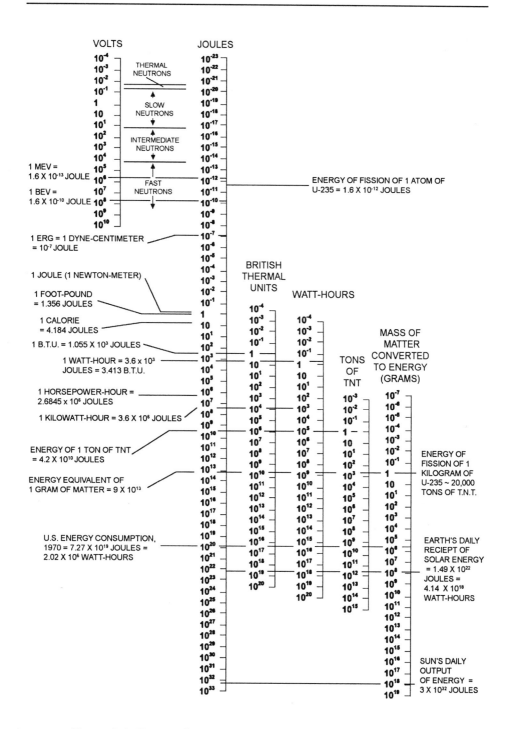

Figure 1-2. Energy Conversion Factors. (Modified after C. Starr, 1971)

Table 1-1. Energy Equivalents and Other Conversion Factors

Unit	Equivalent		Remarks
1 kcal	$= 1.162 \times 10^{-6}$	MWh	
2 Hph (metric)	$= 7.355 \times 10^{-4}$	MWh	
1 BTU	$= 0.293 \times 10^{-6}$	MWh	
1 Q	$= 0.293 \times 10^{12}$	MWh	$1 Q = 10^{18}$ BTU
1 t hard coal	$= 8.134$	MWh	1 kg hard coal equivalent = 7,000 kcal
1 t petroleum products	$= 11$	MWh	varies according to raw material
10^3 m^3 natural gas	$= 10$	MWh	varies according to raw material
1 t U	$= 2.25 \times 10^7$	MWh	200 MeV/U atom
1 t Th	$= 2.30 \times 10^7$	MWh	200 MeV/Th atom
1 t D	$= 6.69 \times 10^7$	MWh	5 MeV/D atom
1 short ton U$_3$O$_3$	$= 0.7693$	t U	
1 kg U$_3$O$_8$	$= 0.8480$	kg U	
1 lb U$_3$O$_8$	$= 0.3847$	kg U	
1 short ton ThO$_2$	$= 0.7972$	1 Th	
1 kg ThO$_2$	$= 0.8788$	kg Th	
1 lb ThO$_2$	$= 0.3986$	kg Th	
1 lb	$= 0.4536$	kg	
1 short ton	$= 0.9072$	t	
1 U.S. barrel (petroleum)	$= 0.1590$	m^3	

Power is the rate at which energy is consumed or generated; for example, Joules are the energy unit, while Watts are Joules/sec, or the power unit.

$$P = \frac{dE}{dt} \qquad\qquad [1\text{-}1]$$

$$E_{year} = \int_0^T P_{inst}\, dt \qquad\qquad [1\text{-}2]$$

and

$$P_{ar} = \frac{1}{T} E_{year} = \frac{1}{T} \int_0^T P_{inst}\, dt \qquad\qquad [1\text{-}3]$$

$$T = 3.15 \times 10^7 \text{ sec} \qquad\qquad [1\text{-}4]$$

Present world energy consumption = 7.1×10^{12} W = 5.3×10^{19} cal/y. The United States alone consumes 1.8×10^{16} kcal/y, which is one-third of the world's total energy consumption. The metabolic consumption per capita is 3100 kcal/day, or 150 Watts. Thus, the United States non-metabolic per capita consumption rate equals 12 kW, which is about 80 times as great as the metabolic consumption.

1.2 TOTAL ENERGY CONSUMPTION

Humans are now consuming fossil energy faster than nature can produce it. We are living off the store of energy of past ages. Fossil fuel is an exhaustible resource, and the energy consumption by humans grows exponentially, as shown in Figure 1-3. This type of growth is inevitably unreasonable. It will double in 18 years even if energy consumption only increases by 4% per year. For an exhaustible resource, the cumulative production and rate production are shown in Figures 1-4 and 1-5.

Nothing in the natural world can grow through too many doubling periods before it runs into some constraint. Eventually all exponential growth curves have to level off, and the growth of energy consumption is no exception. In the long run, energy inputs and outputs for human society must be in a steady state. Because population grows exponentially, our energy consumption is exponential in nature.

$$Q = Q_o \exp(\lambda T)$$
[1-5]

where $e = 2.718$, T = time AD, λ = constant (yr^{-1}).

For example, our present total energy consumption is

$$Q^T = 1.85 \times 10^{-21}\, e^{0.0282T} \text{ (Quad/yr)}$$
[1-6]

The following are some properties from the above exponential equation.

a. **doubling time**, T_2

$$T_2 = \frac{\ln 2}{\lambda} = \frac{0.693}{\lambda}$$
[1-7]

For example,

$$T_2 = \frac{0.693}{0.0282} = 25 \text{ yr}$$

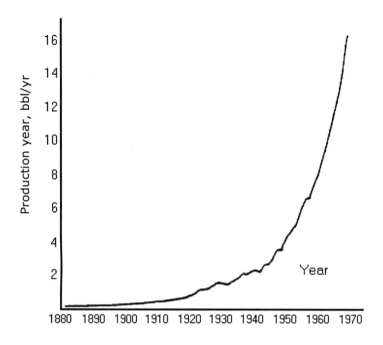

Figure 1-3. World crude oil production (From Hubbert, op. cit., 1974).

b. **yearly fractional increase**, R

$$R = e^\lambda - 1 \qquad\qquad [1\text{-}8]$$

For example,

$$\% \text{ increase} = \left(e^{0.0282} - 1 \right)(100) = 2.86\% \text{ per yr}$$

Derivation of R can be accomplished as follows:

$$R \approx \frac{dQ}{Q\,dt} \qquad\qquad [1\text{-}9]$$

or

$$R = \frac{Q - Q_o}{Q_o t} = \frac{Q}{Q_o} - 1$$

since $t = 1$, or

$$\frac{Q}{Q_o} = R + 1$$

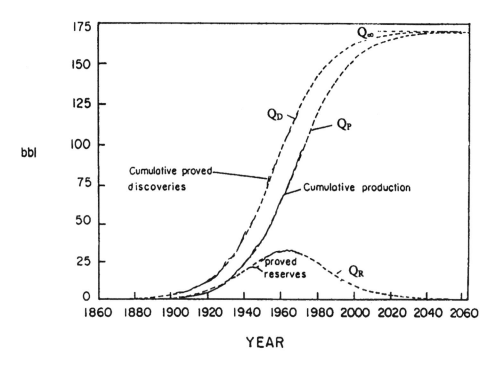

Figure 1-4.   Logistic equations and curves of cumulative production, cumulative discoveries, and proven reserves of crude oil from the contiguous United States 1900-1971 (From Hubbert, op. cit., 1974).

$$\ln \frac{Q}{Q_o} = \ln (R+1)$$

since

$$\ln \frac{Q}{Q_o} = \lambda$$

Thus,

$$\ln (R+1) = \lambda \qquad\qquad\qquad [1\text{-}10]$$

or

$$e^\lambda = R+1$$

or

$$e^\lambda - 1 = R$$

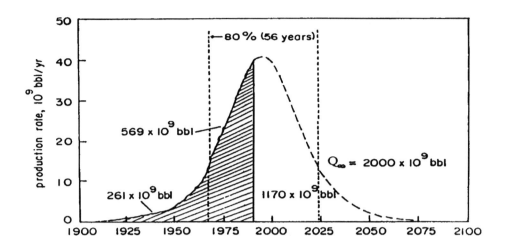

Figure 1-5.   Estimate as of 1972 of complete cycle of world crude oil production (From Hubbert, op. cit., 1974).

c. **crossover**

$$Q_T = 1.85 \times 10^{-21} \, e^{0.0282T} \qquad\qquad [1\text{-}11]$$

For example, gas consumption has become increasingly rapid in recent
years. Gas consumption has the potential to exceed the total energy consumption, and at its current rate it has increased from 6 Q in 1960 to 23 Q in 1980.

Thus,

$$6 = Q_0 \, e^{1960\lambda}$$

$$23 = Q_0 \, e^{1980\lambda}$$

Solving,

$$\lambda = .067 \, y^{-1}$$

$$Q_0 = 1.1 \times 10^{-56}$$

Hence,

$$Q_G = 1.1 \times 10^{-56} \, e^{0.067T}$$

Solving, when $Q_G = Q_T$, $T = 2090$, the crossover will take place.

1.3 FOSSIL ENERGY

The fraction of sunlight (0.1%) used by plants in photosynthesis is converted to chemical energy and stored in the form of carbohydrates such as glucose, sucrose, and starch. The reduced carbon can be reoxidized into carbon dioxide during the process of respiration, providing for the energy needs of biological organisms.

Plants themselves use up about 20% of their carbohydrates for their own energy needs. The remainder, which represents stored photochemical energy, is called the **net primary productivity.**

A small fraction of plant and animal matter, estimated at one part in 10,000, is buried in the earth and removed from contact with atmospheric oxygen. Some of these buried carbon compounds accumulated in deposits and were subjected to high temperatures and pressures in the earth's crust, becoming coal, petroleum, gas, oil shale, and tar sands. These are the **fossil fuels**. In the following sections, we will give some detailed descriptions of each of the fossil fuels.

1.3.1   Petroleum and Gas

Petroleum (rock-oil, derived from Latin "petra," meaning rock or stone, and "oleum", meaning oil) occurs widely in the earth as a gas and a liquid. Gas is a part of petroleum, and petroleum is classified as a mineral according to the U.S. Department of Interior. Chemically, any petroleum is an extremely complex mixture of hydrocarbon compounds, with a minor amount of nitrogen, oxygen, and sulfur impurities. The various fractions can be separated by distillation to determine the overall composition, as shown in Table 1-2. Nearly all petroleum occurs in sediments. These sediments are chiefly of marine origin, and it follows that the contained petroleum is also most likely marine, or related to marine conditions, as shown in Figure 1-6. The process occurs as follows: a small portion of reduced carbon resulting from photosynthesis in the oceans settles to the bottom, where oxidation is negligible. The biological debris is covered by clay and sand particles and forms a compacted organic layer in a porous clay or sandstone. Anaerobic bacteria digest the protein, fat, and carbohydrates, releasing most of the oxygen and nitrogen. As the sediment with the remains, mostly lipid material, becomes buried deeper, the temperature and pressure acting on it rise. Bacterial action decreases, and organic disproportionation reactions are thought to occur through geochemical transformation. Figure 1-8 shows the comparison between reservoir rock and source rock as a function of depth.

There are some other theories regarding the origin of petroleum, such as "inorganic theories," which state that the primary source material for petroleum generation is inorganic. Theories that uphold the inorganic origin of petroleum have few supporters today. Solid proof of an organic origin came about when a C^{13} mass spectroscopy experiment confirmed that the lipid content of marine microorganisms and that of petroleum have very close values, as shown in Figure 1-7.

The most immediate constraint on energy growth is the availability of petroleum and natural gas. The Industrial Revolution was initially fueled by coal, but oil and gas were increasingly substituted because they were cleaner fuels and were transported more easily. Petroleum and gas now constitute nearly 80% of the total energy consumption in the United States.

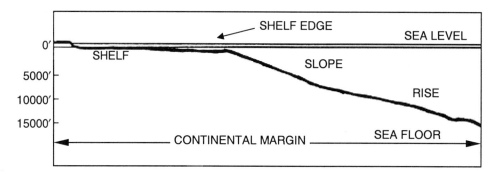

Figure 1-6.   Schematics for continental margins. (Source, T. F. Yen, *Chemistry of Marine Sediments*, Ann Arbor Sci.Pub., 1977, p.4)

Table 1-2.   Petroleum Components

Components	Distillation Temperature (°C)	Structure
Gas	20	$C_P; C_1 - C_4$
Petroleum ether	20 – 60	$C_P; C_2 - C_6$
Ligroin (light naphtha)	60 – 100	$C_P; C_0 - C_7$
Natural gasoline	40 – 205	$C_P, C_N; C_6 - C_{10}$
Kerosene	175 – 325	$C_P, C_N, C_A; C_{12} - C_{18}$
Gas oil	Nonvolatile liquid	$C_P, C_N, C_A; > C_{18}$
Lubricating oil	Nonvolatile semisolid	$C_N, C_A, C_P, X; \sim C_{20}$
Asphaltic bitumens	Nonvolatile	$C_N, C_A, C_P, X; C_{100} - C_{120}$

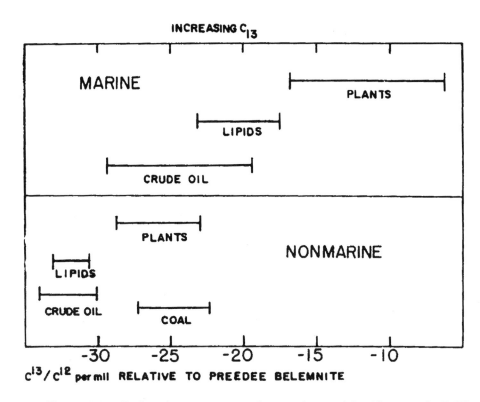

Figure 1-7. Carbon isotope range of natural materials. (Source: S. R Silverman, 1971)

Estimates of the ultimate amount of recoverable liquid petroleum worldwide range from 10,440 to 21,170 Quads (one Quad is equal to 10^{15} Btu). The world oil production until 1983 was approximately 2,800 Quads. In 1982 alone, the world production was about 110 Quads, which is equivalent to 19 billion barrels. Based on the 1982 rate of oil consumption, in the year 2000 our world's reserves of conventional petroleum would have been exhausted. Fortunately, due to conservation measures, that rate has been reduced.

After production, petroleum is transported via pipeline and can be refined or upgraded into a lighter fraction called **gasoline** by the use of a cracking catalyst. Both production and recovery are termed upstream processes, and similarly, both refining and upgrading are termed downstream processes.

Asphalt (or bitumen), the heavy or involatile fractions of petroleum, virtually controls many parameters of the production as well as refining processes. The asphaltic system is comprised of (a) asphaltenes, the dispersed or micellar phase, (b) resins, the peptizing agent (or surfactant), and (c) gas oil, the dispersed phase or intermicellar medium. These fractions can be separated by solvent cuts with appropriate Hildebrand's solubility parameters, as shown in Table 1-3.

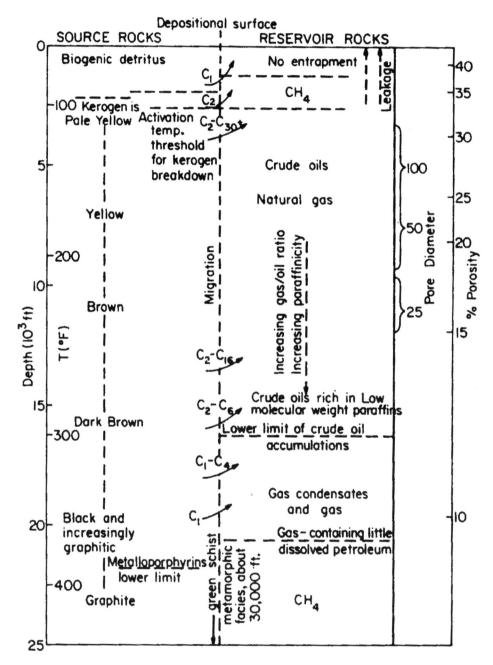

Figure 1-8. Comparison between reservoir and source rock as a function of depth.(Yen, *Energy Sources*, 1973)

Table 1-3. Solvent Fractions of Asphalt and Related Carbonaceous Material

Fraction			Solubility Parameter	
#	Designation	Solubility	δ in hilde-brands	Remarks
1	gas oil	propane soluble	below 6	saturated and aromatic hydrocarbons
2	resin	propane insoluble pentane soluble	6 – 7	combined 1 and 2 are also called maltene or petrolene
3	asphaltene	pentane insoluble benzene soluble	7 – 9	ASTM uses CCl_4 instead of benzene
4	carbene	benzene insoluble CS_2 soluble	9 – 10	ASTM uses CCl_4 instead of benzene
5	carboid	CS_2 insoluble pyridine soluble	10 – 11	combined 4 and 5 are referred to as preasphaltene or asphatol
6	mesophase	pyridine insoluble	above 11	

Volatile-free basis.

1 hildebrand = 2.04 J1/2/cm3/2 = 1 cal1/2/cm3/2.

Because of the flammability of CS_2, pyridine is preferred; fractions 4 and 5 can be combined.

Source: T. F. Yen in *Encyclopedia of Polymer Science and Engineering*, 1990, Wiley, New York.

1.3.2   Oil Shale

Oil shale is diverse fine-grained rock that contains refractory organic material that can be refined into fuels. The soluble fraction, called **bitumen**, constitutes about 20% of this organic material, whereas the remainder exists as an insoluble fraction, **kerogen**. However, most oil shales do contain inorganic minerals accounting for 20–80% of the bulk. All oil shales appear to have been deposited in shallow lakes, marshes, or seas that support a dense algal biota. The latter is a probable source for shale-bound organic precursors.

In Australia, France, and Scotland, oil shales have been the source of products similar to those obtained from petroleum for many years. In the United States, there are over 700 billion barrels of synthetic crude oil spread over the Green River Formation, located in Colorado, Utah, and Wyoming. The total world potential is estimated to be 30 trillion bbl of shale oil.

A typical composition of Green River oil shale is listed in Table 1-4. Kerogen, which can be viewed as a cross-linked organic polymer, constitutes the bulk (approximately 80%) of the available organic material in oil shale. Kerogen itself is 3,200 times richer in biomass than the total available bitumen and petroleum. The liberation of hydrocarbons depends on the degree to which kerogen can be degraded into liquid fuel precursors. Because kerogen is the most abundant organic carbon source in the world, it may be a potential food precursor in an age of food shortages, just as single cell proteins have been.

Table 1-5 shows the elemental and ash analyses of a number of raw fuels.

Table 1-4. Composition of a Typical Green River Oil Shale

General scheme of the oil-shale components		
	Inorganic matrix	Quartz
		Feldspars
		Clays (mainly illite and chlorite)
		Carbonates (calcite and dolomite)
		Pyrite and other minerals
	Bitumens (soluble in CS_2)	
	Kerogens (insoluble in CS_2)	
	(containing U, Fe, V, Ni, Mo)	

Average chemical composition of Green River oil shale, as determined by the writers for several samples from Rifle, Colorado.

FeS_2		0.86%			
$NaAlSi_2O_6 \cdot H_2O$ (analcite)		4.3%			
SiO_2 (quartz)		8.6%			
$KAl_4Si_7AlO_{20}(OH)_4$ (illite) Montmorillonite Muscovite		12.9%			
$KAlSi_3O_8$ (K-feldspar) $NaAlSi_3O_8$–$CaAl_2Si_2O_8$ (plagioclase)		16.4%			
O	22.2%	$CaMg(CO_3)_2$ (dolomite) And calcite	43.1%	Mineral matter 86.2%	Oil shale
Ca	9.5%				
Mg	5.8%				
C	5.6%				
S, N, O	1.28%	Bitumen	2.76%	Organic matter 13.8%	
H	1.42%				
C	11.1%	Kerogen	11.04%		

Source: T.F. Yen and G.V. Chilingarian. *Oil Shale,* Elsevier Sci. Pub. Amsterdam, 1976, p.3

Table 1-5. Elementary and Ash Analyses of Raw Fuels.

Sample	C(%)	H(%)	N(%)	S(%)	O(%)	Mineral and ash(%)
Oil Shale (Colorado)	12.4		0.41	0.04		84.6
Oil Shale (Alaska)	53.9		0.30	1.50		34.2
Kerogen (Green River)	65.7	9.1	2.13	3.15	8.9	10.2
Lignite (Kincaid)	63.1	5.0	0.64	<0.04	21.5	9.3
Bituminous (Lower Freeport)	77.0	5.7	1.42	1.72	8.1	6.6
Anthracite (Pennsylvania)	87.1	3.8	0.60	<0.3	2.6	7.0
Gilsonite (Tarbor vein)	84.5	10.0	2.4	0.53	2.6	0.7
Residuum (Mid-continent)	87.7	6.6	0.90	0.75	1.6	1.3
Asphaltene (Baxterville)	84.5	7.4	0.80	5.60	1.7	0.5

1.3.3 Coal

Coal is composed of the remains of plant matter from the huge, thickly wooded swamps that flourished 250 million years ago during a period of mild, moist climate. Woody plants are made of lignin as well as cellulose and protein as shown in Figure 1-9. **Lignin** is a complex, three-dimensional polymer that contains aromatic groups. While aerobic bacteria rapidly oxidize cellulose when a plant dies, lignin is much more resistant to bacterial action. In swamps, lignin accumulates under water, compacting into a substance called **peat**. Over the geological ages, the peat layers of the primeval swamps were metamorphosed into coal. The first stage of coalification (the formation of coal) is termed **lignite**, where the %C value is 55-75 and the heating value in Btu/lb is 6300-8300.

The next stage is **subbituminous**, where both %C and heating values increase. Next is **bituminous**, which can be divided into high volatile and low volatile types. The last stage can reach **anthracite**, which has a %C value of 87–97. The different stages of coal formation and the corresponding properties can be found in Table 1-6. Figure 1-10 shows the structural representations of bitumen, coal, and kerogen.

The recoverable reserves of coal are estimated to be about 7.4×10^{12} metric tons, with the United States and the former Soviet Union owning about 75% of the supply. Coal is substantially more abundant than oil or gas on a world basis. The size of U.S. coal reserves is also substantially larger than those of oil and gas.

Table 1-6. ASTM Classification of Coals by Rank (in box) and Corresponding Bank Rank Parameters not Used in the ASTM Classication

ASTM Class	ASTM Group	Heating Value Btu/lb (moist, mmf)	Heating Value MJ/kg (moist, mmf)	Agglomerating	R % VM (d, mmf)	R Oil max. (vitrinite)	% Moisture (moist, mmf)
	Peat	3,000 – 4,000[a]	7.0 – 9.3	No	62 – 72	0.2 – 0.4	50 – 95
Lignite	Lignite B	Undefined–6,300[a]	–14.6	No	40 – 65	0.2 – 0.4	45 – 60[b]
	Lignite A	6,300 – 8,300[a]	14.6 – 19.3	No	40 – 65	0.2 – 0.4	31 – 50[b]
Subbituminous	Subbituminous C	8,300 – 9,500[a]	19.3 – 22.1	No	35 – 55	0.3 – 0.7	25 – 38[b]
	Subbituminous B	9,500 – 10,500[a]	22.1 – 24.4	No	35 – 55	0.3 – 0.7	20 – 30[b]
	Subbituminous A	10,500 – 11,500[a]	24.4 – 26.7	No	35 – 55	0.3 – 0.7	18 – 25[a]
Bituminous	High volatile C bit.	10,500 – 13,000[a]	26.7 – 30.2	Yes	35 – 55	0.4 – 0.7	10 – 25[a]
	High volatile B bit.	13,000 – 14,000[a]	30.2 – 32.5	Yes	35 – 50	0.5 – 0.8[c]	5 – 12[b]
	High volatile A bit.	>14,000	>32.5	Yes	31 – 45	0.6 – 1.2[c]	1 – 7[b]
	Med. volatile bit.	>14,000	>32.5	Yes	22 – 31[b]	1.0 – 1.7[c]	<1.5
	Low volatile bit.	>14,000	>32.5	Yes	14 – 22[b]	1.5 – 2.0[c]	<1.5
Anthracitic	Semianthracite	>14,000	>32.5	No	8 – 14[b]	1.8 – 2.6[b]	<1.5
	Anthracite	>14,000	>32.5	No	2 – 8[b]	2.2 – 5.0[b]	0.5 – 2
	Meta-anthracite	>14,000	>32.5	No	<2	>4.5[b]	1 – 3

[a] Air dried   [b] Well-suited for rank discrimination in range indicated.   [c] Moderately well-suited for rank discrimination.

After H.H. Damberger, R.D. Harvey, C.R. Ruch and J. Thomas in *Science and Technology and Coal Utilization*, Plenum Press, 1984.

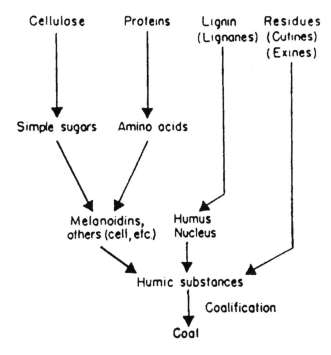

Figure 1-9. Relation of melanoidins to coal as derived from cellulose and protein. [T. F. Yen, ACS Symp. Ser. *18*, 231-266 (1975)]

The major concern with coal is that it is a dirty fuel to burn. Of particular concern is the sulfur dioxide emitted, which causes serious health hazards in urban areas. In addition, being a solid, coal is much less convenient to use than petroleum or natural gas. Using the appropriate chemistry, it is possible to convert coal to liquid or gaseous fuels. The reduced carbon in coal, in which most of its energy content resides, can be transformed into hydrocarbon molecules that are either liquid or gaseous, depending on their molecular weight and structures. A simplified diagram for **interfuel conversion** is shown in Figure 1-11.

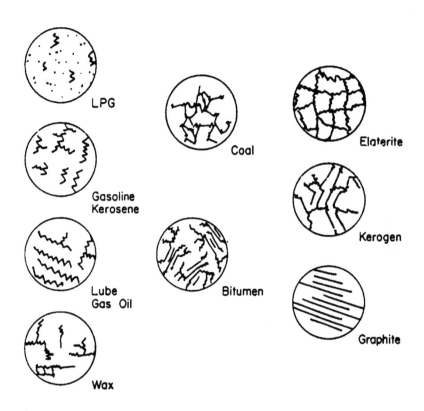

Figure 1-10. Structural representations of bitumen, coal and kerogen, and their relations with LPG (liquified petroleum gas), gasoline, gas oil, and wax. Their interrelations among elaterite and graphite are also indicated. [Source: T. F. Yen, *Chemical Aspects of Interfuel Conversion*, Energy Sources *1(1)* 117-136 (1973)]

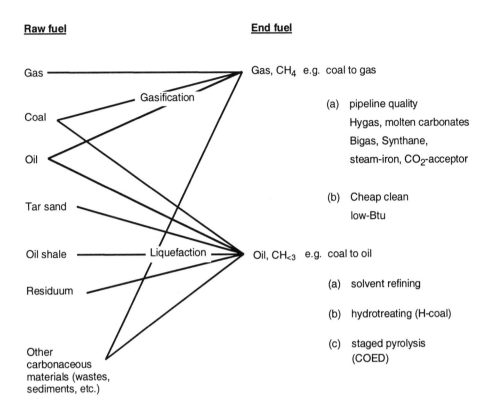

Figure 1-11. Interfuel conversion process

The main problem in converting coal to liquid (coal liquefaction) or gaseous fuel (coal gasification) is finding an efficient way to combine hydrogen with solid carbon, which is a hydrogen-enriched method. The H/C atomic ratio is 4 for methane and about 2 for gasoline, but in coal it ranges from 0.3 to 0.8. Figure 1-12 shows a simplified flow diagram of current coal conversion technologies, including the **Fischer-Tropsch synthesis**.   This process represents the indirect conversion of coal in the synthetic fuel program of the United States Department of Energy.

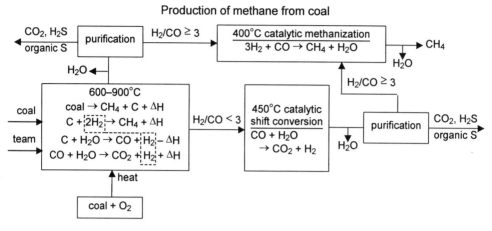

Production of methane from coal

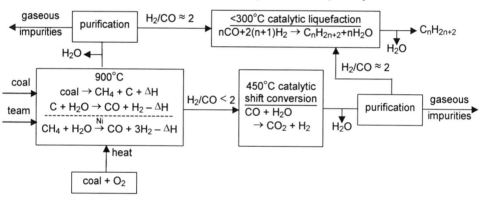

Production of liquid fuel from coal by Fischer–Tropsch synthesis

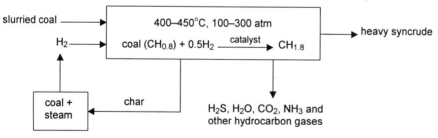

Production of liquid fuel from coal by direct catalytic hydrogenation

Figure 1-12. Coal conversion technology. (Source: Spiro and Stigliani, 1980)

1.3.4   Tar Sands

Tar sands are petroleum deposits and are distinctive in that their bitumen is so viscous that primary production is impossible. The "sands" refer to rock types including limestone, dolomite conglomerate, and shale in addition to consolidated sandstone and unconsolidated sand. Estimates of reserves range from 2.5 to 6 trillion barrels of oil equivalence. The world's most well-known sand deposit is Athabasca, in Canada. In Alberta, the sands of the McMurray Formation of the lower Cretaceous age contain one billion barrels of oil in place. Next to this in size and recognition is the Orinoco heavy oil belt in eastern Venezuela. The biggest deposits of sand appear to be lodged in deltaic sediments. Mechanisms for turning migrating oil into heavy residue are still under debate, including bacterial degradation and meteoric water washing. Usually, these bitumen will have API gravities ranging from −10 to +16 degrees.

$$°\text{API} = \frac{141.5}{\text{specific gravity at } 60°\text{F}} - 131.5 \qquad [1\text{-}12]$$

Viscosities range from 10,000 to more than one million centipoise (cp). Therefore, the bitumen cannot readily move through the carrying rock or the associated inorganic materials. The general methods for providing fuel from tar sands in-place are combustion and the injection of hot water or steam. Both methods are based on petroleum **enhanced oil recovery** (EOR) of the tertiary recovery schemes. The in-situ combustion method for tar sand recovery always encounters problems in formation permeability. Currently, commercial production of tar sands includes strip mining of the deposit and extraction with hot water (Suncor and Syncrude). One of the serious environmental problems is the gigantic volume of the tailing ponds containing the oily residues and the abrasive, quartz-like inorganic particulates that currently cannot be disposed of.

1.4 RESOURCES MODELS

Of the available evaluations of our energy and resources on Earth, there have been many attempts to point out that all forms of resources are limited. This will affect the suppliers as well as the users; it also has an impact on government policies. In the following sections, we will discuss four such types of models.

1.4.1 Macroscopic Approach by King Hubbert

The essential feature of this model is based on the record of past discoveries and production history records.

$$Q\,(\text{resource}) = \int_{t_1}^{t_2} P\,(\text{Production})\,dt \qquad [1\text{-}13]$$

Initially, there is exponential growth and leveling off (refer to Figs. 1-3 and 1-4).

$$P = P_0 e^{\lambda t}$$

$$Q = \int_0^T e^{\lambda t}\, dt = \frac{P_0}{\lambda} e^{\lambda t}\Big|_0^t = \frac{P_0}{\lambda}\left(e^{\lambda t} - 1\right) \tag{1-14}$$

At the point of leveling off, the maximum production (P_m) is reached, suggesting a bell-shaped curve that can easily be represented by a probability curve.

$$Q_\infty = \int_{-\infty}^{\infty} P\, dt = \int_{-\infty}^{\infty} P_m \exp\left[-\frac{1}{2}\left(\frac{t - t_m}{\sigma}\right)^2\right] dt$$

$$Q_\infty = \sqrt{2\pi}\,\sigma\, P_m$$

$$P_0 = P_m \exp\left[\frac{1}{2}\left(\frac{t_m}{\sigma}\right)^2\right]$$

or

$$t_m = \sigma\sqrt{2 \ln \frac{P_m}{P_0}} \tag{1-15}$$

where σ is standard deviation and t_m is the maximum time for production (notice also $t = 0$ for the preceding calculation). According to King Hubbert, energy resource and the proved reserves are equivalent as shown in Figure 1-13. If Q_D is from discovery and Q_P is from production, then $Q_D - Q_P = Q_R$ (resources). The rate of the increase of resources or approved reserves can be simply represented by a differential curve, as shown in Figure 1-14.

$$Q = \int_0^t \left(\frac{dQ}{dt}\right) dt = \text{resource} \tag{1-16}$$

$$\frac{dQ_D}{dt} - \frac{dQ_P}{dt} = \frac{dQ_R}{dt} \tag{1-17}$$

When Q_R is maximum,

$$\frac{dQ_R}{dt} = 0 \tag{1-18}$$

then

$$\frac{dQ_D}{dt} = \frac{dQ_P}{dt} \tag{1-19}$$

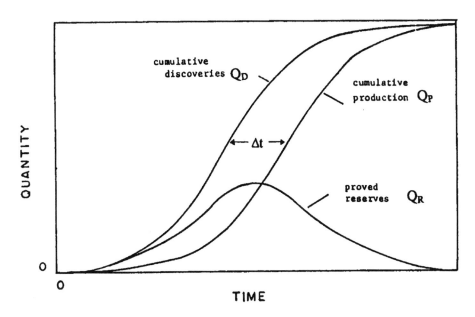

Figure 1-13. Variation of proved reserves Q_R, cumulative production Q_p, and cumulative proved discoveries Q_D, during a complete cycle of petroleum production. (Source: M.K. Hubbert, Energy Resources: A Report to the Committee on Natural Resources, Nat. Acad. Sci., Nat. Res. Counci. Publ. 1000-D, 1962: and Hubbert, op. cit., 1974)

but

$$Q_D = \frac{Q_\infty}{1 + ae^{-bt}}$$

Q_∞ is ultimate discovery

$$t = 0, \quad Q_D = \frac{Q_\infty}{1 + a}$$

$$t = \infty, \quad Q_D = Q_\infty$$

$$\ln\left(\frac{Q_\infty}{Q_D} - 1\right) = \ln a - bt \qquad [1\text{-}20]$$

a and b can be evaluated from

$$\ln\left(\frac{Q_\infty}{Q_D} - 1\right) \text{ vs. } t \text{ plot}$$

In this regard, the fossil utilization is finite, only occurring as a short time interval when one considers a ±5000 year period. The fossil fuel use exploited by mankind is only a blink of time (Figure 1-15). Long-term reserves may come from nuclear fusion or solar power as indicated in Figure 1-16.

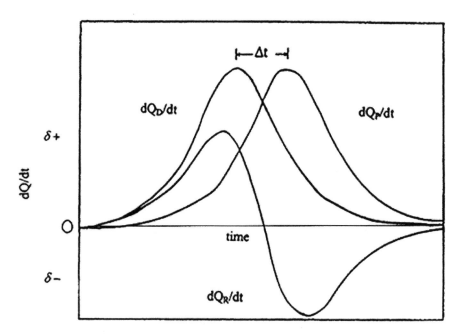

Figure 1-14. Variation of rates of production, of proved discovery, and of rate of increase of proved reserves of crude or natural gas during a complete production cycle. (After Hubbert, op. cit., 1962 and 1974)

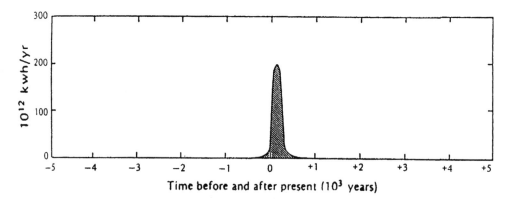

Figure 1-15. Epoch of fossil-fuel exploitation in perspective of human history from 5,000 years in the past to 5,000 years in the future. (Modified from Hubbert, 1962)

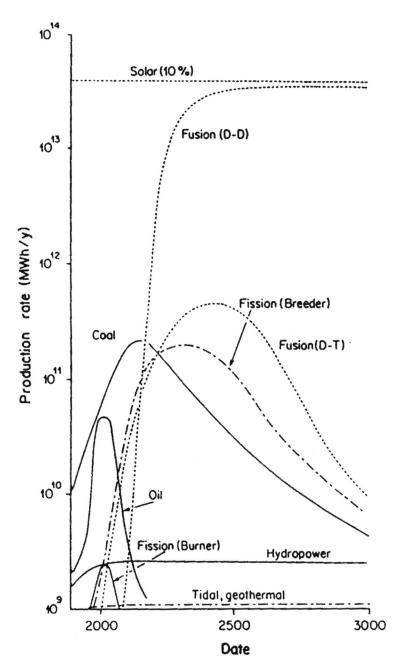

Figure 1-16. Possible fuel consumption levels and projected availability of reserves.

1.4.2   McKelvey's Microanalytical Model

This model is based on cost and geological factors by a stepwise analysis of the deposit domains in a specific region. The analytical methods involve statistical approximations including a regressional analysis.

$$L = \frac{REI}{P} \qquad\qquad [1\text{-}21]$$

Here:

L = living standard

R = resource

E = energy

I = ingenuity (socioeconomic, political, and technological)

P = capita

$$\frac{E}{P} = a + b\frac{G}{P} \qquad\qquad [1\text{-}22]$$

G = GNP

$$E_t = \frac{E_m}{g} + E_s \qquad\qquad [1\text{-}23]$$

Here:

E_t = total energy required

E_m = mining energy

E_s = smelting and refining energy

g = grade

Therefore, **resource** depends on an increasing degree of both **geological assurance** (measured, indicated, inferred, hypothetical, speculative, and so on) and **economic feasibility** (submarginal, paramarginal, and so on), as shown in Figure 1-17. For each mineral there is a separate mineralogical barrier and threshold above crustal abundance, as shown here:

Metal	Percentage threshold above crustal abundance
Fe	5
Al	8.13
Cu	0.007

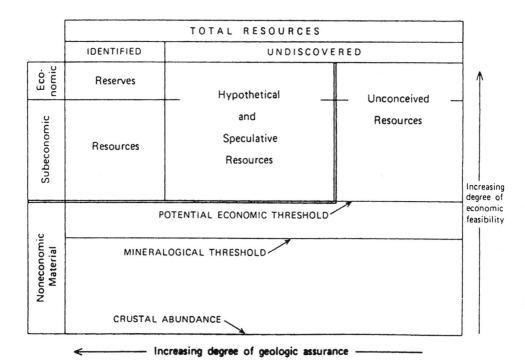

Figure 1-17.   The relation of resources to non-economic mineral materials. (Source: Brobst, 1979)

Such a barrier is illustrated by Figure 1-18 for the copper ore. If the ore grade is below 0.007%, it is not economical to recover.

1.4.3   Dynamic Model of Meadows

Dennis Meadows' model concluded that reserves are finite, that usage rate will create substitution technology, and that recycling is minor.   He suggested a static reserve life index (s) and an exponential reserve life index (e), where

$$e = \frac{\ln (Rs + 1)}{R} \qquad [1\text{-}24]$$

where R = annual fractional growth rate.

	s (yr)	e (yr)
Coal	900	110
Fe	400	86
Al	175	60
Ni	140	55
Petroleum	70	35
Cu	40	26
Ag	20	16
Pb	15	12
Hg	13	11

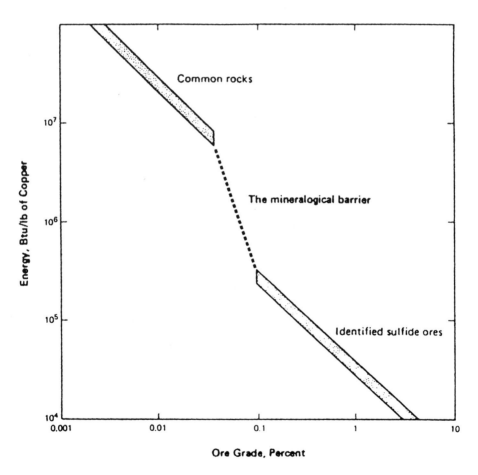

Figure 1-18.   The mineralogical barrier.

All resources are limited from 1980 to 2050 as shown in Figure 1-19.

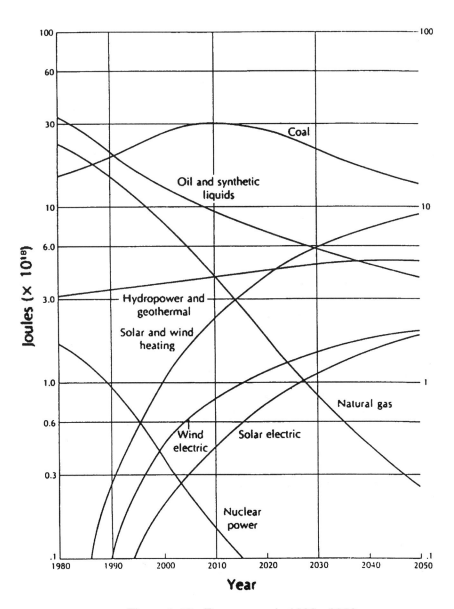

Figure 1-19. Energy supply 1980 - 2050.

1.4.4   Odum Ecological Model

Howard Odum proposed that the energy network is similar to electrical circuitry, as shown in Figure 1-20. All calculations can be performed as they would be in systematic engineering. The essential points are as follows:

- Energy flows in the complex forest ecosystem, which includes all humans, such that

$$n = 1 + \frac{\left(\dfrac{\log E}{NPP}\right)}{\log \eta}$$
[1-25]

where

n = trophic level

E = energy

NPP = net primary production

η = ecological efficiency

- Man's diverse work substitutes ecosystem variety in an agricultural system.
- Fossil fuel subsidized agriculture is a colonial member of technological society with maximum solar conversion.
- Energy is the organizing principle from which all economic, social, and political values are derived.
- Maximum power is in the middle of high rate and low efficiency, and vice versa for all competitors.

Finally, a diagram can illustrate the fossil energy flow in our ecosystem, as shown in Figure 1-21.

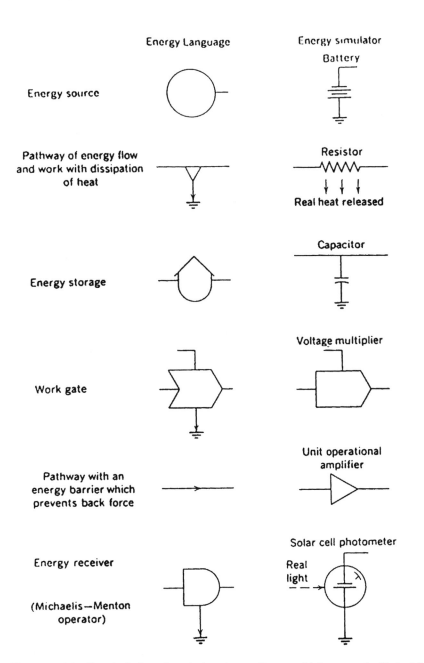

Figure 1-20. Symbols for electric hardware items which are substituted for parts of the energy network diagram in making an energy simulator. (Source: Odum, 1976, Ref. 1-9)

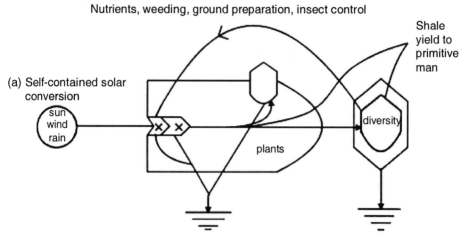

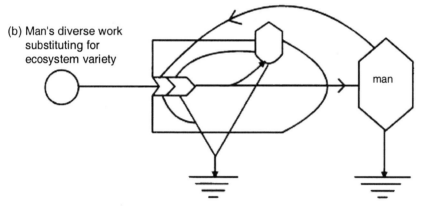

Figure 1-21.   Forest ecosystem.

1.5 ENERGY CONSERVATION AND ITS CHEMICAL PERSPECTIVE

Following the 1973 Arab oil embargo, the United States developed technologies based on new and alternative sources. At the same time, social, economic, and political realizations had exerted influence on the public towards a **soft path** on energy policy instead of developing "hard" technology to accelerate the depletion of nonrenewable resources, as seen in the past. As defined by Lovins, the soft path consists of the following alternatives:

- relying heavily on renewable energy sources such as solar, and biomass

- being diverse and tailored for maximum effectiveness under specific circumstances

- being flexible and accessible to most of the public

- being matched in both geographic distribution and scale to prominent end use needs

- having a good agreement between energy quality and end use

In 1991, the Union of Concerned Scientists and other conservation organizations united and issued three other scenarios (soft path) other than the U.S. government analysis, which anticipates a one-time increase of energy consumption by 2030 based on 1988. Even the market driven scenario of a GNP 2.1% annual growth is better. The softest scenario is the climate stabilization scenario, which is designed to achieve a 25% reduction of CO_2 emulsions by 2005 and a 50% reduction by 2030. Regardless, since the selection of the soft path, many technologies have been developed for the conservation of energy. They will be discussed in the following sections.

1.5.1 Intra, Inter and Combined Fuel Conversions

Intrafuel conversion signifies that a given raw fuel can produce more than one type of end-use fuel. For example, highly volatile bituminous B coal can be used for liquefaction as well as gasification; thus, it can produce coal gas as well as coal liquid. The optimum yield of either product will depend on the type of raw coal that is fed to a given process. Furthermore, direct conversion or indirect conversion (such as Fischer-Tropsch Synthesis) can be evaluated (refer to Fig. 1-10). **Interfuel conversion** is the use of two or more raw fuels as feedstocks to derive one type of end fuel. Petroleum has been used with coal for a liquefaction process. The use of spent nuclear fuel as a catalyst for petroleum refining is another example. The **combined fuel conversion** is intended for the end use for two or more usages from one type of fuel. A good example is **cogeneration** in which both electricity and space heating can be achieved. Magnetohydrodynamics is another example where coal is thermally decomposed to obtain energy; in the meantime, iodine vapor is conditioned in the magnetic field to increase the current to yield more electricity. The combined cycles for geothermal energy production to increase the efficiency is another example.

1.5.2   Enhanced Oil Recovery (EOR) and Ultimate Oil Recovery (UOR)

When a petroleum oil field is discovered, the natural stored energy (expansion of natural gas or volatile component) assisted by pumping will produce oil. This stage is usually referred to as a **primary recovery**. As the energy is depleted, production declines and water is injected to the reservoir and this is called **secondary recovery**. When the water-to-oil production ratio of the field approaches an economic limit of operation, the net profit diminishes because the difference between the value of the produced oil and the cost of water injection and treatment becomes too narrow. The **enhanced oil recovery (EOR)** or the **tertiary recovery** then begins. The combined total oil production of primary and secondary recovery is generally less than 40% of the original oil in place. The target setting for EOR is an additional 10–15%. In most cases, the oil well is shut down and about 50% of the oil can never be recovered.

Current EOR technology involves the following:

- **surfactant flooding** — reduction of interfacial tension assistance in emulsification

- **polymer flooding** — addition of water-soluble polymers such as polyacrylamide or polysaccharide to increase the viscosity of water for mobility control

- **miscible flooding** — use of CO_2 for oil swelling

- **steam or fire flooding** — thermal recovery to improve sweep efficiency

- **alkaline flooding** — injection of alkali to and from the in situ surfactant with the acid portion of the oil components

For recovery, either one or a combination of the preceding processes is used in the field. A good example is **microbial enhanced oil recovery (MEOR)**, which uses injected bacteria and nutrients for the production of surfactants, polymers, gases, solvents, and biomass simultaneously for recovery. The bacteria also can be used as a selective plugging agent for great heterogeneous distributions of the reservoir racks in propensity and permeability. **Ultimate oil recovery** (UOR) is the employment of the final physical, chemical, and microbiological techniques for the recovery of the remaining residual oil in the reservoir. This technique has not been developed. At any rate, the EOR process is essential for the duration of the 21st century, as shown in Figure 1-22.

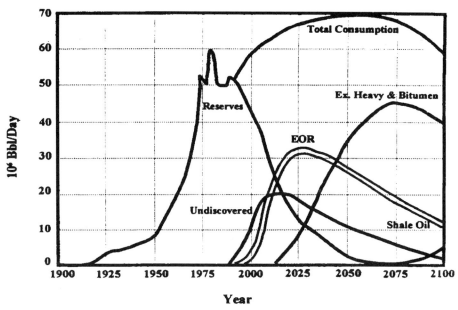

Figure 1-22. World Crude Oil Supply from 1900 to 2100. (Source: from DOE, *Energy Outlook to the Year 2000 an Overview*, 1990, p.3)

1.5.3 Substitution Technology

Because waste can be looked upon as a resource, recycling becomes important. The problem is that based on the constraints of materials, some resources cannot be recycled indefinitely; for example, paper can only be recycled up to four times — unless new technology is invented, such as ink that will fade after a fixed duration so that paper can be reused. The consideration of material is essential to proper recycling. This theory will be illustrated in the following example. In many communities, the use of Styrofoam food containers has been banned because the material is nonbiodegradable. Yet, in a careful study of a styrofoam cup versus a paper cup on a per cup basis, the styrofoam cup has less environmental impact as compared to a paper cup. This is due to the weight of the required raw material (polyfoam vs. paper cup, 1.59g vs. 10.1g). As resources deplete, especially for metals, new technology or substitution should be in place. Figure 1-23 demonstrates that chromium usage requires substitution. Recycling of metals plays an important role because it only requires a small fraction of energy compared to the energy used to extract and manufacture the metals from ores. Typically, the reprocessing of aluminum cans of soft drinks only requires 5% of the energy compared to the processing from aluminum ore. Some metals are higher; for example, steel requires 48%.

Due to the shortage and deprivation of nonrenewable resources, new substitution technology will be initiated. For example, carbon fibers will replace steel. In substitution, model economics becomes important. For example, the market share, F, of the introduction of new technology will rise exponentially in the beginning and reach a saturation lag phase.

$$\frac{dF}{dt} = \alpha F(1-F) = \alpha F - \alpha F^2 \qquad \text{[1-26]}$$

Upon integration,

$$F(t) = \frac{1}{\{1 + \exp[-(\alpha t + c)]\}} \qquad \text{[1-27]}$$

α determines the speed of the substitution process, and c is the integration constant. This case is for two competing commodities. From system dynamics, it can also be stated as:

$$\frac{dF}{dt} = \frac{\alpha F(1-F)^2}{[\gamma F + (1-F)]} \qquad \text{[1-28]}$$

Now γ is time dependent to exogenous restrictions.

1.5.4   Environmentally-Benign Processing (Green Process)

In the 90s, energy efficiency is applicable to most industrial processes. For example, automobile energy intensities in the United States (mJ per vehicle per km) and fuel economy (L/100 km) have improved from 1970 to 1990. Improvement will be made by cutting the loss of efficiency due to aerodynamic drag and breaking. An effort will also be made to reduce the weight by (possibly) using carbon fiber composites (see Figure 1-24 for the "greener" vehicles). Research and development are centered on novel process designs and process improvement that would reduce the potential for environmental release. Also, for fossil fuel production, feedstock substitutions, and alternative synthetic and separation procedures with efficient catalysts, a new process that would minimize by-products formation and reduce waste at the source will be developed. Some areas are under consideration.

- more highly selective catalysts
- low temperature and pressure operations
- low-energy separation techniques
- synthesis that bypass toxic feedstock and solvents
- substitution of halogenated solvents

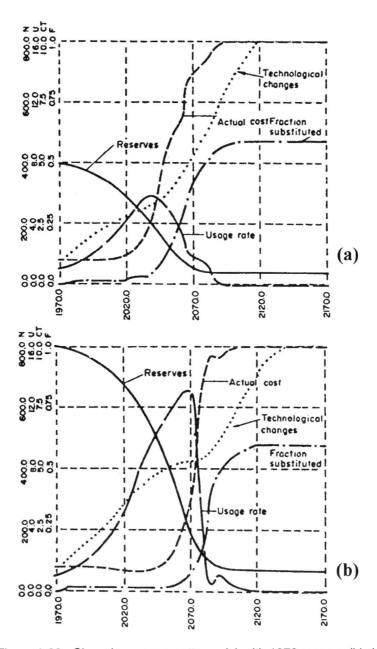

Figure 1-23. Chromium usage patterns (a) with 1970 reserve (b) double the reserve U (usage), N (reserves), C (cost), T (technology), F (fraction substituted). (After Meadows et al., Ref. 1-6)

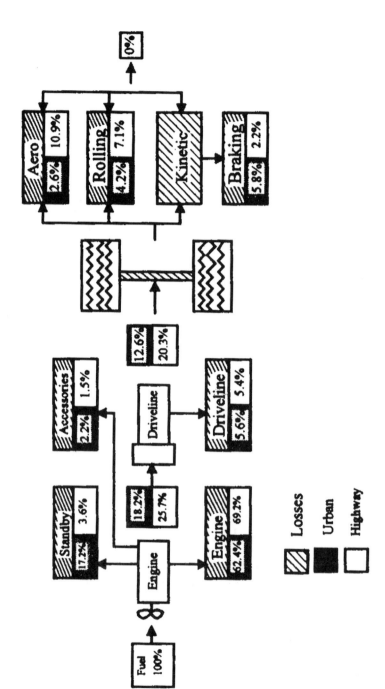

Figure 1-24.   Energy losses in automobile transport point to strategies for improving fuel efficiencies in automobiles. (Source: D.L. Illman (1994). Auto-makers move toward new generation of "greener" vehicles. *Chemical and Engineering News* 72(31): 8-16. Copyright 1994 by American Chemical Society. Reprinted with permission)

REFERENCES

1-1   T. F. Yen and G.V. Chilingarian (eds), *Oil Shale*, Elsevier Science, New York, 1976.

1-2   T. F. Yen, *Science and Technology of Oil Shale*, Ann Arbor Science, Ann Arbor, Michigan, 1976.

1-3   E. C. Donaldson, G.V. Chiligarian, and T.F. Yen, *Enhanced Oil Recovery, I, Fundamentals and Analyses*, Elsevier Science, New York, 1985.

1-4   R. E. Zimm, "Nuclear Energy Resources and Long-Term Energy Requirement," Angewandte Chem. 10, 1-19 (1971).

1-5   K. A. D. Inglis, *Energy, from Surplus to Scarcity?* Applied Science, Barking, Essex, UK, 1974.

1-6   D. H. Meadows, D. L. Meadows, J. Randers, and W. W. Behrens III, *The Limit to Growth: A Report for the Club of Rome's Project on the Predicament of Mankind*, Universe Books, New York, 1972.

1-7   Readings from Scientific American, *Chemistry in the Environment*, W.H. Freeman, San Francisco, California, 1973.

1-8   H. T. Odum, *Environment, Power, and Society*, John Wiley, New York, 1970.

1-9   H. T. Odum and E. C. Odum, *Energy Basis for Man and Nature*, McGraw Hill, New York, 1976.

1-10 C. A. S. Hall, C. J. Cleveland, and R. Kaufmann, *Energy and Resource Quality*, Wiley-Interscience, New York, 1986.

1-11 J. T. McMullan, R. Morgan, and Murray, *Energy Resources and Supply*, Wiley, London, 1976.

1-12 A. B. Lovin, *Soft Energetics: Toward a Durable Peace*, Harper and Row, New York, 1979.

1-13 Union of Concerned Scientists Alliance to Save Energy, American Council for An Energy Efficient Economy, Natural Resources Defense Council, American's Energy Choices, *Investing in a Strong Economy and a Clean Environment*, Union of Concerned Scientists, Cambridge, Massachusetts, 1991.

1-14 M. B. Hocking, "Paper Versus Polystyrene: A Complex Choice," Science, 251, 504-505, 1991.

1-15 T. F. Yen, "Chemical Aspects of Interfuel Conversion," Energy Sources, 117-136, 1973.

1-16 E. C. Donaldson, G. V. Chilingarian, and T. F. Yen, *Enhanced Oil Recovery, II, Process and Operation*, Elsevier Science, Amsterdam, 1989.

1-17 T. F. Yen, *Microbial Enhanced Oil Recovery: Principal and Practice*, CRC Press, Boca Raton, Florida, 1990.

1-18 M. K. Hubbert, Resources and Man, W.H. Freeman, San Francisco, California, 1973.

1-19 M. Christian and H. M. Groscarth, "Modeling Dynamic Substitution Process in Energy Supply Systems," Energy Sources, 17, 295-311 (1995).

1-20 Y. P. Hsia and T. F. Yen, "Evaluation of Coal Liquefaction Efficiency Based on Various Ranks," Energy Sources, 3, 46-53 (1976).

1-21 *The Institue of Fuel, Fuel and the Environment*, London, 1973.

1-22 L. H. Keith, *Energy and Environmental Chemistry, Fossil Fuels*, Ann Arbor Science, Ann Arbor, Michigan, 1982.

1-23 W. J. Mitsch, R. K. Rasade, R. W. Bosserman, and J. A. Dillon, Jr. *Energetics and Systems*, Ann Arbor Science, Ann Arbor, Michigan, 1982.

1-24 R. L. Seale and R. A. Sierka, *Energy Needs and the Environment*, University of Arizona Press, Tucson, Arizona, 1973.

1-25 J. K. Jacques, J. B. LeSourd and J. M. Ruiz, *Modern Applied Energy Conservation*, Ellis Horwood, Chichester, England, 1988.

1-26 W. J. Mitsch, R. K. Rasade, R. W. Bosserman and J. A. Dillon, Jr. *Energetics and Systems*, Ann Arbor Science, Ann Arbor, MI, 1982.

1-27 T. L. Shaw, D. E. Lennard, and P. M. S. Jones, *Policy and Development of Energy Resources*, Wiley, Chichester, England, 1984.

1-28 F. Benn, J. Edewor, and C. McAuliffe, *Production and Utilization of Synthetic Fuels--An Energy Economic Study*, Halsted, New York, 1981.

1-29 D. Trantolo and D. Wise, *Energy Recovery from Lignin, Peat and Lower Rank Coals*, Elsevier, Amsterdam, 1989.

1-30 E. M. Goodger, *Alternative Fuels: Chemical Energy Resources*, Wiley, New York, 1980.

1-31 E. W. Erikson and L. Waverman, *The Energy Question, An International Failure of Policy*, Vol. 1, *The Word*, Vol. 2. North America, University of Toronto Press, Toronto, 1974.

1-32 R. K. Hessley, J. W. Reasoner, and J. T. Riley, *Coal Science, An Introduction to Chemistry, Technology, and Utilization*, Wiley, New York, 1986.

1-33 R. F. Naill, *Managing the Energy Transition, a System Dynamics Search for Alternatives to Oil and Gas*, Ballinger, Cambridge, Massachusetts, 1977.

1-34 A. B. Lovins, *Soft Energy Patterns, Toward a Durable Peace*, Ballinger, Cambridge, Massachusetts, 1977.

1-35 B. R. Cooper and W. A. Ellingson, *The Science and Technology of Coal and Coal Utilization*, Plenum, New York, 1984.

1-36 G. V. Chillingarian and T. F. Yen, *Bitumen Asphalts and Tar Sands*, Elsevier, Amsterdam, 1978.

PROBLEM SET

1. Our energy consumption can be described as follows:

   $$Q = Q_0 \exp(\lambda T)$$
   Where $\lambda = 0.035$ per year.

   a) Define and calculate the doubling time T_2
   b) Also calculate the yearly fractional increase R

2. a) Calculate the API gravity of water
   b) If an oil has $-3°$ API, what is the specific gravity at $60°$ F

3. Calculate the following for the daily output of the Hoover Dam
   a) Power in Joules/sec
   b) Energy in Joules

   (Given power of Hoover Dam = 4.3×10^9 KW)

CHAPTER 2

ALTERNATIVE ENERGY SOURCES

*I*n the preceding chapter, we focused our attention on fossil energy — petroleum, coal, oil shale, and tar sand. There are a number of alternatives to fossil fuels (such as solar, wind, tidal, geothermal, chemical, and nuclear energy) that are also important in helping us meet our energy demands. These are called **alternative energy sources**.

There are many different ways to classify different fuel types, and at this time there are quite a few types that warrant our attention. The first category is the improvement of conventional energy; a good illustration being enhanced oil recovery (EOR) or enhanced gas recovery (EGR). Magnetohydrodynamics (MHD), the most efficient use of coal, also belongs to this group. The next category is the interfuel refining and intrafuel optimization. Both categories are largely related to fossil fuels although the second will include nuclear energy. The third category is **non-renewable energy** such as nuclear and geothermal energy. Then follow **supplementary fuels** (auxiliary fuels) such as solar, tidal, wind, and ocean currency. The last important category is **chemical fuels** such as hydrogen and methanol. Hydrogen is often called "ecofuel," signifying that it is compatible with the environment. The supplementary fuels and chemical fuels are **renewable energy**.

In this chapter, we will first spend some time discussing the thermal pollution problem, and then move to the main topic, the primary source — solar energy. Finally, we will look at other alternative sources of energy such as wind, tidal, geothermal, and chemical power.

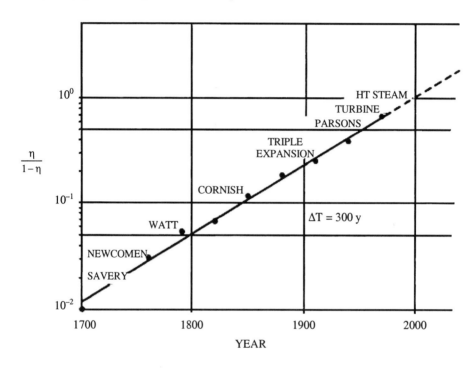

Figure 2-1.   Efficiency of prime movers.

2.1 WASTE HEAT

Even if we could be supplied with an unlimited amount of energy, there would still be difficulties associated with the ever-increasing use of energy. One general problem under the heading of thermal pollution is the disposal of all the heat that accompanies energy utilization. All energy generation, no matter what form it may take for the moment, eventually appears as a **heat burden** on the environment. A power station that generates electricity uses heat supplied by either the burning of fossil or nuclear fuels to produce the steam that runs a turbine. Even if everything works with perfect efficiency, however, only a fraction of the heat can be converted to useful work. Throughout history, humans have tried their best to develop better heat engines as shown in Figure 2-1, but ultimate efficiency is still limited by their nature. These limits, expressed by the laws of thermodynamics, cannot be circumvented by clever engineering tricks. They represent the best we could do if everything went as well as possible. As illustrated in Figure 2-2, for a steam electric power plant, heat is delivered to a boiler at a temperature of T_2, and after driving the turbine, the steam is condensed at a lower temperature of T_1. The maximum efficiency, or the ratio of work output to heat input, is given by the temperature difference, $T_2 - T_1$, divided by the input temperature, T_2. This ratio is a consequence of the second law of thermodynamics and is a fundamental property of nature.

The first law of thermodynamics tells us that energy is conserved, so that

$$Q_s = Q_r + W \qquad\qquad [2\text{-}1]$$

where

Q_s = energy supplied to the heat engine

Q_r = heat rejected (waste heat) to the environment

W = quantity of work produced

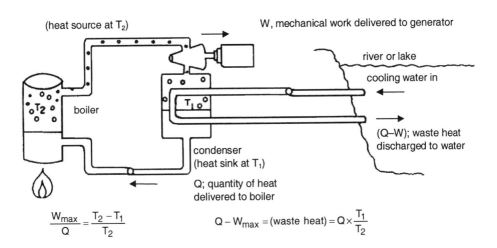

Figure 2-2. Maximum work and waste heat from steam electric power plant.

A diagram of the energy flow is

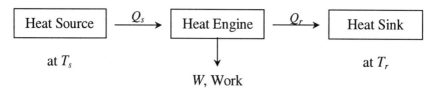

The ideal efficiency of the thermal engine is defined as

$$\eta = (W_{max}/Q_s) = (Q_s - Q_r)/Q_s = (T_s - T_r)/T_s \qquad [2\text{-}2]$$

So

$$W = \eta Q_s \qquad [2\text{-}3]$$

$$Q_r = (1 - \eta)\, Q_s \qquad [2\text{-}4]$$

Similarly, the rate can be written as $Q_r = (1 - \eta)Q_s$. The waste heat produced increases rapidly with decreasing efficiency as shown in Figure 2-3. If we also take the stack losses into consideration as shown in Figure 2-4, then

$$Q_r/W = (1 - \eta - s)/\eta \qquad [2\text{-}5]$$

Table 2-1 shows the thermal efficiencies of heat engines using different energy sources. A plant operating at 40% efficiency produces 4 calories of useful work and 6 calories of waste heat for every 10 calories of fuel burned. This production means that for every calorie of useful work produced, 1.5 calories of waste heat have to be carried away.

Small portions of the waste heat from a fossil fuel plant are transferred to the environment through the hot gases emitted from the smokestack. A large fraction — 85% — is transferred via the condenser, which cools and condenses the hot steam. In a nuclear plant that has no smokestack, essentially all of the waste heat is rejected to the surroundings via the cooling water, except in the chemonuclear reactor concept. For large, modern, power stations, the volume of water required is enormous. It is estimated that 100,000 ft^3/min of cooling water is required in a 1000 mega-watt power plant. Already more than 10% of the total water stream flow in the United States is used for this purpose. It is becoming increasingly difficult to find sites where this much water is available. Moreover, there are the ecological consequences of the cooling water stream flow to consider. For one thing, fish or other marine life can be sucked into the intake and killed. There is also concern that the increase in the temperature of the discharge water, as shown in Figure 2-5, may be harmful to aquatic life.

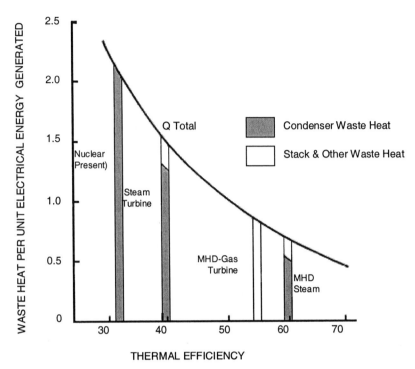

Figure 2-3.   Overall thermal efficiency of different electrical generating systems. The waste heat produced increases rapidly with decreasing efficiency. (Source: Richard J. Rosa, Avco Corp.)

Table 2-1. Thermal Efficiencies of the Heat Engines Using Different Energy Sources

Energy Source	Q_r/W	η
Fossil fuel	1.75 – 1.6	0.36 – 0.33
Nuclear	2.33 – 2.0	0.30 – 0.33
Geothermal	4.2	0.19
Solar farm	2.33	0.30
Fusion	1.0	0.50

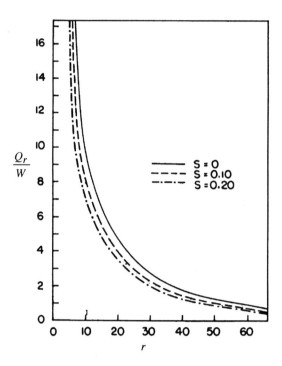

Figure 2-4.   The ratio of heat rejection to useful work as a function of overall plant efficiency. The effect of heat rejected in terms of stack loss (S) is also shown.

[Example 2-1]   Considering a coal plant for the generation of 1400 MW of power, calculate the required cooling water flow if the condenser temperature change is 20°F. Assume 85% of the waste heat is carried off by the condenser cooling water.

From Equation 2-4,

$$\dot{Q}_r = r(1-\eta)\dot{Q}_s \tag{a}$$

Here, $\dot{Q} = \dfrac{dQ}{dt}$ and r is the plant efficiency.

If one Btu raises the temperature of one lb of water by one °F, then the flow rate of water,

$$Q = \frac{\dot{Q}_r}{\Delta T} \tag{b}$$

If we can express the power generated, P as KW, then from Equation 2-3, we have

$$P = \dot{W} = \eta \dot{Q}_s \qquad \text{(c)}$$

Because 3413 Btu = 1 KWH, after substitution we get

$$\dot{Q}_r = \frac{3413r(1-\eta)P}{\eta} \quad \frac{\text{Btu}}{\text{hr}} \qquad \text{(d)}$$

If we express by Q as cfs, then dividing by 3600 (sec/hr) / 62.4 (lb/ft^3)

$$Q = \frac{0.0152\, r\, (1-\eta)\, P}{\eta\, \Delta T} \quad \text{cfs}$$

Because $r = 0.85$ and $\eta = 0.35$

$$Q = \frac{(0.0152)(0.85)(0.65)(1.4 \times 10^9)}{(0.35)(20)} = 1.68 \times 10^6 \quad \text{cfs}$$

Measures have been taken to reduce the demand for cooling water. Dry cooling towers, using air as a coolant, have been adopted in some power plants. The disadvantages are more noise and less efficiency. There are also ways to use the waste heat from electrical plants purposefully, which also reduces cooling water consumption. One way is to simply use the hot water outflow to heat residential or commercial buildings. Another possibility is to combine electrical generation with the production of industrial process heat. The latter is called **cogeneration**, which has been mentioned in section 1.5.1, and is the process of combining the water heat from electrical generation with space heating. Research has also been done for applying waste hot water to heat fish ponds. It was found that the ratio of the weight that the fish gained to the weight of feed food increases from 25% (temperature of the pond at 2°C) to 75% (temperature at 30°C). Most of the waste heat product from fossil fuel is of **low grade** (below 120°C) and fluctuates considerably. Low grade waste heat can be used to defrost fruits and plants to extend the growing season.

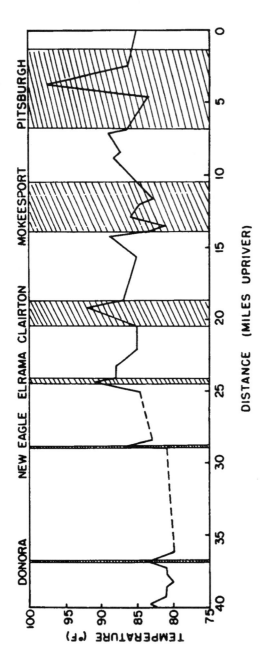

Figure 2-5. Water temperatures can become very high, particularly in summer along rivers with concentrated industry. The chart shows the temperature of the Monongahela River measured in August, along a 40-mile stretch upriver from its confluence with the Ohio.

2.2 PLANK'S LAW AND ABSORPTION

Every object is a source of **radiant energy**. The radiation arises from the electron motions of the atoms and molecules of the substance. These motions represent the heat energy the object possesses. The hotter it is, the more extensive and higher in frequency the motion becomes. A heated body emits energy in the form of **electromagnetic waves**. This energy is radiated in all directions, and upon falling on a second body, is partially absorbed, partially reflected, and partially transmitted, as indicated in Figure 2-6. The fraction of the incident radiation absorbed is known as the **absorptivity (a)** and the fraction reflected, the **reflectivity** of the body. The reflectivity is also called **albedo (A)**, which varies with surfaces, as shown in Table 2-2. The amount transmitted will therefore depend on the following two properties:

$$I = Ab + R + T$$

$$a = (1-A) \qquad\qquad [2\text{-}6]$$

where I is incident energy, Ab is absorbed energy, R is reflected energy, and T is transmitted energy.

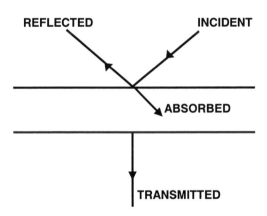

Figure 2-6.   Reflection, adsorption, & transmission of radiation.

As previously mentioned, electromagnetic waves are emitted from hot surfaces — the higher the temperature of the surface, the greater the radiation energy flux leaving it, and the higher the frequencies of the departing electromagnetic waves. The way in which the energy in black body emissions is distributed over the wavelength spectrum can be explained by **Plank's quantum theory**. The emissive power of a black surface is

$$P(T) = \frac{2\pi\, hc^2}{\left[\lambda^5\left(\exp\left(\frac{hc}{\lambda kT}\right)\right) - 1\right]}$$ [2-7]

where

P = power density in W/m^3

λ = wavelength

T = absolute temperature

c = velocity of light

h = Plank's constant

k = Boltzmann constant

Table 2-2.   Reflectivity or "Albedo" of Various Surfaces

Surface	% Reflected
Clouds (stratus) <500 ft thick	5 - 63
500 – 1000 ft thick	31 – 75
1000 – 2000 ft thick	59 – 84
average of all types and thicknesses	50 – 55
Concrete	17 – 27
Crops, green	5 – 15
Forest, green	5 – 10
Meadows, green	10 – 20
Ploughed field, moist	14 – 17
Road, black top	5 – 10
Sand, white	34 – 40
Snow, fresh fallen	75 – 90
Snow, old	45 – 70
Soil, dark	5 – 15
Soil, light (or desert)	25 – 30
Water	8*

*Typical value for water surface, but the reflectivity increases sharply from less than 5% when the sun's altitude above the horizon is greater than 30°, to more than 90% when the altitude is less than 3°. The roughness of the sea surface also affects the albedo somewhat.

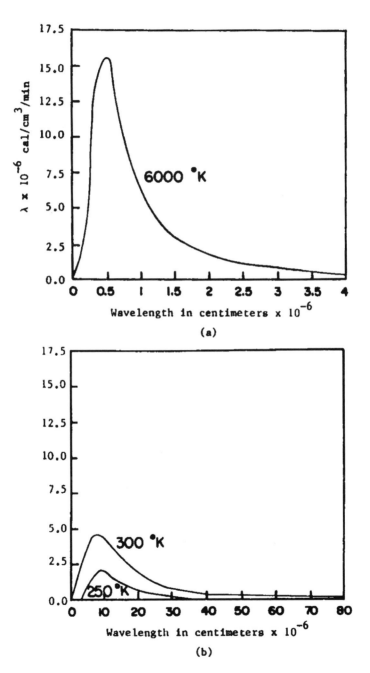

Figure 2-7.   Black-body emission of (a) a hot body such as the sun, and (b) a cool body such as the earth.

A comparison of the black body emission of a hot body and a cold body is shown in Figure 2-7. For any temperature, the distribution curve has a maximum value. The value of λ for which this maximum occurs may be found by determining $(dP/d\lambda)_T$ and finding what value of λ will make the derivative zero.

$$\left(\frac{\partial P}{\partial \lambda}\right)_T = 0 \tag{2-8}$$

Let

$$x = \frac{hc}{\lambda kT}$$

$$P = \frac{2\pi k^5 T^5}{h^4 c^3} \frac{x^5}{e^x - 1} \tag{2-9}$$

$$\ln P = \ln \frac{2\pi k^5 T^5}{h^4 c^3} + 5 \ln x - \ln\left(e^x - 1\right)$$

then

$$\left(\frac{\partial \ln P}{\partial x}\right)_T = 0 + \frac{5}{x} \cdot \frac{1}{e^x - 1} \frac{\partial(e^x - 1)}{\partial x} = \frac{5}{x} - \frac{e^x}{e^x - 1}$$

$$= \frac{5}{x} - (1 + e^{-x} + e^{-2x} + e^{-3x} + \ldots) = \frac{5}{x} - 1 - e^{-x} = 0 \tag{2-10}$$

$$x + x e^{-x} = 5, \quad x = 4.9651 = \frac{hc}{\lambda kT} \tag{2-11}$$

The result is that

$$\lambda T = 0.288 \text{ cm-degree K} \tag{2-12}$$

which is **Wien's displacement law.** Equation [2-12] can predict the wavelength of a black body; for example,

$$\lambda_{(sun)} = \frac{2880}{6000} = 0.48 \, \mu m,$$

$$\lambda_{(earth)} = \frac{2880}{273} = 11 \, \mu m$$

The calculated value is in agreement with the earth's spectrum shown in Figure 2-7.

2.3 STEFAN-BOLTZMANN LAW AND EMISSIVITY

The rate at which a hot body radiates energy is proportional to the fourth power of its absolute temperature and this relationship is known as the **Stefan-Boltzmann Law**.

$$P = \sigma T^4 \qquad\qquad [2\text{-}13]$$

where

T = absolute temperature

σ = Stefan-Boltzmann constant which has the value $5.67{\times}10^{-8}$ W/m^{2}K^4

The solar flux incident on the earth, the **solar constant (S)**, can be estimated by equating the following two equations, [2-14] and [2-15]:

$$\text{Power radiated by Sun} = 4\pi R_S^2 \sigma T_S^4 \text{ (based on black body)} \qquad [2\text{-}14]$$

where $R_S = 6.95 \times 10^8$m (radius of Sun)

$$\text{Power radiated by Sun} = 4\pi R_{SE}^2 S \text{ (based on } S \text{ from the earth)} \qquad [2\text{-}15]$$

where $R_{SE} = 1.49 \times 10^{11}$m (earth-sun distance)

Thus,

$$4\pi R_S^2 \sigma T_S^4 = 4\pi R_{SE}^2 S \qquad\qquad [2\text{-}16]$$

or

$$T_S = \left(\frac{R_{SE}}{R_S}\right)^{\frac{1}{2}} \left(\frac{S}{\sigma}\right)^{\frac{1}{4}} \qquad\qquad [2\text{-}16A]$$

In the preceding equation, if the average temperature of the sun is 6000K, then S can be calculated as 1600 W/m^2. This value is equivalent to 2 cal/cm^2-min or 2 langleys/min. A simple schematic for those values can be found in Figure 2-8.

A **black body** is defined as one that absorbs all the radiation falling on it; that is, its absorptivity = 1. The **emissivity, ε,** of a body is a dimensionless factor related to the rate of thermal radiation from a particular surface. It varies with surface temperature, roughness, and color and is an important parameter for indicating the energy trapping ability, as shown in Table 2-3. The emissivity of the earth can be estimated by the following calculation, which sets Equation [2-17] equal to Equation [2-18]:

$$\text{Radiation energy absorbed by the earth} = (1-A)\pi R_E^2 S \qquad [2\text{-}17]$$

(depending on earth's cross-sectional area)

$$\text{Energy emitted by the earth outward} = (4\pi R_E^2)\,\varepsilon\,(\sigma T_E^4) \qquad [2\text{-}18]$$

At equilibrium $(1 - A)\,\pi R_E^2 S = (4\pi R_E^2)\,\varepsilon\,(\sigma T_E^4)$, then

$$(1 - A)S = 4\varepsilon\sigma T_E^4 \qquad [2\text{-}19]$$

Taking $A = 0.35$ and $T_E = 286°K$ (13°C), we obtain $\varepsilon = 0.6$.

By slightly modifying equation [2-19], one can easily estimate the effect of human utilization of energy on the earth's temperature.

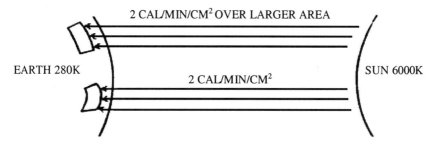

Figure 2-8. The energy flux from the sun arriving at the earth's surface.

Table 2-3. Emissivity of Surfaces Near T = 300K

Material		$\varepsilon\,(T \approx 300K)$
Aluminum	Polished	0.04
	Rough plate	0.06
	Oxidized	0.15
Cast iron		0.50
Sheet steel		0.70
Wood, black lacquer, white enamel, plaster, roofing paper		0.90
Porcelain, marble, brick, glass rubber, water		0.94

[**Example 2-2**] If the earth's average absolute temperature, T, is given by a balance between the absorbed solar energy, s(1-A)/4, and black body radiation, σT^4, what would be the direction and magnitude of the temperature change of
  a) the albedo increased by 1%, and
  b) power that is generated from fossil or nuclear fuels at a rate equivalent to 1% of the incidental solar energy

$$\frac{S(1-A)}{4}=\sigma T^4$$

$$T=\left[\frac{S(1-A)}{4\sigma}\right]^{\frac{1}{4}}$$

$$=\left[\frac{2\,\text{cal/min/cm}^2\,(1-0.4)}{4\left(1.43\times10^{-12}\,\text{cal/sec/cm}^2\,/\,\text{K}\times60\,\text{sec/min}\right)}\right]^{\frac{1}{4}}=244\,\text{K}$$

a) $$\frac{dA}{A}=0.01\,,\quad\frac{A}{1-A}=\frac{0.4}{0.6}=\frac{2}{3}$$

$$dT=\left(\frac{-1}{4}\right)\left(\frac{dA}{A}\right)\left(\frac{A}{1-A}\right)(T)=-0.25(0.01)(0.667)(244)=-0.4\,\text{K}$$

b) $$\frac{dS}{S}=0.01$$

$$\frac{dT}{T}=\frac{1}{4}\left[\frac{S(1-A)}{4\sigma}\right]^{-1}\left(\frac{1-A}{4\sigma}\right)dS=\frac{1}{4}\frac{dS}{S}$$

$$dT=\frac{1}{4}\left(\frac{dS}{S}\right)T=\frac{1}{4}(0.01)(244\,\text{K})=+0.6\,\text{K}$$

As we have seen, the sun provides us with a large input of energy every day. Solar energy is our largest and most lasting energy resource as shown in Table 2-4. To maintain a steady state, the earth must dispose of the energy that it receives from the sun. In section 2.2, we calculated that the earth emits a spectrum of energy with a peak wavelength of about 10,000 nm. If the heat influx to the earth increases or decreases, the outflux will adjust accordingly, and the average temperature of the earth will increase or decrease. It has been estimated that human consumption of energy in the form of fossil or nuclear fuels will equal the solar heat flux within another 14 doubling periods, which at the current rate of increase corresponds to about 320 years. Using Equation [2-19], it is easy to calculate that the average temperature would then increase by 19% or 46°C. Long before that, severe effects, including the melting of the polar ice caps, would be felt.

Human activity could also upset the heat balance by inadvertently changing the albedo. For example, the clearing of forest for agricultural land use increases the albedo from about 10% to 15%. If erosion sets in, leading to dust bowl and eventually desert conditions, the albedo is further increased to 30% (refer to Table 2-2).

Table 2-4. Estimated World Energy Resources $Q = 10^{15}$ Btu

Resource	Recoverable Quantity	Energy in Q
Oil	2500 billion barrels	15
Natural gas	12000 trillion cubic feet	12
Oil shale	2000 billion barrels	12
Tar sand	300 billion barrels	2
Coal	7600 billion tons	190 - 231
Uranium	1.3 million tons	100
Lithium	0.67 million tons	230
Deuterium	50×10^6 million tons	12×10^6
Photosynthesis on earth		1 Q/year
Solar energy on earth		4000 Q/year
Total solar output		13×10^{13} Q/year

2.4 Solar Technology

Solar energy is an immense resource. Enough sunlight falls yearly on each square meter of earth to equal the energy content of 420 pounds of high-grade bituminous coal. But it is also characterized by low energy density and intermittency. The most convenient and straightforward application of solar energy is the heating of buildings and water. These so-called passive energy systems often involve architectural designs that enhance the absorption of solar energy. Active energy systems require mechanical power such as pumps to circulate air, water, or other fluids from solar collectors to heat sinks (for storage). In addition to these applications, there are two more difficult methods for generating electricity from sunlight. One is **solar thermal electric conversion (STEC)**, which indirectly transforms solar energy into electrical energy; the other, **photovoltaic energy conversion**, is a method of direct transformation.

2.4.1 Solar Thermal Electric Conversion

Solar thermal electric conversion (STEC) uses collected heat to run a boiler in a steam generator, as shown in Figure 2-9. To achieve sufficiently high temperatures for efficient operation, the sunlight must be focused from many collection units, as shown in Figure 2-10.

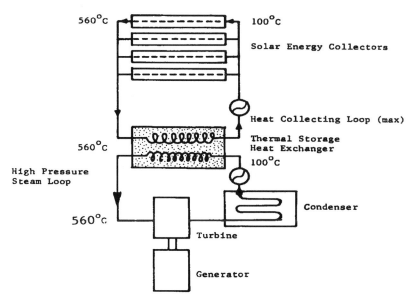

Figure 2-9.   Basic solar energy system using thermal conversion.

Approximately 90% of the solar spectrum is at wavelengths shorter than 1.3 μm, and the spectrum of the escaping infrared radiation, even at 900K, overlaps it very little, as shown in Figure 2-11. Thus, the selective surface for the collectors must be black for wavelengths shorter than 1.3 μm and mirror-like for longer wavelengths. Different materials have different absorption spectra. Figure 2-12 shows that for various gases, oxygen and ozone have strong absorptions of radiation at wavelengths shorter than 0.3 μm. Figure 2-13 shows the calculated optical performance of a thin-film stack for an absorber panel. The panel reflects most of the radiation for wavelengths greater than 1 μm, the reflectance value is close to unity, and the panel absorbs most of the radiation in shorter wave lengths. So far, silicon is the best material for the absorption, as shown in Figure 2-14.

The following is an example of how to find the temperature of a solar panel:

$$\text{Absorption by panel} = (1-A)S\ (A_p) \qquad\qquad [2\text{-}20]$$

$$\text{Radiation by panel} = \sigma T_p^{\,4}\varepsilon\ (A_p) \qquad\qquad [2\text{-}21]$$

where A_p = area of panel and T_p = panel temperature.
At equilibrium

$$\left(1-A\right)S = aS = \sigma\ 4T_p^{\,4}\varepsilon \qquad\qquad [2\text{-}22]$$

$$T_p = \left(\frac{aS}{4\sigma\varepsilon}\right)^{\frac{1}{4}} \qquad\qquad [2\text{-}23]$$

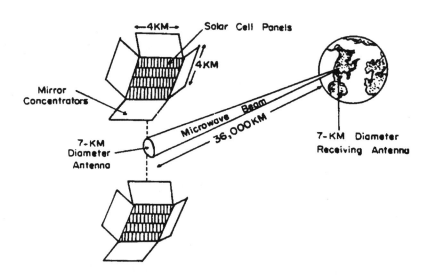

Figure 2-10. A schematic drawing showing the approximate solar cell panel and antenna dimensions required, with present technology, to produce 10,000 MW of electrical power on earth using solar cells in synchronous earth orbit.

Therefore, the properties of a collector must be

a) transparent to visible and UV
b) opaque to IR
c) high a and low ε values

Assume a = 80%, ε = 50%, S = 1000 W/m², then T_p = 125°C. This temperature is too low. At 30°C, which is ambient, $\eta = (T_H - T_L)/T_H = 25.9\%$. There are two ways, using new film and a focusing mirror, that the panel temperature can be increased. In the case of a **new film**, for example, Al coated with CdTe, a = 90%, ε = 3.9%, and the calculated T_p = 526°C. For a **focusing mirror** of 1000:1, a = 90%, ε = 1 %, and the calculated value is T_p = 6000°C. For a solar farm, the temperature should reach 560°C.

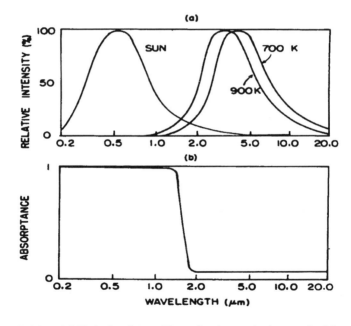

Figure 2-11. (a) Relative intensities of solar emission and of thermal re-radiation for two absorber temperatures. Approximately 90% of the solar spectrum is at wavelengths shorter than 1.3μm, and the spectrum of the escaping infrared radiation, even at 900k, overlaps it very little. Thus, the selective surface must be black for wavelengths shorter than 1.3μm and mirror like for longer wavelengths. (b) Spectral characteristic of an ideal selective absorber.

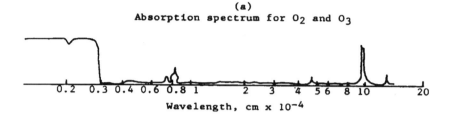

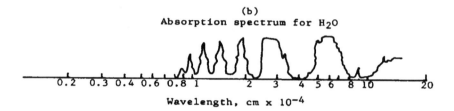

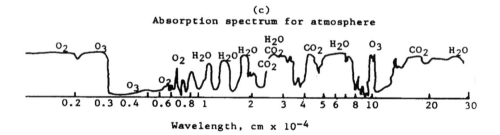

Figure 2-12.   Absorption of radiation at various wavelengths by (a) O_2 and O_3; (b) H_2O; and (c) the principal absorbing gases.

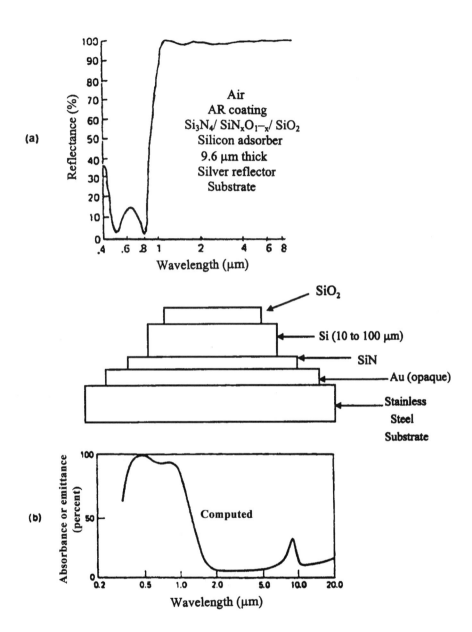

Figure 2-13. (a) Calculated optical performance of that film stack for absorber panel. (b) A possible frequency-selective solar absorption stack and its computed absorption curve. (From A. B. Meinel and M. P. Meinel, "Physics Looks at Solar Energy," Physics Today, Feb1972)

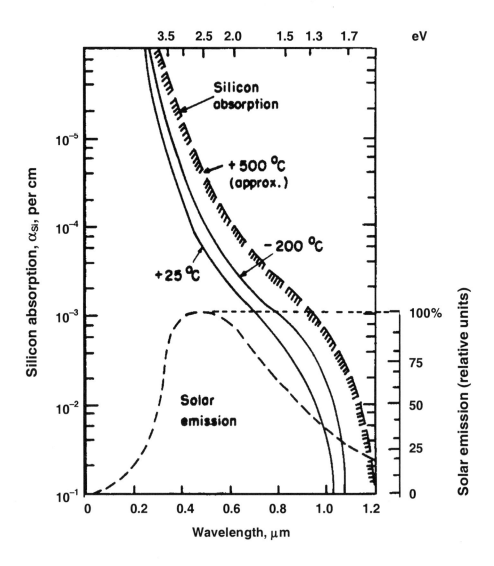

Figure 2-14.   Overlay of silicon absorption and solar emission spectra, indicating optimum thickness of 10μm for the silicon layer.

2.4.2   Photovoltaic Devices

Photovoltaic devices convert light to electricity by means of a nonthermal process, using solar cells comprised of semiconductors such as silicon. The process works at ambient temperatures and produces no by-products other than heat from sunlight, which is not converted to electricity. Figure 2-14 indicates an overlay of silicon absorption and solar emission spectra. Silicon solar cells, developed for the space program, are the most successful current design. This design consists of a sandwich of p-type (positive) silicon semiconductors that conduct positive charges and n-type (negative) silicon semiconductors that conduct negative charges. A charge separation is developed across the junction between them, as shown in Figure 2-15.

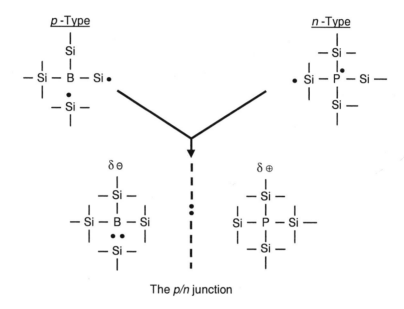

The *p/n* junction

Figure 2-15. Origin of barrier field.

There are net positive and negative charges, and the field created is similar to a chemical complexation, as shown in Figure 2-16. If crystalline silicon is doped with boron atoms, a bonded negative charge results (B⁻) as well as free positive holes; on the other hand, if crystalline silicon is doped with arsenic atoms, a bonded positive charge results (As⁺) as well as free negative electrons. When a p-type semiconductor is joined with an n-type (p-n junction), the free charges move away from the interface because of repulsion by the fixed charges. This movement results in a separation of charge and creates a potential across the interface. If a photoelectron strikes the p-type semiconductor, it will cause the charge to accelerate across the interface before it combines with a hole. The silicon cell produces electricity reliably, but is quite expensive because high-grade crystalline silicon is required.

Solar energy is a time-dependent energy resource. A means of storing energy is needed to provide an even and continuous supply. There are various methods of storing energy, such as the storage of compressed air in caverns, mechanical energy storage in flywheels, and direct electrical storage in large superconducting magnets. Materials that undergo a change of phase in a suitable temperature range may also be useful for energy storage. Table 2-5 lists some common materials used for heat storage. Batteries are also good devices for energy storage, representing chemical storage of electrical energy in a portable form. Table 2-6 lists characteristics for various rechargeable batteries. Although the chemical battery is very efficient in converting electrical energy to chemical energy and back, it is a heavy and expensive storage medium. The lithium-polymer battery is especially suitable for the future because it can easily be packaged in a "credit card" form, shown in Figure 2-17. Table 2-7 lists some common energy densities in storage. Figure 2-18 illustrates the specific power and energy of various sources.

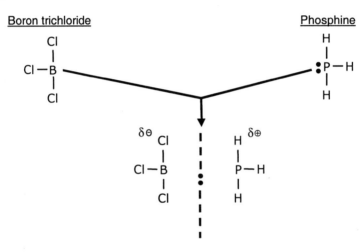

Figure 2-16.   Field creation similar to chemical complexation.

Table 2-5.   Heat Storage in Phase Change

Compound	Transition Temperature (°K)	Heat (J/g)
$CaCl_2 \cdot 6H_2O$	302 – 312	174
$Na_2CO_3 \cdot 10H_2O$	305 – 309	267
$Na_2HPO_4 \cdot 12H_2O$	309	265
$Ca(NO_3)_2 \cdot 4H_2O$	313 – 315	209
$Na_2SO_4 \cdot 10H_2O$	305	241
$Na_2S_2O_2 \cdot 5H_2O$	322 – 324	209

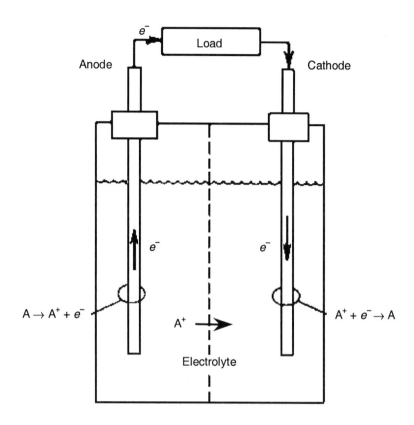

Figure 2-17. General diagram of an electrochemical cell.

There are several points regarding pollution by solar energy that are worth mentioning, because utilizing solar energy is definitely not pollution-free. High temperatures near the solar panel may cause unfavorable air turbulence, resulting in the death of flying birds. So-called "optical pollution" of the solar farm is also felt by some people. In addition, toxic materials are generated in manufacturing the photovoltaic systems.

Table 2-6. Summary of Current Battery Characteristics

	Specific Power (watts per kg)	Energy Density (watt-hours per L)	Specific Energy (watt-hours per kg)	Life (years)	Cycle Life[a] (80%DOD)	Ultimate Cost ($ per kWh)
BATTERY GOALS[b]						
Mid-term goal	150	135	80	5	600	<150
Long-term goal	400	300	200	10	1,000	<100
CURRENT BATTERY STATUS						
Lead-acid	67 – 138[c]	50 – 82	18 – 56	2 – 3	450 – 1,000	70 – 100
Nickel-iron	70 – 132	60 – 115	39 – 70	na	440 – 2,000	160 – 300
Nickel-cadmium	100 – 200	60 – 115	33 – 70	na	1,500 – 2,000	300
Nickel-metal hydride	200	152 – 215	54 – 80	10	1,000	200
Sodium-sulfur	90 – 130	76 – 120	80 – 140	na	250 – 600	100+
Sodium-nickel chloride	150	160	100	5	600	>350
Lithium-polymer	100	100 – 120	150	na	300	50 – 500

a cycle life = number of discharges in battery lifetime at 80% depth of discharge.

b USABC figures.

c CARB estimates 170 watts per kg.

na = not available.

Source: "The Keys to the Car: Electric and Hydrogen Vehicles for the 21st Century," World Resources Institute.

Table 2-7. Energy Density in Storage

	Chemical Energy kcal/C	Electric-Mechanical Energy watt-hr/lb (20% heat efficiency)
Gasoline	11.0	1,150
Lipid	9.3	
Methanol	5.2	550
Ammonia	4.8	510
Carbohydrate	4.1	
Protein	4.1	
Sodium-sulfur battery		385
Conceptual super flywheel		200
Lead acid battery		85
Super fly wheel		40
Rubber band		1

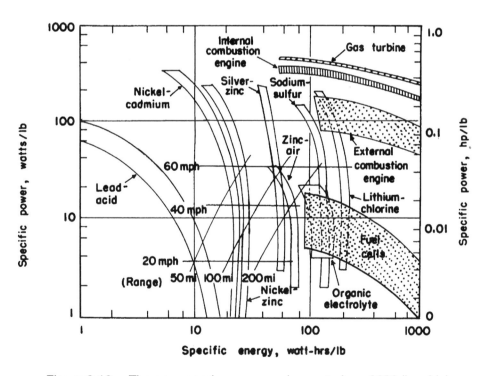

Figure 2-18. The power and energy requirements for a 2000-lb vehicle utilizing a 500-lb motive power source for steady driving. Power and energy are assumed to be at the output of the device. After Ref. (8).

2.5 MISCELLANEOUS ENERGY SOURCES

This section is used to illustrate some of the supplementary energy sources such as wind and tidal waves. Furthermore, chemical fuels such as alcohol and hydrogen will be discussed under miscellaneous energy sources.

2.5.1 Wind Power

Wind power has a yearly energy potential of four times the current United States consumption. Windmills have a long history of use throughout the world, and there have been many technical advances in the design of windmills and wind turbines. It seems quite likely that wind power will be an additional economical source of electricity in the near future. The energy density, E, is in J/m^3.

While the energy E' is $E' = (1/2) MU^2$ (in J), or $(1/2) \rho A X U^2$, the energy density, E, is

$$E = \frac{1}{2}\rho U^2 \quad \text{in} \quad \left(\frac{J}{m^3}\right) \qquad [2\text{-}24]$$

where $\rho = \dfrac{M}{V}$, $V = \dfrac{RT}{P}$, U is velocity of wind, ρ is density and V is volume

The power density, P, is (power can be expressed as $P_w = dE'/dt = (1/2) \rho A U^3$ in (W))

$$P = EU = \frac{1}{2}\rho U^3 \quad \text{in (W/ m}^2) \qquad [2\text{-}25]$$

Assuming velocities (U) and pressures (p) according to Figure 2-18 and Figure 2-19

U, p at entrance
U_t, p_t at turbine
U_c, p_c at exit

The axial force is

$$F = A\, p_t = \frac{1}{2}\rho A (U^2 - U_c^2) \text{ in Newtons} \qquad [2\text{-}26]$$

which is derived from Bernoulli's equation.

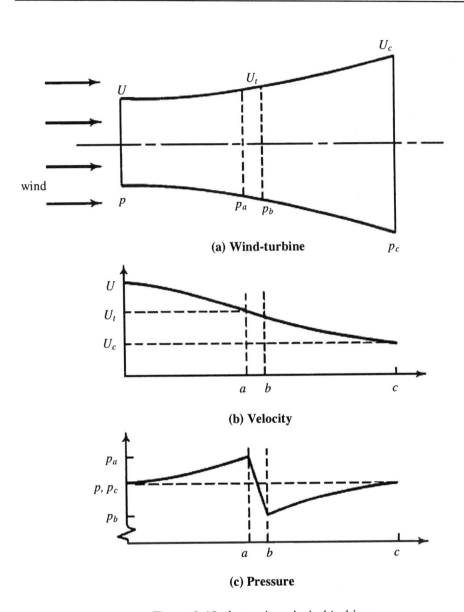

(a) Wind-turbine

(b) Velocity

(c) Pressure

Figure 2-19. An enclosed wind-turbine.

Also, the flow rate,

$$Q = \rho A U_t \quad \text{in (kg/s)} \tag{2-27}$$

$$U_t = \frac{1}{2}(U + U_c) \tag{2-28}$$

$$F = QU - QU_c = \rho A U_t (U - U_c) \quad \text{in (kg - }\frac{m}{s^2}) \tag{2-29}$$

Thus,

$$\rho A U_t (U - U_c) = \frac{1}{2} \rho A (U^2 - U_c^2) = \frac{1}{2} \rho A (U - U_c)(U + U_c) \qquad \text{[2-30]}$$

Also,

$$P = \frac{1}{2} Q U^2 - \frac{1}{2} Q U_c^2 = \frac{1}{2} \rho A U_t (U^2 - U_c^2) = \frac{1}{4} \rho A (U + U_c)(U^2 - U_c^2) \quad \text{[2-31]}$$

$$\frac{dP}{dU_c} = (-3U_c + U)(U_c + U) = 0 \qquad \text{[2-32]}$$

or

$$U_c = \frac{U}{3} \qquad \text{[2-32a]}$$

For maximum power density,

$$P_{max} = \frac{8}{27} \rho A U^3 \qquad \text{[2-33]}$$

where P is in Watts.

Because entrance wind flux is $1/2 \, \rho A U^3$, P_{max} = 16/27 = 59.3% of wind speed.

Furthermore, P is power density in W/m², and

$$P(\text{W/m}^2)\, A(\text{m}^2) = \qquad E(\text{J/m}^3) \qquad U(\text{m/s}) \qquad A(\text{m}^2) \qquad \text{[2-34]}$$

$$\downarrow \qquad\qquad \downarrow \qquad\qquad \downarrow$$

physical   wind   sail

\   /

Economic

where E is the physical limit, and U and A are economic limits. Thus, for a windmill with four sails of total area A = 80 m², at a wind velocity of 10m/sec (and thus a kinetic energy density, E, of 50 J/m³), the maximum wattage is P_w = EUA is 50×10×80 or 40 kilowatts. In this manner, the wind energy is controlled by physical and economic constraints. The maximum power that can be obtained by wind power as a function of wind speed is shown in Figure 2-20. At present, wind power can only be used as a backup energy source.

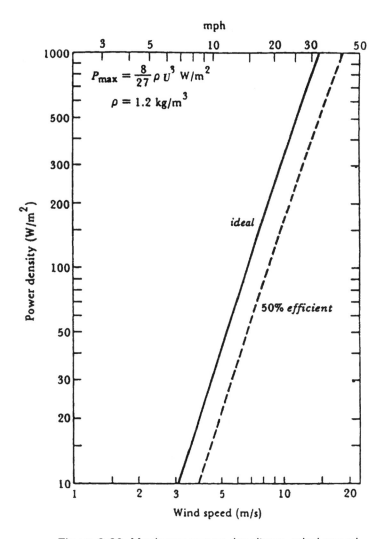

Figure 2-20. Maximum power density vs. wind speed.

2.5.2 Tidal Power

The source of **tidal** energy is the gravitational energy of the earth-moon-sun system. Tidal power projects utilize the differences in height between low and high tides. The efficiency of a tidal power station can be quite high; at La Rance in France, 25% of the theoretically obtainable power is generated as electricity. In the United States, the plant at Passamaquoddy Bay near the Canadian – U.S. border is also successful. Still, the total available tidal energy is only a small fraction of the world's energy needs. Figure 2-21 shows a schematic diagram of a tidal power dam and turbine.

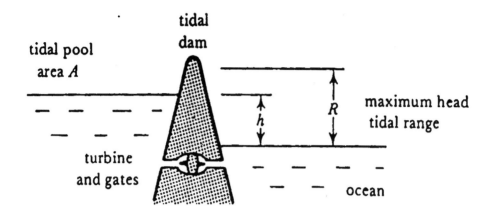

Figure 2-21. A tidal power dam and turbine.

$$dE = g\rho h dV \qquad\qquad [2\text{-}35]$$

where h= head and R= tidal range.

$$E = \int_{R}^{0} dE = - g\rho A \int_{R}^{0} h dh = \frac{1}{2} g\rho \; AR^{2}$$

$$\therefore E_{period} = g\rho \; AR^{2} \qquad\qquad [2\text{-}36]$$

$$dV = - A dh \qquad\qquad [2\text{-}37]$$

Assuming $R = 5$m, then $E = .245$ TJ/km^2, and dividing by the tidal period $= 12$ h 25 min, then $P = 5.48$ MW/km^2. An expression for average power can be obtained, $P_{Ave} = 0.219 \; R^2$ MW/km^2, which is determined by the tidal range.

Tidal energy does affect coastal waters by markedly changing the habitats of fish.

2.5.3   Geothermal Power

Geothermal energy is another source being discussed nowadays. It has been successfully used on a small scale in a number of volcanic areas around the world. Figure 2-22 gives schematic diagrams of open and closed system geothermal power-plant cycles. Geothermal energy has substantial advantages for high-power energy production; that is, geothermal energy, unlike solar energy, can generate power continuously. In utilizing natural systems, pollution problems that occur are the disposal of the condensed steam and the impurities extracted from it. Also, gaseous emissions, particularly hydrogen sulfide, often accompanied the production. The modern approach to geothermal power is based on the fact that between 10 and 15 km under the earth's crust, the rock temperature is several hundred degrees Celsius. This temperature is high enough to produce steam and generate power efficiently. However, there are problems associated with such a heat exchange rate, which need to be solved before the rate is commercially viable. A severe environmental problem is linked to the elimination of boron, fluorine, and some toxic metals in the waste sludge.

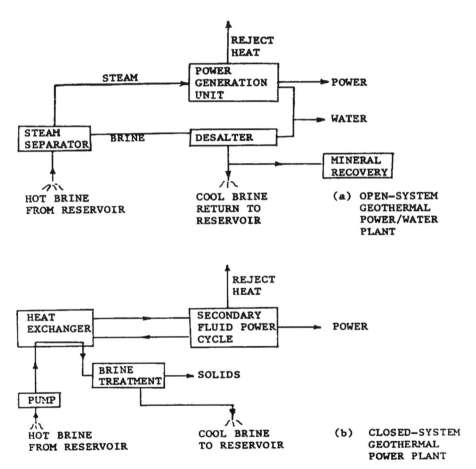

Figure 2-22. Open- and closed-system geothermal power-plant cycles.

2.5.4   Hydrogen Fuel – "Eco-energy"

Hydrogen is portable enough that it can be transported through the existing natural gas system, and certainly it is clean enough that it can produce water upon combustion. For catalytic burners, it can be flameless even for central heating to have adjustable accomplished humidity of the circulated air. As a fuel, it can be suitable for aircraft with M > 1. It can even transform edible food such as *Hydrogenomonas* yeast. Hydrogen can also be easily stored as metal hydrides or intermetallic compounds. Hydrogen can be produced by electrolysis or direct photolysis, but the best way is by successive thermal decompositions through mineral cycles as long as there is **high quality waste heat** (about 730°C) available. One such cycle is illustrated here, with the schematics shown in Figure 2-23.

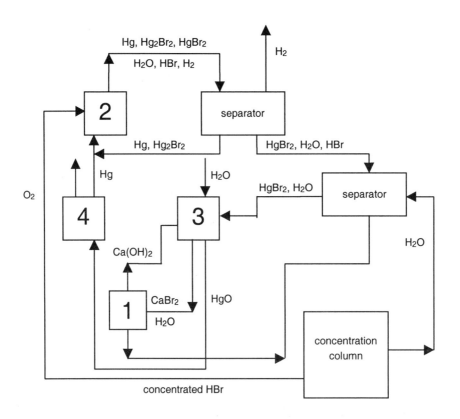

Figure 2-23. DeBeni-Marchetti's Mark 1-cycle schematics. The numerals indicate the various reactions occurring. They are summarized in the text.

This scheme is termed the **De Beni-Marchetti's Mark-1 Cycle**, and proceeds as follows:

1. Water splitting, $CaBr_2 + 2H_2O \rightarrow Ca(OH)_2 + 2HBr$ at 730°C
2. Hydrogen switch, $2HBr_2 + Hg \rightarrow HgBr_2 + H_2$ at 250°C
3. Oxygen shift, $HgBr_2 + Ca(OH)_2 \rightarrow CaBr_2 + H_2O + HgO$ at 100°C
4. Oxygen switch, $HgO \rightarrow Hg + \frac{1}{2}O_2$ at 600°C

The net balance: $H_2O \rightarrow H_2 + \frac{1}{2}O_2$

All the chemicals are in a closed system. The only net result is water splitting to produce oxygen and hydrogen. The waste heat for nuclear power is well suited; for example to the **chemonuclear reactor** concept. In this manner the petrochemical stocks such as CO and H_2 can be produced from the undesirable CO_2 gas, as shown in Figure 2-24.

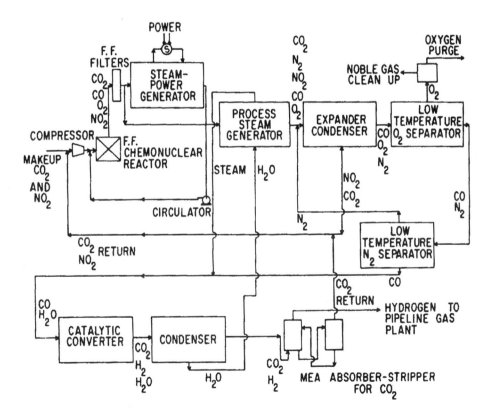

Figure 2-24. Fission-fragment chemonuclear reactor for production of CO and H_2 from CO_2 and for power production.

2.5.5 Fuel Cell

For energy conversion fuel cells are gradually becoming more important, and will be a major device in the future. It can be used as a stationary electric power plant or motor power in vehicles, on-board electric power for marine vessels, and space ships. Fuel cells operate, in principle, like a battery. However, fuel cells do not run down or require recharging. The fuel cell is able to produce electrical energy and heat as long as fuel is supplied. A fuel cell consists of two electrodes that sandwich an electrolyte. Oxygen passes over one electrode and hydrogen passes over the other. This generates electricity, water, and heat, as indicated by Figure 2-25. The electrolyte shown in the figure not only transports dissolved reactants to the electrode, but also conducts ionic charges between the electrodes and thereby completes the cell electric circuit. It also provides a physical barrier to prevent the fuel cell and oxidant gas streams from directly mixing.

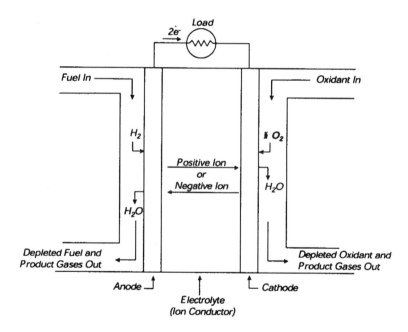

Figure 2-25. Schematic of an Individual Fuel Cell. (After Fuel Cell Handbook, 1998)

There are a variety of fuel cells, and each are in different stages of development. They can be classified using many categories, depending on the combination of fuel and oxidant, whether the fuel is processed outside or inside the fuel cell (known as external reforming or internal reforming), the type of electrolyte, the temperature of operation, and whether the reactants are fed to the cell by internal of external manifolds. The most common classification of fuel cells is by the type of electrolyte used in the cells. This classification includes the following:

- **Polymer electrolyte fuel cell (PEFC):** An ion exchange membrane, fluorinated sulfonic acid polymer or similar polymers, is used as the electrolyte in the PEFC because of its excellent proton conducting. There is little corrosion concern in the fuel cell, since the only liquid in the cell is water. The concern is that the byproduct water does not evaporate faster than it is produced because the membrane must be hydrated. PEFC operating temperature then is below 120°C because of restrictions of the polymer membrane, and an H_2-rich gas with minimal or no CO is used in order help with water balance.

- **Alkaline fuel cell (AFC):** The electrolyte used in this fuel cell, KOH, can be used to operate at a higher or lower temperature. If the concentration is 85 wt% KOH, the fuel cell can be operated at ~250°C, and if the concentration is less, 35-50 wt%, the fuel cell is operated at <120°C. The electrolyte is retained in a matrix, usually asbestos, and a wide range of electrocatalysts can be used, including Ni, Ag, metal oxides, spinels, and noble metals. The fuel supply is restricted to non-reactive constituents, except for hydrogen. CO is a poison, and CO_2 will react with KOH to form K_2CO_3, altering the electrolyte.

- **Phosphoric acid fuel cell (PAFC):** 100% concentrated phosphoric acid is used as the electrolyte in PAFC, which can operate from 150 - 220°C. At lower operating temperatures, phosphoric acid is poor at conducting ions, and CO poisoning of the Pt electrocatalyst in the anode becomes severe. Concentrated phosphoric acid is relatively stable, compared to other common acids, and thus PAFC can operate at the high end of the acid temperature range (100 to 220°C). The use of acid at 100% concentration minimizes water vapor pressure, so cell water management is relatively simple. The universal matrix to contain the acid is silicon carbide, and electrocatalyst in the anode and cathode is Pt.

- **Molten carbonate fuel cell (MCFC):** A combination of alkali carbonates is usually used as the electrolyte in this fuel cell. The electrolyte is retained in a ceramic matrix of $LiAlO_2$. At the operating temperature of 600 - 700°C, the alkali carbonates form a highly conductive molten salt, and carbonate ions provide ionic conduction. Due to the high operating temperature, Ni in the anode and nickel oxide in the cathode are adequate to promote reaction.

- **Solid oxide fuel cell (SOFC):** A solid, non-porous metal oxide, usually Y_2O_3-stabilized ZrO_2, is used as the electrolyte. The cell operates at 650 - 1000°C, where ionic conduction by oxygen ions takes place. The anode is usually $Co\text{-}ZrO_2$ or $Ni\text{-}ZrO_2$, and the cathode is Sr-doped LaMnO3.

For the above listed fuel cell types, the last two, MCFC and SOFC, are operated at higher temperatures. This has the advantage because hydrogen is usually used for the fuel input of the cell reaction. If we supply with fossil fuel or natural gas, certainly high temperature is required to reform the original fuel mixture into the hydrogen-rich component so that the fuel cell can be operable. Both internal and external reforming processes favor high temperatures, so this would be the preferred method for the fuel cell.

There is one type of fuel cell called the **direct methanol fuel cell (DMFC)**, which is similar to the PEFC. This type of fuel cell behaves the same as PEFC, with the anode side drawing oxygen from the liquid methanol, eliminating the use of fuel reformer. Although the efficiency is lower, it can operate at 50 – 100°C, making this type of fuel cell useful for household applications such as laptops, cell phones, and other domestic purposes.

REFERENCES

2-1.   A. P. Kapitza, "Physics and the Energy Problem," *New Scientists*, 7, 10 October (1976).

2-2.   R. Wilson and W. Jones, *Energy, Ecology, and the Environment*, Academic Press, New York, 1974.

2-3.   T. L. Brown, *Energy and the Environment*, Charles E. Merrill, Columbus, Ohio, 1971.

2-4.   J. A. Duffie and W. A. Beckman, *Solar Energy Thermal Process*, Wiley, New York, 1974.

2-5.   M. Steinberg, "A Review of Nuclear Sources of Non-Fossil Chemical Fuels," *Energy Sources 1*, 17-29 (1973).

2-6.   J. Edmonds and J. M. Reilly, *Global Energy—Assessing the Future*, Oxford University Press, New York, 1985.

2-7   C. Starr, "Energy and Power," *Scientific American 225*, 37-49 (1971).

2-8   G. L. Johnson, *Wind Energy Systems*, Prentice-Hall, Englewood Cliffs, New Jersey, 1985.

2-9   J. H. Krenz, *Energy Conversion and Utilization*, Allyn and Bacon, Boston, Massachusetts, 1976.

2-10.   S. S. Lee and S. Sengupta, *Waste Heat Management and Utilization*, Vol. 1-3, Hemisphere, Washington DC, 1979.

2-11.   J. T. McMullan, R. Morgan, and R. B. Murray, *Energy Resources and Supply*, Wiley, London, 1976.

2-12.   J. J. Kroushuar and R. A. Ristinen, *Energy and Problems of a Technical Society*, revised edition, Wiley, New York, 1988.

2-13.   J. R. Williams, *Solar Energy: Technology and Applications*, Ann Arbor Science, Ann Arbor, Michigan, 1974.

2-14.   S. W. Angrist, *Direct Energy Conversion*, 3rd ed., Allyn and Bacon, Boston, Michigan, 1976.

2-15.   R. C. Bailie, *Energy Conversion Engineering*, Addison and Wesley, Reading, Michigan, 1978.

2-16.   S. Sengupta and S. S. Lee, *Waste Heat: Utilization and Management*, Hemisphere, Washington D.C., 1983.

2-17.   H. P. Garg, *Advances in Solar Energy Technology*, Vol. 3, *Heating, Agricultural and Photovoltaic Application of Solar Energy*, Reidel, Dordrecht, Holland, 1987.

2-18.   H. Yiincii, E. Paykoc and Y. Yener, *Solar Energy Utilization*, Martinus Nijhoff, Dordrecht, Holland, 1987.

2-19.   J. S. Hsieh, *Solar Energy Engineering*, Prentice-Hall, Englewood Cliffs, New Jersey, 1986.

2-20.   A. deVos, *Endoreversible Thermodynamics of Solar Energy Conversion*, Oxford University Press, New York, 1992.

2-21.   J. Schmid and W. Palz, *European Wind Energy Technology*, Reidel, Drodrecht, Holland, 1986.

2-22.   R. J. Goldstick and A. Thumann, *The Waste Heat Recovery Handbook*, Fairmont Press, Atlanta, Georgia, 1983.

2-23.   J. H. Krenz, *Energy: From Opulence to Sufficiency*, Praeger, New York, 1980.

2-24.   P. Auer, *Advances in Energy Systems and Technology*, Vol. 1, Academic, New York, 1978.

2-25.   T. F. Spiro and W. M. Stigliani, *Environmental Issues in Chemical Perspective*, State University of New York Press, Albany, New York, 1980.

PROBLEM SET

1.   A device generates energy of 5600 Joules to the heat engine, and the heat rejected to the environment is 3920 Joules. What could be the energy source of the heat engine? (Use Table 2-1).

2.   If the absorptivity of a planet is 0.6 and emissivity is 0.7, what is the planet's average absolute temperature T_E ?

3.   The solar plant is used to collect heat to operate a steam generator. Calculate the panel temperature based on a = 0.45 and $\varepsilon = 0.75$.

NUCLEAR POWER

*N*uclear energy is a by-product of weapons research. The development of nuclear reactors began in earnest in 1942. In 1953, the United States launched the Atoms for Peace program with the intention of harnessing the destructive potential of nuclear power and dedicating it to supply energy to the world. The first unit of electricity from nuclear energy was experimentally generated in 1953. From commercial introduction in 1960, nuclear power expanded rapidly throughout the early 1970s. Nuclear power is presently the most highly developed alternative to energy supplied by coal. This chapter will begin with a summary of applicable nuclear phenomena, followed by a bird's-eye view on the types of nuclear power reactors that have been developed. The fuel cycles of nuclear energy will be briefly discussed. Finally, the generation of nuclear waste, its impact on the environment, and management of the waste, including separation and treatment technologies will be addressed.

3.1 RADIOACTIVITY

Nuclear energy originally resides in the nuclei of **atoms**, the smallest particles of chemical elements. **Nuclei** are made of collections of protons and neutrons called **nucleons,** which are held together by strong nuclear forces. **Protons** are positively charged, while **neutrons** are neutral. The number of protons, P, determines the number of negatively charged **electrons** that surround the nucleus, which in turn determines the chemical properties of the element. An **element** in the periodic table could be characterized by $_Z Symbol^A$, in which A is the number of nucleons in the nuclei, and Z corresponds to the **atomic number** and is equal to the number of electrons around the nucleus of a neutral atom. This symbol is commonly referred to as **nuclide**. The number of neutrons, N, is equal to the atomic weight, A, expressed as the nearest whole number less the number of protons: $N = A - Z$. **Isotopes** of the elements, such as $_{20}Ca^{40}$ and $_{20}Ca^{42}$, have the same number of protons but a different number of neutrons. **Isotones** such as $_{20}Ca^{40}$ and $_{19}K^{39}$ are elements that have same number of neutrons. Elements with the same number of nucleons, such as $_{20}Ca^{40}$ and $_{18}Ar^{40}$, are called **isobars**. **Isodiaspheres** are elements with same N - Z values, such as $_{20}Ca^{40}$ and $_{19}K^{38}$. A summary is shown here:

$_Z Element^A$

A = mass no. = no. of nucleons in nuclei

Z = atomic no. = no. of protons

$A - Z = N$ = no. of neutrons

Isotope — same Z, $_{20}Ca^{40}$, $_{20}Ca^{42}$ (on vertical lines of Fig. 3-1)

Isotone — same N, $_{20}Ca^{40}$, $_{19}K^{39}$ (on horizontal lines of Fig. 3-1)

Isobar — same A, $_{20}Ca^{40}$, $_{18}A^{40}$

Isodiasphere — same N - Z = A - 2Z, $_{20}Ca^{40}$, $_{19}K^{38}$ (on 45° line of Fig3-1)

A = 2Z proton and neutron are equal, for example, $_1H^2$, $_2He^4$, $_3Li^6$, $_5B^{10}$, etc.

Isomer — same Z, A, N; differ only in half-life

Nuclear forces increase as the number of nucleons increases, yet so does the electrostatic repulsion among the positively charged protons. To a degree, the stability of a nucleus can be empirically related to the ratio of the number of neutrons to the number of protons it contains: N/Z. The values of N/Z range from 1 to 1.59. For stable light nuclei, N/Z is approximately 1 (Figure 3-1).

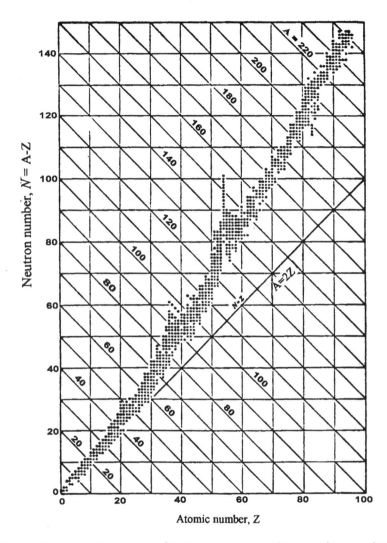

Figure 3-1. Nuclidic chart. Stable, negatron emitter, positron emitter (or electron capture), and alpha emitter are illustrated where a nuclide has more than one mode of disintegration, the predominant mode has been used. The sloping lines represent constant A values (isobars).

The highest nuclear stability is associated with atoms of intermediate atomic weight in the vicinity of the element iron. For heavier elements, the repulsive forces among protons gradually become more important, and elements heavier than bismuth, with 83 protons, are unstable. The shakedown of an unstable nucleus to a more stable form results in the natural process of **nuclear decay** and the ejection of particles or radiation from the nucleus. Some of these heavy elements can also undergo **spontaneous fission**, in which they are split into two daughter atoms of intermediate atomic weights, with the release of a great deal of energy.

3.1.1   Nuclear Binding Energy

The mass of a nucleus is less than the sum of its separate constituent particles, protons and neutrons. The difference is called **mass defect**. The mass is lost in forming the nucleus and appears as energy, known as the **binding energy**. The amount of binding energy can be related to the mass defect by Einstein's famous energy formula, $E = mc^2$, where c is the velocity of light in a vacuum. It has been calculated that 1 **a.m.u. (atomic mass unit)** is equivalent to 931.14 MeV (million electron volts). Binding energy per nucleon is greatest near the nucleus of iron (A = 56). The nuclei in this region are the most strongly bound of all, or, in nuclear terminology, the most stable. Most of the energy goes off as gamma rays, and a small amount becomes the kinetic energy of the atom's recoil.

[Example 3-1]   Compute the binding energy of $_{25}Mn^{55}$ if the mass measured by mass spectrum is 54.9558 a.m.u.

This nuclide has 30 neutrons, 25 protons, and 25 electrons.

Total mass = 30(1.00898) + 25(1.00758) + 25(0.00055)= 55.47265 a.m.u.

Mass defect = 55.47265 − 54.9558 = 0.517 a.m.u. = 0.517(931.14) = 481 MeV

Mass defect per nucleon = 481/(30 + 25) = 8.7 MeV

3.1.2   Nuclear Reactions

Early radioactive materials workers were cognizant of the presence of only one form of radiation, and its properties were similar to those of X rays. There are actually several different types of nuclear reactions. The following symbols are used: $_2\alpha^4$ (α-ray here $_2\alpha^4$ is equivalent to $_2He^4$), $_0\gamma^0$ (γ-ray), $_{-1}\beta^0$ (β-ray), $_{-1}e^0$ (electron), $_1\beta^0$ (positron), $_0n^1$ (neutron), $_1p^1$ (proton), $_1d^2$ (deutron), and $_1t^3$ (triton). In general, six types of nuclear reactions are presented here:

(1) **Alpha Emission** (α-decay)

$$_Z X^A \rightarrow {}_{Z\text{-}2} X^{A\text{-}4} + {}_2\alpha^4 \qquad \text{[3-1]}$$

For example,

$$_{88} Ra^{226} \rightarrow {}_{86} Rn^{222} + {}_2\alpha^4 \qquad \text{[3-2]}$$

Alpha radiation is not true electromagnetic radiation as are light and X rays; it consists of particles of matter. **Alpha particles** are actually doubly charged ions of helium with a mass of 4 ($_2He^4$). Although they are propelled from the nuclei of atoms at velocities of about 10% the speed of light, they do not travel much more than 10 cm in air at room temperature. They ionize half the atoms in their path. If an alpha emitter is ingested into the body (the likeliest route is via inhalation of dust particles that carry radioisotopes), it produces a high density of localized damage, with an appreciable potential for cancer induction.

(2) **Beta Emission** (β-decay)

$$_Z X^A \rightarrow {}_{Z\text{+}1} X^A + {}_{\text{-}1}\beta^0 \qquad \text{[3-3]}$$

For example,

$$_{15} P^{32} \rightarrow {}_{16} S^{32} + {}_{\text{-}1}\beta^0 \qquad \text{[3-4]}$$

(3) **Beta-Positron Emission**

$$_Z X^A \rightarrow {}_{Z\text{-}1} X^A + {}_1\beta^0 \qquad \text{[3-5]}$$

For example,

$$_{53} I^{121} \rightarrow {}_{52} Te^{121} + {}_1\beta^0 \qquad \text{[3-6]}$$

A **beta ray** is actually an energetic electron. For an element having a large N/P ratio, neutrons tend to be converted to protons by releasing $_{\text{-}1}\beta^0$. On the other hand, if the N/P ratio is small, protons have a tendency to become neutrons by releasing $_1\beta^0$. Because of the low mass, the ionizing power of beta radiation is much weaker than that of alpha radiation.

(4) **Electron Capture**

$$_Z X^A + {}_{\text{-}1}e^0 \rightarrow {}_{Z\text{-}1} X^A \qquad \text{[3-7]}$$

For example,

$$_{26} Fe^{55} + {}_{\text{-}1}e^0 \rightarrow {}_{25} Mn^{55} \qquad \text{[3-8]}$$

One element can be bombarded by high energetic electrons and transformed into another element.

(5) **Neutron Capture**

$$_{Z}X^{A} + {_{0}n^{1}} \rightarrow {_{Z}X^{A+1}}$$

$$_{28}Ni^{56} + {_{0}n^{1}} \rightarrow {_{28}Ni^{57}}$$

(6) **Proton Release** (α-bombardment)

$$_{Z}X^{A} + {_{2}\alpha^{4}} \rightarrow {_{Z+1}X^{A+3}} + {_{1}p^{1}} \tag{3-9}$$

for example

$$_{7}N^{14} + {_{2}\alpha^{4}} \rightarrow {_{8}O^{17}} + {_{1}p^{1}} \tag{3-10}$$

Protons can be released as a result of bombardment of alpha particles. The shorthand notation for the preceding reaction is $N^{14}(\alpha,p)O^{17}$ — usually the Z is omitted and sometimes the A can be written on the left side, for example N^{14} or ^{14}N.

For upper layers of the atmosphere, the following reactions are important:

$$N^{14} \, (\gamma, \beta) \, C^{14}$$

$$N^{14} \, (n, t) \, C^{12}$$

$$N^{14} \, (n, p) \, C^{14}$$

Can you compute the power necessary for the following nuclear reaction? What do you name this reaction?

$$O^{16} \, (\gamma, \alpha) \, C^{12}$$

We tentatively name it the Christ equation! The energy can change water to wine, a transmutation of O^{16} in water to C^{12} in alcohol. In addition to the preceding reaction, there are some reactions used to produce neutrons as shown in Table 3-1. The particles that are generated from the above reactions – that is, alpha particles, beta particles, neutrons, and protons – could induce nuclear reactions by bombarding other atoms. Gamma emission is a high-energy photon, often released as energy by nuclear transformations or shifts of orbital electrons.

Table 3-1. Nuclear Reactions Used to Produce Neutrons

Incident radiation	Reaction	Lowest Neutron energy, Mev
Alpha particles	^{9}Be (α,n) ^{12}C	5.71
	^{11}B (α,n) ^{14}N	
	^{7}Li (α,n) ^{10}B	
Deutrons	^{2}H (d,n) ^{3}He	2.45
	^{3}H (d,n) ^{4}He	14.05
	^{9}Be (d,n) ^{10}B	
Protons	T (p,n) ^{3}He	1.19
	^{9}Be (p,n) ^{9}B	
	^{7}Li (p,n) ^{7}Be	1.88
Photons	D (γ,n) H	0.3
	^{9}Be (γ,n) ^{8}Be	0.16
	^{238}U (γ,n) ^{237}U	6.0
	^{2}H (γ,n) ^{1}H	0.1

3.1.3 Radioactive Decay

A **radioactive element** emits radiation spontaneously. Radioactive decomposition is a true unimolecular reaction and its rate is not affected by external circumstances such as temperature, pressure, or even chemical combination. If we start with a certain amount of a radioactive element and arrange to measure its activity as a function of time, we find that the activity steadily diminishes. The quality of the radiation — that is, the energy of each alpha, beta, or gamma ray — does not change, but the quantity decreases.

The decomposition of a radioactive element is a true first-order reaction. In such a reaction the rate of decomposition is directly proportional to the amount of undecayed material. If the initial concentration is N_o at time $t = 0$, and if at some later time, t, the concentration has fallen to N, the following relationship is valid:

$$\frac{dN}{dt} = -\lambda N$$

$$N = N_o \exp(-\lambda t) \qquad [3\text{-}11]$$

where λ is the decay constant. For radioactive substances, it is customary to express decomposition rates in terms of **half-life** ($t_{1/2}$), or the time required for the amount of substance to decrease to half its initial value. Then

$$\lambda = \frac{0.693}{t_{\frac{1}{2}}} \qquad \text{[3-12]}$$

For example, the half-life of I^{131} is 8 days, and the decay constant $= (0.693)/8 = 0.087$ day^{-1}. Half-lives of the radioactive elements vary from fractions of a second to around 10^{12} years.

For successive decays,

$$N_1 \xrightarrow{\lambda_1} N_2 \xrightarrow{\lambda_2} N_3 \qquad \text{[3-13]}$$
$$t_{\frac{1}{2}}\,(1) \qquad\quad t_{\frac{1}{2}}\,(2)$$

$$\frac{dN_1}{dt} = -\lambda_1 N_1, \quad \frac{dN_2}{dt} = \lambda_1 N_1 - \lambda_2 N_2 \qquad \text{[3-14]}$$

Solving,

$$N_2 = \frac{\lambda_1 N_{01}}{\lambda_2 - \lambda_1}\left(e^{-\lambda_1 t} - e^{-\lambda_2 t}\right) \qquad \text{[3-15]}$$

If $\lambda_2 \gg \lambda_1$,

$$\lambda_2 N_2 \approx \lambda_1 N_{01} e^{-\lambda_1 t} = \lambda_1 N_1$$

[Example 3-2]   The nuclide $_{34}Se^{70}$ decays by beta emission to $_{33}As^{70}$ and again $_{33}As^{70}$ decays in the same manner to $_{32}Ge^{70}$. The decay constant for the selenium nucleide is 0.0158 min^{-1} and for the arsenic nuclide is 0.0133 min^{-1}. Assuming the original $_{34}Se^{70}$ is 10^{10} atoms, how many atoms of $_{33}As^{70}$ are formed in 10 minutes?

From the problem we can write the following:

$$_{34}Se^{70} \rightarrow {}_{33}As^{70} + {}_{1}\beta^{0}$$

$$_{33}As^{70} \rightarrow {}_{32}Ge^{70} + {}_{1}\beta^{0}$$

Writing out the kinetics, we get

$$\frac{dN_{As}}{dt} = \lambda_{Se} N_{Se} - \lambda_{As} N_{As} \qquad \text{(a)}$$

and

$$-\frac{dN_{Se}}{dt} = \lambda_{Se} N_{Se} \tag{b}$$

Integrating (b), we get

$$\int_{N_{Se}}^{N_0} \frac{dN_{Se}}{N_{SE}} = \int_{t}^{0} -\lambda_{Se}\, dt$$

or

$$N_{Se} = N_0 \exp(-\lambda_{Se} t) \tag{c}$$

Substituting (c) to (a), we get

$$\frac{dN_{As}}{dt} + \lambda_{As} N_{As} = \lambda_{Se} N_0 \exp(-\lambda_{Se} t) \tag{d}$$

In order to solve (d), we assume

$$a(x) = \lambda_{As}, \ Q(x) = \lambda_{Se} N_0 \exp(-\lambda_{Se} t)$$

Then,

$$e^{\int a(x)dx} = \exp\left(\int \lambda_{As}\, dt\right) = \exp(\lambda_{As} t)$$

Thus,

$$\int Q(x) e^{\int a(x)dx}\, dx = \int \lambda_{Se} N_0 \exp(-\lambda_{Se} t)\exp(\lambda_{As} t)\, dt$$

$$= \frac{\lambda_{Se} N_0}{\lambda_{As} - \lambda_{Se}} \exp[(\lambda_{As} - \lambda_{Se})t]$$

After integration,

$$N_{As} \exp(\lambda_{As} t) = \frac{\lambda_{Se} N_0}{\lambda_{As} - \lambda_{Se}} \exp[(\lambda_{As} - \lambda_{Se})t] + c \tag{e}$$

When $t = 0$, $N_{As} = 0$, we can evaluate the constant c,

$$c = \frac{-\lambda_{Se} N_0}{\lambda_{As} - \lambda_{Se}}$$

Substituting into (e),

$$N_{As} = \frac{\lambda_{Se} N_0}{\lambda_{As} - \lambda_{Se}} \left[\exp(-\lambda_{Se} t) - \exp(-\lambda_{As} t) \right]$$

$$= \frac{0.0158 N_0}{(0.0133 - 0.0158)} \left(e^{-0.158} - e^{-0.133} \right) = 0.136 N_0$$

Because $N_o = 10^{10}$, $N_{As} = 1.36 \times 10^9$ atoms.

3.1.4 Units of Radioactivity

The unit of radioactivity is the **curie**. It was formerly considered to be the number of disintegrations occurring per second in 1 gram of pure radium. Now it is defined to be a fixed value of 3.7×10^{10} disintegrations per second as the standard curie (Ci). The curie is a fairly large unit; therefore, smaller units such as pico-curie (pCi) are more commonly used (1 pCi = 2.2 disintegrations/min). Derivation of the basic unit is shown here:

$$\text{Ci (Curie)} - 1g \text{ of } Ra^{226} \text{ (t½} - 1590 \text{ yr)}$$

$$\lambda = \frac{0.693}{1590} \text{ yr}^{-1} = 1.38 \times 10^{-11} \text{ sec}^{-1}$$

$$-\frac{dN}{dt} = \lambda N = \lambda \frac{1}{226} \left(6.02 \times 10^{23} \right) = 3.7 \text{ x } 10^{10} \text{ disintegration/sec.}$$

One disintegration/sec (dis/sec) is called one Bq (**Becquerel**), and some common units are summarized in Table 3-2.

The **roentgen (r)** is a unit of gamma or X ray radiation intensity. It is of value in the study of the biological effects of radiation that result from ionization induced within cells by the radiation. One roentgen is defined as the amount of gamma or X ray radiation that will produce 1 electrostatic unit (esu) of electricity in 1 cubic centimeter of dry air at standard conditions.

Examples include mCi, μCi, nCi, etc.,

$$1 pCi = 0.037 \text{ dis/sec} = 2.2 \text{ dis/min} = 37 \text{ m Bq}$$

With the advent of atomic energy involving exposure to neutrons, protons, and alpha and beta particles, which all have effects on living tissues, it has become necessary to have other means of expressing ionization produced in cells. Three methods of expression have been used.

Table 3-2.   Radiation Quantities and Units

Quantity	SI Units	Equivalents
Activity (Becquerel)	Bq	1 curie = 3.7×10^{10} Bq
		1 pCi = 0.037 Bq
Concentration	Bq m^{-3}	1 pCi l^{-1} = 37 Bq l^{-1}
Equilibrium Equivalent	EEC$_{222}$	1 WL = 3,740 Bq m^{-3}
Concentration	EEC$_{220}$	1 WL = 276 Bq m^{-3}
Absorbed dose (Gray)	Gy	1 Gy = 100 rad
Dose equivalent (Sievert)	Sv	1 J kg^{-1} = 100 rem
Working Level	WL	1 WL = 1.3×10^5 MeV l^{-1}
Working Level Month	WLM	1 WLM = WL (hours/ 170)
Potential Alpha Energy Concentration (no longer in use).	PAEC	1 PAEC = 1 J m^{-3}

The **roentgen-equivalent-physical (rep)** is defined as the quantity of radiation which produces ionization in 1 gram of human tissue equivalent to that produced in air by 1 roentgen (33.8 ergs of energy).   The **roentgen-absorption-dose (rad)** is a unit of radiation corresponding to an energy absorption of 100 ergs per gram of any medium.   One **gray** (Gy) is equivalent to 100 rads. It was found that 1 roentgen is approximately equivalent to 100 ergs/g of tissue.   The term **roentgen-equivalent man (rem)** has been specially developed for man.   It corresponds to the amount of radiation that will produce an energy dissipation in the human body that is biologically equivalent to 1 roentgen of radiation of X rays or approximately 100 ergs/g.   One **sievert** (Sv) is equivalent to 100 rems.   The recommended **Maximum Permissible Dose (MPD)** for radiation workers is 5 rem/year and for nonradiation workers 0.5 rem/yr. Table 3-3 shows the dosages of some common exposures.   A summary is expressed here:

R (Roentgen) is defined as the quantity of radiation that can produce 1 stat coulomb (1/3 x 10^{-9} coulombs) of ionized charge in 1 cm^3 of dry air at standard conditions (34 eV for each collision).

$$E = (34eV)N = 34\frac{\left(\frac{1}{3} \times 10^{-9}\right)}{\left(1.6 \times 10^{-19}\right)} = 7.083 \times 10^{10} \text{ eV/cm}^3 = 1.133 \times 10^{-8} \text{ J/cm}^3$$

$$= 8.76 \times 10^{-3} \text{ J/Kg} = 87.6 \text{ ergs/g of air}$$

rep (Roentgen-equivalent-physical)

$$E = 93 \text{ ergs/cm}^3 \text{ in tissue}$$

rad (radiation dosage)

$$E = 100 \text{ ergs/g in any media}$$

rem (Roentgen - equivalent - man)

$E = (\text{rad})(\text{QF})$ (QF is quality factor for γ, β = 1, and QF for n, neutron = 3)

for example, Natural background ~ 120 mrem/yr

Table 3-3. Levels of Exposure

Exposures	Dosage (milli rem)
Chest X ray	10-2000
Watching color T.V.	0.002
Flight from New York to San Francisco	1

3.1.5   Natural Radiation

Most of the presently existing atoms on earth are stable, but some are radioactive. **Naturally occurring radioactive materials** (NORM) generally contain radionuclides found in nature. The largest terrestrial sources of radiation are ^{40}K, ^{238}U, ^{235}U, and ^{232}Th. The last three are alpha emitters, while ^{40}K is a beta emitter. ^{232}Th and ^{238}U are both quite abundant in the earth's crust. ^{235}U, which is the naturally occurring isotope that can undergo fission, has become scarce. The total energy emission from ^{40}K in the earth's crust is estimated to be 4×10^{12}W. The heat generated by the decay of these elements has resulted in a much slower cooling of the earth than would have otherwise been the case.

Unstable isotopes can also be created by the interaction of stable nuclei with neutrons or with high-energy particles. As shown in Table 3-4, many radioactive materials are generated inside the nuclear reactors by the bombardment of neutrons or protons, making it difficult to manage radioactive waste. Bombardment of the earth with radiation from outer space also converts some stable atoms to radioactive ones. A good example is the production of ^{14}C from ^{14}N by cosmic rays, which are charged particles entering the earth's atmosphere at high velocities. All living forms are kept uniformly radioactive with ^{14}C because atmospheric carbon, which contains this produced material, is the primary source of carbon for life; this is the basis of the **radiocarbon dating method**. Radioactive changes involve the transmutation of one element into another, such as the radioactive decay of U to stable Pb. Such a change cannot occur in one step, and many intermediate steps are involved. In addition, many intermediate products are generated, as shown in Figure 3-2. Natural radiation is rather high in some parts of the world, for example, 0.2–2 roentgen/yr in Morrode Ferro, Brazil and Kerala India registered 2.6 R/yr. The nuclei present are $_{19}K^{40}$, $_{37}Rb^{87}$, $_{75}Re^{187}$.

The radioactivity occuring in nature are:

$$U^{238} \xrightarrow{\alpha} Th^{234} \xrightarrow{\beta^-} Pa^{234} \xrightarrow{\beta^-} U^{234} \xrightarrow{\alpha} Th^{230} \xrightarrow{\alpha} Ra^{226} \xrightarrow{\alpha}$$

$$Rn^{222} \xrightarrow{\alpha} Po^{218} \xrightarrow{\alpha} Pb^{214} \xrightarrow{\beta^-} Pb^{214} \xrightarrow{\beta^-} Po^{214} \xrightarrow{\beta^-} Po^{214} \xrightarrow{\alpha} Pb^{210}$$

The sequence of fallout can be expressed as follows:

In bone,

$$_{38}Sr^{90} \xrightarrow[28.1y]{\beta^-} {}_{39}Y^{90} \xrightarrow[64\ hr]{\beta^-} {}_{40}Zr^{90}$$

In muscle,

$$_{55}Cs^{137} \xrightarrow[30\ y]{\beta^-} {}_{56}Ba^{137}$$

and in thyroid,

$$_{53}I^{131} \xrightarrow[8\ d]{\beta^-} {}_{54}Xe^{131}$$

Currently, there are no federal regulations (U.S.) concerning NORM which include the uranium and other radioactive elements in coal ash, oil, gas production wastes, etc.

Table 3-4. Typical Activation Products that appear in Wastes from Power Reactors

Radionuclide	Half-life	Mechanisms of formation		Activated source
Light-water-cooled reactors				
^{16}N	7 sec	^{16}O	(n,p) ^{16}N	water
^{17}N	4 sec	^{17}O	(n,p) ^{17}N	water
^{19}O	30 sec	^{18}O	(n,p) ^{18}O	water
^{3}H (tritium)[b]	12.0 yrs	^{6}Li	(n,α) ^{3}H	lithium hydroxide
^{16}F	1.8 hr	^{18}O	(p,n) ^{18}F	water
^{36}Na	15.0 hr	^{28}Na	(n,y) ^{24}Na	sodium in water
		^{27}Al	(n,α) ^{34}Na	aluminum
^{28}Al	2.0 min	^{17}Al	(n,y) ^{28}Al	aluminum
^{31}Si	2.8 hr	^{31}P	(n,p) ^{31}Si	water treatment with phosphates
^{41}Ar	1.8 hr	^{40}Ar	(n,y) ^{41}Ar	air
^{53}Mn	300 days	^{53}Fe	(n,p) ^{54}Mn	steel
^{56}Mn	2.6	^{55}Mn	(n,y) ^{56}Mn	steel
		^{50}Fe	(n,p) ^{54}Mn	steel
^{55}Fe[b]	29 yrs	^{54}Fe	(n,y) ^{55}Fe	steel
^{59}Fe[b]	45.0 days	^{58}Fe	(n,y) ^{59}Fe	steel
		^{58}Co	(n,p) ^{59}Fe	steel
^{18}Co[b]	71.0 days	^{57}Co	(n,y) ^{38}Co	steel and stellite
		^{18}Ni	(n,p) ^{58}Co	steel
^{60}Co	5.2 yrs	^{18}Co	(n,y) ^{44}Co	steel
^{44}Cu	12.8 hr	^{61}Cu	(n,y) ^{64}Cu	17-4 ph steel
^{64}Cu	4.6 min	^{65}Cu	(n,y) ^{66}Cu	and Cu-Ni alloy
^{51}Cr[b]	27.0 days	^{50}Cr	(n,y) ^{51}Cr	steel
^{96}Zr	65 days	^{94}Zr	(n,y) ^{94}Zr	zirconium fuel structure
^{192}Hf	40 days	^{190}Hf	(n,y) ^{13}Hf	halfnium control rods
^{182}Ta[b]	111.0 days	^{181}Ta	(n,y) ^{182}Ta	steel
^{187}W	24.0 hrs	^{186}W	(n,y) ^{187}W	stellite
sodium-cooled reactors				
^{41}Ar	1.8 hr	^{40}Ar	(n,y) ^{41}Ar	impurity in nitrogen
^{21}Na[b]	15.0 hr	^{23}Na	(n,y) ^{24}Na	sodium
^{22}Na[b]	2.6 yr	^{23}Na	(n,2n) ^{22}Na	sodium
^{86}Ru[b]	19.5 days	^{85}Rb	(n,y) ^{84}Rb	collant impurities
^{121}Sb[b]	60.0 days	^{127}I	(n,α) ^{124}Sb	coolant impurities

[a]Reprinted from *Radioactive Waste Handling in the Nuclear Power Industry*, Electric Institute, 750 3rd Ave., N.Y., 1960.

[b]Isotopes likely to be most prevalent and that can be important in waste-handling operations.

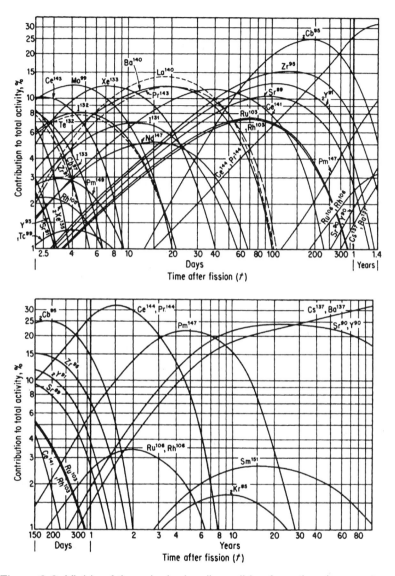

Figure 3-2. Yields of the principal radionuclides from the slow-neutron fission of ^{235}U. (Hunter and Ballou, 1951)

3.1.6 Nuclear Fission

The principle is illustrated here:

Bombardment with a neutron

$$_{92}U^{235} + {_0}n^1 \rightarrow {_{92}}U^{236} + {_0}\gamma^0 \quad \text{(Absorption)}$$

Fragmentation

$$_{92}U^{235} + {_0}n^1 \rightarrow {_{56}}Ba^{142} + {_{36}}Kr^{91} + 3{_0}n^1$$

$$\text{or } _{92}U^{235} + {_0}n^1 \rightarrow {_{57}}La^{139} + {_{40}}Zr^{95} + 5{_{-1}}\beta^0 + 2{_0}n^1$$

In general, 2.5 neutrons are gained from the two fragments or

$$_{92}U^{235} + {_0}n^1 \rightarrow {_z}X^A + {_{z'}}Y^{A'} + [(z + z') - 92]\,{_{-1}}\beta^0 + [236 - (A + A')]\,{_0}n^1 + \text{energy}$$

Also, the principle of a breeder reaction is discussed here:

$$_{92}U^{238} + {_0}n^1 \rightarrow {_{92}}U^{239} + {_0}\gamma^0 \xrightarrow{\ \beta^-\ } {_{93}}Np^{239} \xrightarrow{\ \beta^-\ } {_{94}}Pu^{239}$$

$$_{90}Th^{232} + {_0}n^1 \rightarrow {_{90}}Th^{233} + {_0}\gamma^0 \xrightarrow{\ \beta^-\ } {_{91}}Pa^{233} \xrightarrow{\ \beta^-\ } {_{92}}U^{233}$$

but for the liquid metal-cooled fast breeder (LMFBR), the following is valid:

$$_{11}Na^{23} + {_0}n^1 \rightarrow {_{11}}Na^{24} + {_0}\gamma^0 \xrightarrow{\ \beta^-\ } {_{12}}Mg^{24}$$

3.1.7   Nuclear Fusion

In contrast to fission, where a large atom can be split into a smaller atom resulting in the release of energy, **fusion** is when light atoms are combined so that energy can be released. The principle of nuclear fusion is outlined here:

Hydrogen burning — Sun

$$H^1 \ (p, \ \beta) \ H^2$$

$$H^2 \ (p, \ \gamma) \ He^3$$

$$He^3 \ (p, \ \beta) \ He^4$$

$$4p \rightarrow \alpha + 2\beta + \gamma$$

D-D and D-T reaction

$$H^2 \ (d, \ n) \ He^3$$

$$H^2 \ (d, \ p) \ H^3$$

$$H^3 \ (d, \ n) \ He^4$$

$$He^4 \ (d, \ p) \ He^4$$

$$6d \rightarrow 2\alpha + 2p + 2n, \text{ or } 3d \rightarrow \alpha + p + n$$

Li Breeding (from seawater)

$$H^3 \, (d, n) \, He^4$$

$$\text{either } Li^6 \, (n, \alpha) \, H^3$$

$$\text{or } Li^7 \, (n, \frac{\alpha}{n}) \, H^3$$

$$d + Li^6 = 2\alpha$$

or

$$d + Li^7 = 2\alpha + n$$

3.2 NUCLEAR REACTORS

In using nuclear energy, heat is again generated to produce steam to drive a steam turbine to generate electricity. Due to the size and complexity of "burners" for either nuclear fission or nuclear fusion, they can only be used in the largest factories or for electricity generation. Here we will focus on electricity generation. The fusion of two or more light atomic nuclei to form the nucleus of a heavier element generally produces more energy than the fission of heavy elements. Research conducted to produce a **controlled fusion reaction** for power production has proven unsuccessful up to the present, so the discussion here will be focused on nuclear fission reactors.

In general, there are two types of fission reactors. One is called a **thermal type reactor** and the other a **breeder type reactor**. Three main types of thermal reactors have been developed: the **boiling water reactor (BWR)**, the **pressurized water reactor (PWR)**, and the **high temperature gas-cooled reactor (HTGR)**, as shown in Figure 3-3. The first two are commonly called **light water reactors (LWR)**. The breeder type reactors are more advanced systems of which there are three main types: the **liquid metal fast breeder reactor (LMFBR)**, the **gas-cooled fast breeder reactor (GCFBR)**, and the **molten salt breeder reactor (MSBR)**.

PRESSURIZED WATER REACTOR (PWR)   BOILING WATER REACTOR (BWR)

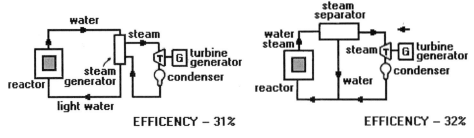

EFFICENCY – 31%   EFFICENCY – 32%

HIGH TEMPERATURE GAS REACTOR (HTGR)

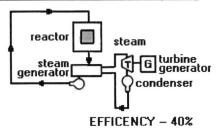

EFFICENCY – 40%

Figure 3-3.   Basic thermal reactor concepts.

3.2.1   Thermal Reactors

The major difference between the two types of light water reactors is that the PWR circulates highly pressurized water (2200 psi) through the system, whereas the BWR uses boiling water instead. The BWR concept was developed because of the desirable reduction in pressure and plant costs by eliminating the large temperature drops and the expense associated with the steam generators of the PWR. Also, BWR pressure vessels may have thinner walls. Disadvantages arise mainly from radioactivity in the cooling water and in the steam circulating through the turbine and other parts of the heat-power loop, thus potentiating large radioactive releases to the atmosphere. The efficiency of both types of reactors is pretty close: 31% for PWR and 32% for BWR.

The lower part of Figure 3-4 shows the operating principles of a pressurized light water reactor. The reactor itself consists of a number of metal tubes containing pellets of uranium or uranium oxide. The reactor is controlled by adding material that absorbs (captures) neutrons and prevents more than one neutron per fission from causing further fissions.

Such a material could be boron or cadmium, a solid that can be conveniently placed on a control rod. Control rods are lowered automatically among the fuel rods to a level that adjusts the neutron flux so that the chain reaction is maintained but does not run out of control. Surrounding the fuel and control rods is a bath of water, which acts both as a **coolant** to carry away the energy generated in the fission reaction and also as a **moderator** — that is, a substance that slows down the neutrons to increase the fission probability. Neutrons, released by the fission, travel at such high velocities that, in the absence of a moderator, they would escape before inducing further fission. The water circulates through a heat exchange that generates steam from a secondary coolant. This steam is then used to drive a turbine to generate electricity.

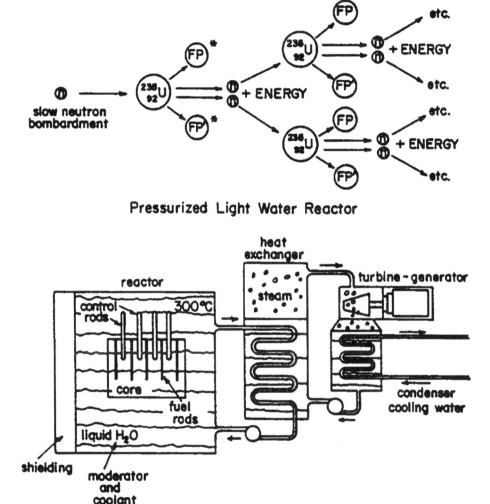

Figure 3-4. Chain reaction induced with slow (thermal) neutrons; pressurized light water reactor. FP and FP' are fission products. (following Spiro and Sriglianti, 1980)

The reactor components are summarized below:

Fuel	initial fissile material U^{235}, Pu^{239}
	fertile material Th^{232}, U^{238}
Moderator	H_2O, D_2O, C, Be
Coolant	H_2O, air (Ar, Kr), CO_2, He, Na, Bi, Na-K,
	organic (polyphenylenes)
Reflector	Be, C
Shielding	Concrete (Si, Na, Ca), steel, Pb, H_2O (P, Cl)
Control Rod	Cd, B, Hf, Zr
Structure	Al, steel (Co, Ni, Cr, W, Ta, Mn, Cu)

The HTGR, using helium instead of water to transfer heat, permits the use of graphite and other materials at elevated temperatures. These materials provide thermal and mechanical stability and allow the transfer of heat at temperatures up to 850°C. This feature affords higher heat power efficiency (40%), the option of steam or gas turbines for power production, and the use of heat for chemical processing.

3.2.2   Breeder Reactors

Although U-235 represents an extremely concentrated form of energy, there is not a great deal of it present in the world. While U-235 is the only naturally occurring fissionable isotope, isotopes of other heavy elements can also undergo induced fission. Of particular interest is plutonium-239, which can be made by neutron bombardment of uranium-238, the abundant form of uranium. As shown in Figure 3-5 it is feasible to convert inactive uranium-238 to fissionable plutonium-239, thereby converting all the uranium to nuclear fuel.

The breeder reactor is designed to extract power from the fission of either U-235 or Pu-239, but at the same time to produce plutonium from U-238 at a fast rate so that more fuel is produced than is consumed. The breeder reactor usually uses **fast neutrons** so that the surrounding material captures few neutrons, and the number of neutrons per fission is greater. The heat-transfer material in a fast-neutron reactor cannot be water, for that would slow the neutrons down. It must therefore be either a gas or another liquid. Molten sodium is usually chosen because it does not slow down the neutrons, has good heat transfer properties, and is a liquid over a wide temperature range, 98–808°C. In general the fast neutron breeder technology is called a generation III system.

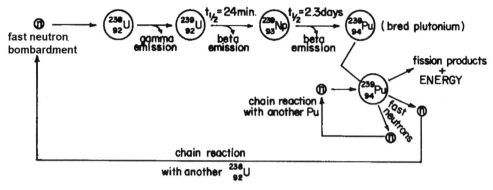

Liquid Sodium Breeder Reactor

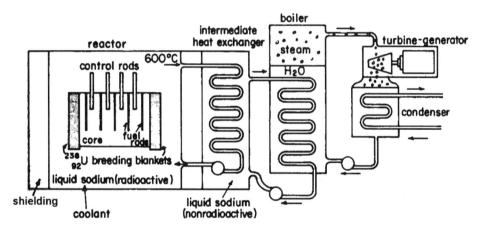

Figure 3-5. Breeder reaction with fast neutrons and liquid sodium breeder reactor. (Following Spiro and Sriglianti, 1980)

The use of breeder reactors would stretch the supply of uranium fuel by at least a factor of 50. However, the technology of breeder reactors is far more complex and is still in the process of being developed.

3.3 NUCLEAR FUEL CYCLE

The amount of energy concentrated in uranium is vastly greater than that contained in coal or oil. One gram of uranium-235 is equivalent to about 2 metric tons of high-grade coal or 12 barrels of oil. But is it a cheap energy source to use? As discussed before, natural U-235 is in dilute form and needs to be concentrated for reactors. A lot of energy has to be spent in the **separative work (SW)**, which can be represented as a separation factor, α where:

$$\alpha = \frac{\left(\dfrac{N'}{1-N'}\right)}{\left(\dfrac{N}{1-N}\right)} = \frac{R'}{R}$$

where N = concentration fraction of desired, N' = undesired, and R = abundance.

From the separation factor, the following terms are used in nuclear science and technology.

Process difference , ε, $\varepsilon = \alpha - 1$

$$\alpha \approx \left(\frac{m_2}{m_1}\right)^{\frac{1}{2}}$$

Separation power $\propto \varepsilon^2$

Separation potential, $V(N)$, $V(N) = (2N-1)\ln\left(\dfrac{N}{1-N}\right) + AN + B$

Separative work, SW, $SW = m_p\, V(N_p) + m_w\, V(N_w) - m_f\, V(N_f)$

Where the subscripts indicate p – products; w – wastes; f – feeds

A **unit of separative work (SWU)** refers to the energy required to double the U-235 content in 1 kg of uranium (output). The use of gas diffusion, centrifugation, or the recently-developed laser technology still requires a lot of energy to separate U-235. Table 3-5 lists the energy consumption in some of the nuclear centers around the world. The electricity used in Oak Ridge, Tennessee, for example, is approximately 6,000 million watts per year, the sum of the total electricity consumption of Philadelphia, San Francisco, and Denver. A fuel cycle for the enrichment of U-235 is shown in Figure 3-6.

Table 3-5. Energy Consumption in Some of the World's Nuclear Centers

Location	10^6 SWU/ yr.
Oak Ridge, TN	4.7
Portsmouth, OH	5.2
Paducah, KY	7.3
Capenhurst, (U.K.)	0.4
Pierrelatte (FRANCE)	0.2

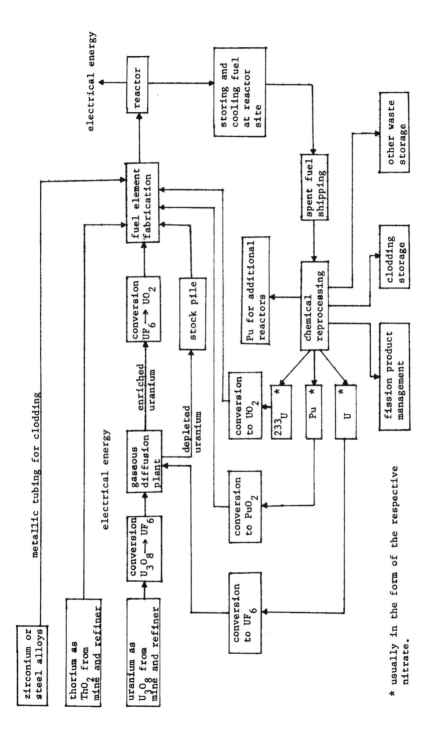

Figure 3-6. Schematic representation of the nuclear-fuel cycle. Each box represents a stage in the processing of materials from their raw state to their reuse or management as radioactive waste.

3.3.1 Nuclear Accidents

At Hiroshima and Nagasaki, A-bombs with yields of approximately 20 kT were used, the explosions spreading about half a mile in the middle of each city. An H-bomb with a yield of 1 MT can destroy everything and everyone within 7 miles. The intercontinental missiles of the Cold War were equipped with over 3-MT of thermonuclear warheads. Nuclear power is not only used for destruction, it also allows for peaceful use. Some of the examples for the peaceful use of nuclear energy are presented here.

In the past, Project Handcar (yield 10 kT) was designed for the exploration of carbonate rocks at the Nevada testing sites. Project Gas Buggy (yield 26 kT), which was supported by the El Paso Natural Gas Co. was used for stimulation of natural gas in the San Juan Basin. Project Rulison (yield 40 kT) was supported by the CER Geonuclear Corp. and explored natural gas in Garfield, Colorado. Project Branco, which was for the exploration of oil shale, never took place. There were also some other practical uses including the construction of harbors and waterways, blockcave mining, and aggregate production.

According to the Rasmussen Report, the risk of major accidents or calamities for nuclear reactors is extremely rare when compared with other accidents. Nevertheless, accidents like those in the Three Mile Island and Chernobyl incidents do occur. From WASH 740, the maximum hypothetical accident of a 500 MW (t) reactor in a population center of 1 M and a population density of 500/ miles^2 will result in 3,400 deaths, 43,000 injuries, and property damage of $7B. Modern reactor technology has been built based on the safety principles of redundancy, separation, and diversity to avoid a core meltdown due to a **loss of coolant accident** (LOCA). To prepare for this event, modern reactors are provided with the **emergency core-cooling system** (ECCS) to allow rapid shutdown of reactor transient.

The transportation of radioactive material, including wastes, is of some concern. In the past, thousands of rail shipments containing radioactive materials resulted in 26 shipping accidents. None of them caused overexposure or injury to the public. The most serious accident involved the storage of radioactive waste. For example, in the 74 million gallons of high-level waste stored at Hanford operation 10, out of 149 underground tanks developed leaks which resulted in 227,400 gal to ground. For the 17 million gallons stored at Savannah River site, four tanks developed leaks. All of these leaks will contaminate the soil and surface water, as shown in Figure 3-7. For example, the spreading of tritium contamination in concentrations above safe drinking-water standards has increased steadily from 1964 to 1993 in the Hanford site, as shown in Figure 3-8.

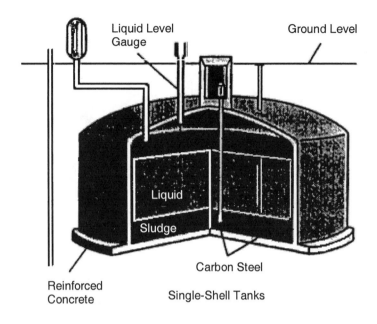

- 149 tanks constructed 1943-64
- ~210 m³ to 3,800 m³ capacity (55 kgal to 1 Mgal)
- Bottom of tanks at least 50 m (150 feet) above groundwater
- No waste added to tanks since 1980
- Tanks currently contain: ~136,800 m³ (36 Mgal) of salt cake, suldge and liquid ~555×10¹⁶ Bq (150 MCi)
- 67 are assumed to have leaked ~3,800 m³ (~1 Mgal)

Figure 3-7. A cutaway view of a typical single-shell tank in the Hanford Site.

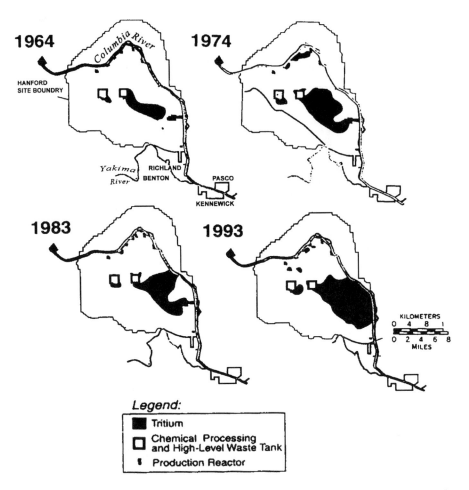

Figure 3-8.   Spreading tritium contamination at the Hanford Site in Washington. The shaded areas on these maps show how tritium contamination
in concentrations above safe drinking-water standards have spread over
time. (Source: US DOE, "Closing the Circle on the Splitting of the Atoms,
1996)

3.4 RADIOACTIVE NUCLEAR WASTE

As the utilization of nuclear power expands, more radioactive wastes are generated either from the separative work of U-235 or from fission reactions in the
reactors. Table 3-6 gives the amount of radioactive wastes as a result of the expanding use of nuclear power in the United States.

Table 3-6. Radioactive Wastes as a Function of Expanding U.S. Nuclear Power[*]

	Calendar Year		
	1970	1980	2000
Installed nuclear capacity, MW	11,000	95,000	734,000
Volume high-level liquid waste[a,b]			
Annual production, gal/yr	23,000	510,000	3,400,000
Accumulated volume, gal[c]	45,000	2,400,000	39,000,000
Accumulated fission products, megacuries[b]			
Sr^{90}	15	750	10,800
Kr^{35}	1.2	90	1,160
H^3	0.04	3	36
Total for all fission products	1,200	44,000	860,000
Accumulated fission products, tons	16	388	5,350

[*]Source: Snow, 1967

[a]Based on 100 gallons of high-level acid waste per 10,000 thermal megawatt days (MWd) irradiation.

[b]Assumes 3-yr lag between dates of power generation and waste production.

[c]Assumes wastes all accumulated as liquids.

Spent reactor fuels are usually reprocessed. This process involves cutting up the old fuel rods, chemically extracting the uranium and plutonium from the fission products, and preparing the products for radioactive waste disposal. Maintenance and safety problems are much more severe than those in the operation of a nuclear reactor, and the possibilities for the accidental release of radioactivity are much greater. Transportation of nuclear fuels between stations also has a great potential for environmental hazard.

Although the radiation released from nuclear reactors in normal operation is quite low, the reactors themselves contain a large quantity of intense radioactivity. Concerns that the system might fail in an actual crisis situation were raised considerably after the Three Mile Island accident (discussed in the previous section titled "Nuclear Accidents"). Although there are several barriers inside the reactor to keep all product activity under control, a small amount of fission product gases (krypton, xenon, and iodine) does leak through pinhole leaks to the primary coolant circuit. They are usually released to the environment through a stack, along with hot air to make them rise and disperse. Nevertheless, these gases are radioactive, as shown in Table 3-7.

Table 3-7. An Example of Annual Discharge of Radioactive Waste (Noble Gases)

Country	Facility	Discharge Limit (Ci/a)	Activity Released (Ci/ a)			
			1969	1970	1971	1972
FED. REP. GERMANY	VAK	8.8×10^4	1,750	3,340	2,455	—
	AVR	51.6	18	32	27	30
	KRB	1.9×10^6	11,400	7,350	6,780	11,105
	MZFR[a]	3×10^3	—	—	526	955
	KWL	3.1×10^6	166,000	114,000	9,000	5,300[b]
	KWO	8×10^4	5,560	7,700	1,456	3,202
	KWW	3.2×10^4	—	—	—	594
	KKS	6.1×10^4	—	—	3	2,445
	BR-3		—	26,680[c]	—	252
	Chinon	4×10^5 [d]	12,300	8,085	4,225	11,515
	SENA	2.5×10^6 [d]	—	3	4,500	31,342
	EL-4	4×10^5 [d]	46	72	53,810	144,450[e]
	St-Laurent-des-Eaux	4×10^5 [d]	1,900	305	3,425	3,863
	Bugey	4×10^5 [d]	—	—	—	841
	Latina	5×10^3	1,500	2,500	2,470	3,660
	Garigliano	6.3×10^5	140,000	275,000	640,000	290,000
	Trino	5×10^5	—	19	585	1,031
	Dodewaard	3×10^5	—	~3,000	~3,000	8,400
UNITED KINGDOM	CEGB Power Stations: gaseous activity not measured routinely [f]					

[a]In addition, MZFR discharged the following amounts of tritium into the atmosphere: in 1969 1300 Ci; in 1970 1190 Ci; in 1971 1130 Ci; in 1972 542 Ci. The discharge limit is 4000 Ci/ a.

[b]Shut-down of power station for 8 months.

[c]Exceptional discharge due to reactor operation experimentally with faulty fuel elements.

[d]At this discharge level, assuming an atmospheric dilution factor of 1.5×10^{-5} s/m3 and a probability of over 20% that the wind is blowing in one direction, the maximum concentration in air at ground level is equal to the MPC to population in air.

[e]In addition, EL-4 discharged 83 Ci of tritium into the atmosphere.

[f]In the CEGB power stations, gas activity (Ar-41) is not systematically measured. Occasional measurements have shown that at the Bradwell, Hinkley Point, and Trawsfynydd power stations the annual discharge is approximately 40,000 Ci, 200,000 Ci and 130,000 Ci respectively.

3.4.1 Nuclear Waste and Waste Types

Any activity that provides radioactive materials generates radioactive waste. The waste will include byproducts of mining, nuclear power generation, and various processes in industry. The defense industry in particular can use a gas, liquid, or solid form and can remain radioactive for a few hours, months, or over thousands of years. Currently in the United States alone, a minimum of over 45,300 sites handle radioactive material; of these sites a single complex may have as high as 1,500 contaminated sites. During the past 50 years of nuclear weapon production, vast quantities of hazardous materials and radionuclides have been released into the air, groundwater, surface water, sediments and soils, as well as vegetation and wildlife. Data and description of vast amounts of buried waste within the contaminated pits, ponds, and lagoons state that the migration of the contaminants to water are presently unknown, except for a small number of examples.

Nuclear wastes can be categorized into **high-level waste** (HLW) and spent nuclear fuel, **transuramic** (TRU) **waste**, **low-level waste** (LLW), and **mixed waste**. The HLW includes fissionable product elements, active nuclide and their daughter elements. The temperature of these wastes can reach up to 930°C and take 5 years for cooling. The half-life of the element and its daughter compound can be very long-lived (refer to Fig. 3-2). Much of HLW was generated by defense activities and stored in underground storage tanks at the Savannah River site, the Idaho National Engineering Laboratory, and the Hanford site. Spent nuclear fuel from commercial utilities has to be reprocessed to recover materials for defense purposes. The remainder is handled similarly as HLW in the sites. TRU is a waste contaminated with alpha-emitting radionuclides, with an atomic number greater than 92 (U), a half-life greater than 20 years, and in concentrations greater than 100 nCi per gram of waste.

The LLW can have a radioactivity level that is less than 10μCi and up to a 1 m Ci/L level, and can be discarded after dilution with water.

Usually, if the dilution extent reaches the **decontamination factor** (DF), then it can be considered clean (DF is the ratio of impurity concentration relative to the desired product before processing to that concentration after processing. Mixed waste results when radioactive wastes are also contaminated with hazardous wastes). Mixed waste also contains hazardous materials with HLW and LLW.

As indicated by Table 3-6, the increase in HLW is exponential. The amount of HLW currently in storage (in m^3) has been summarized by U.S. DOE in 1991 as:

Savannah River Site	132,000
Hanford Site	254,000
Idaho Chemical Processing Site	12,000
West Valley Demonstration Project	1,230

3.4.2 Biological Concentration of Radioactive Elements

Some substances will be biologically concentrated. Of those potentially emitted by power stations, iodine, strontium, radium, plutonium, and cesium are probably the most important. For example, iodine can fall on grasslands and be eaten by cows; it will then be concentrated in milk. The milk, when ingested by humans, will lead to radioactive iodine concentrated in the thyroid. Figure 3-9 illustrates the $^{129}I/^{127}I$ atom ratio in various materials showing the change in the pathway from air to food. These indicate the risk to the biosphere and especially to mankind.

Microbial processes for nuclear waste treatment can result in:

- the removal, recovery, and stabilization of radionuclides,

- the biodegradation of organic constituents to binolous products, and

- the overall reduction of the volumes of waste for disposal under proper conditions.

Microorganisms bring about dissolution or immobilization of radionuclides and toxic metals by one or more of the following mechanisms:

- redox reactions that affect solubility

- changes in pH and Rh that affect the valence or ionic states

- solubilization and leaching of certain elements by metabolites or decomposition products such as organic acid metabolites, dielates, or specific sequestering agents

- volatilization due to alkylation (biomethylation)

- immobilization leading to formation of stable minerals or bioaccumulation by microbial biomass

In a flow-through bioreactor, Pseudomonas aeroginosa immobilized within a matrix of calcium alginate can reduce uranium levels in a simulated wastewater from 10 ppm to 6.8 ppb over 350 column volumes before breakthrough occurs and the uranium can be recovered by addition of dilute HNO_3 of 5 column volumes. Biosorption is unique for removal of trace uranium in wastewater because of its high solubility in water.

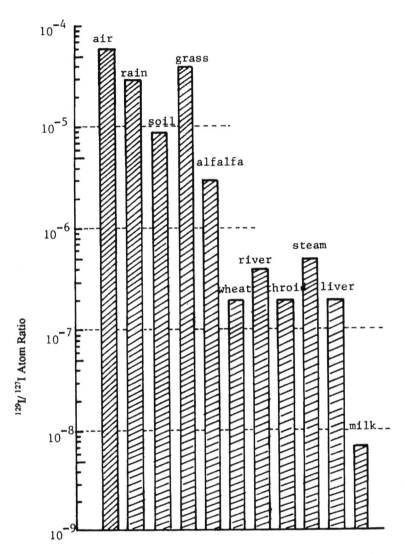

Figure 3-9. $^{129}I/^{127}I$ in various materials showing the change in the pathway from air to food.

3.5 NUCLEAR WASTE DISPOSAL AND TREATMENT TECHNOLOGY

The cost estimate for nuclear waste cleanup in the United States has been rising rapidly. It has increased from \$200 B to \$500 B, and in 1995, the estimate for the total cleanup costs could reach 1 trillion dollars and take 30 years to complete! This is a serious matter as much of the treatment technology is handicapped by the limited information available about the contamination profile. For example, U.S. DOE has been identified as particularly intractable for the following:

- groundwater contamination at almost all sites
- plutonium in soil (for example at Rocky Flats and Mound Plant)
- silos containing uranium processing residuals at Fernald
- single-shell tanks containing HLW at Hanford
- buried TRU waste at INEL

Disposal of radioactive wastes is difficult because there is no way to render them nonradioactive. They must be kept isolated from the biosphere until they decay to the background level, requiring about 10 half-lives and perhaps isolation for hundreds of years. Figure 3-10 shows the taxonomy of nuclear waste disposal options. Possible options are storage in salt vaults, geologically secure strata of the earth, and granite layers.

Recently, after significant institutional and political maneuvers, the Yucca Mountain Project in the United States is proceeding with the characterization of the volcanic tuft for development as a repository. Historically, nuclear repositories have always brought about controversy. One of the most famous cases is at the salt dome in Lyons, Kansas. Allowing 3 years for licensing and 6 years for construction, operations for Yucca Mountain are scheduled to begin by 2010. The repository would have a capacity of 70,000 metric tons of uranium (MTU), which is equivalent to nuclear waste that contains only a fraction of current stockpiles.

Much of the treatment and cleanup technology of nuclear waste remains in conceptual and research phases only. The following section gives a brief account of the efforts, except the biological ones, which were covered in Section 3.4.2.

3.5.1   Transmutation

This concept involves the elemental changes of the waste constituents into elements with shorter radioactive lives or even into nonradioactive elements through nuclear reactions by a reactor or an accelerator. Actinides burning (refer to Figure 3-10), also called waste partitioning-transmutation (P-T), is an advanced method for radioactive waste management based on the idea of destroying the most toxic components in the waste. It consists of:

- selective removal for the most toxic radionuclides from HLW

- conversion of these radionuclides into less toxic radioactive elements and/or stable elements.

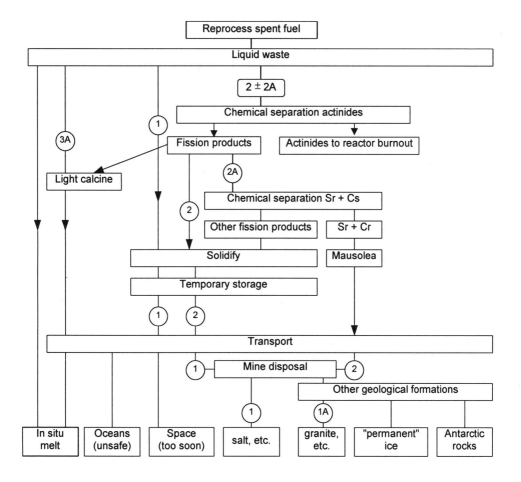

Figure 3-10.   Taxonomy of nuclear waste disposal options.

Transmutation of key fission products of long-lived Tc^{99} and I^{129} is feasible as a thermal reactor. For example, both the long-lived radioactive nuclides are converted by neutron capture and beta decay into stable products as shown here:

$$Tc^{99} + n \rightarrow Tc^{100} \rightarrow Ru^{100} + n \rightarrow Ru^{101} + n \rightarrow Ru^{102}$$

$$(165h)\beta \text{ stable} \qquad \text{stable} \qquad \text{stable}$$

$$I^{129} + n \rightarrow I^{130} \rightarrow Xe^{130} + n \rightarrow Xe^{131} + n \rightarrow Xe^{132}$$

$$(12.4\,h)\beta \quad \text{stable} \qquad \text{stable} \qquad \text{stable}$$

$$I^{127} + n \rightarrow I^{128} \rightarrow Xe^{128} + n \rightarrow Xe^{129} + n \rightarrow Xe^{130}$$

$$\text{stable} \quad (25m)\beta \; \text{stable} \qquad \text{stable} \qquad \text{stable}$$

The main product from TRU is Pu^{239}, which can fission well by neutron capture to Pu^{240} and end with high mass isotopes of Pu, Am, and Cm. With complexity and several competing routes, the anticipated goal is difficult to achieve. The extent of the depletion of TRU by transmutation is measured by the depletion ratio $\chi(t)$.

$$\chi(t) = \frac{\text{total amount of TRU supplied}}{\text{TRU in transmuter and its waste}}$$

3.5.2   Polymer Extraction

This method uses water-soluble chelating polymers containing multiple hydroxamine acid functional groups to retain certain metal ions of interest while the unbound metal ions are removed with the bulk of the aqueous solution as they permeate by membrane ultrafiltration. For actinides, selective retention of americium (III) and plutonium (III) from dilute solution high in salt content has been achieved. Chelators also have been used to bind tetravalent plutonium and thorium over trivalent ions for the TRU wastes. Chelating ligands include polyhydroxamates, bis(acylpyrazdones), and malonamides. Functionalized adsorption particles (1-15μ) with a magnetic core and polymer coating of adequate ligands have been developed for in situ groundwater treatment when a magnetic filter "wall" is installed.

3.5.3   Aqueous Extraction for Separation

Two major aqueous processes have been developed. The **PUREX** (plutonium-uranium extraction) process separates U(VI) and Pu(IV) from fission product species in nitric acid solution by solvent extraction with tributyl phosphate (TBP). Americium, curium, and neptuniun (in most cases) remain in waste stream. **TRUEX** (transuranic extraction) is a solvent extraction procedure that can efficiently separate TRU elements (for example Np, Pu, Am and Cm) from aqueous nitrate or chloride containing wastes. The key extractant is actyl (phenyl) N,N-diisobutylcarbamoyl-methylphosphine oxide (CMPO) (refer to Fig. 3-9). This is combined with TBP and a diluent such as a normal paraffinic hydrocarbon (C_{12}-C_{14} mixture) or a nonflammable chlorocarbon (tetrachloroethylene) to formulate the TRUEX solvent.

There are also other processes developed for removal of TRU that are more efficient. For example, the use of DMDBTDMA (shown in Figure 3-11), for mostly the Am(III). If the TRU constant can be lowered to below 100 nCi/g of solid, the waste can be classified as non-TRU. If ^{137}Cs and ^{90}Sr can also be reduced to an acceptable level, then decontamination is complete. Many extraction processes also employ coextractions of both actinides and lanthanides with HDEHP (refer to Figure 3-11), and, with the help of DTPA, have been successfully attempted.

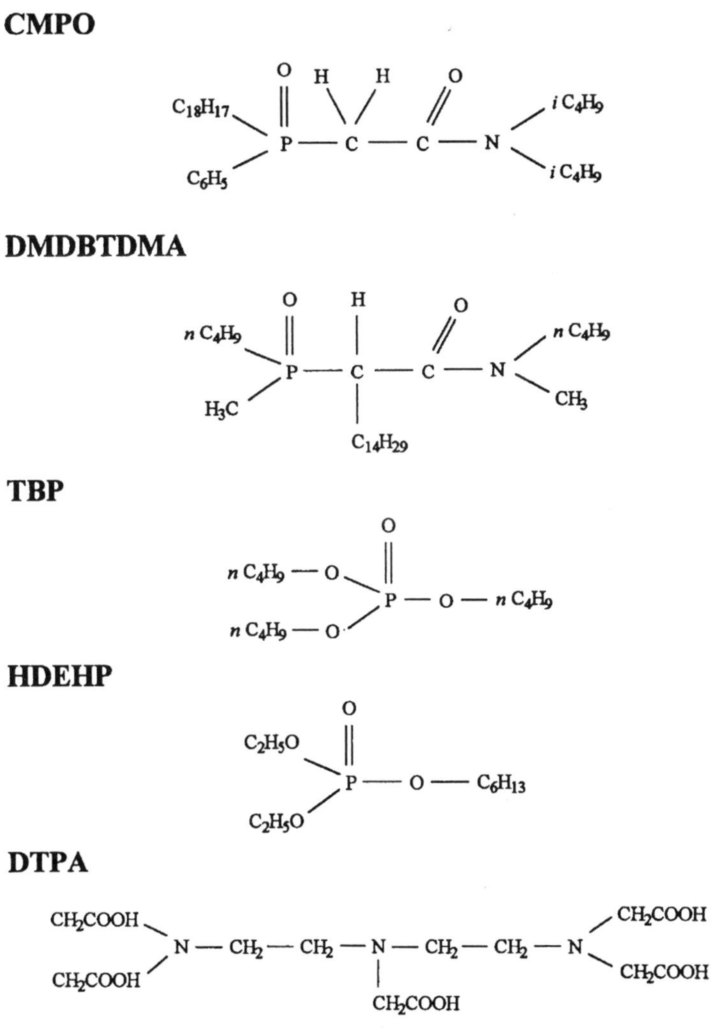

CMPO

DMDBTDMA

TBP

HDEHP

DTPA

Figure 3-11. Chemicals for solvent extraction of radioactive metals in nuclear wastes.

REFERENCES

3-1.   J. Edmondo and J. M. Reilly, *Global Energy-Assessing the Future*, Oxford University Press., New York, 1985.

3-2.   N. C. Rasmussen, *Reactor Safety Study, An Assessment of Accident Risks in U.S. Commercial Nuclear Plants*, WASH-1400, U.S. Nuclear Regulatory Commission, 1975.

3-3.   D. Bodansky, *Nuclear Energy: Principles, Practices and Prospects*, AIP, Woodbury, New York, 1996.

3-4.   R. Wilson and W. J. Jones, *Energy, Ecology, and the Environment*, Academic Press, New York, 1974.

3-5.   J. J. Duderstadt, *Nuclear Power*, Marcel Dekker, New York, 1979.

3-6.   S. S. Penner (ed.), *Energy, Vol. III, Nuclear Energy and Energy Policies*, Addison-Wesley, London, 1976.

3-7.   M. W. Firebaugh and M. T. Ohanian, *An Acceptable Future Nuclear Energy System*, Oak Ridge Associated Universities, Oak Ridge, Tennessee, 1980.

3-8.   D. Faude, M. Helm, and W. Weisz, *Fusion and Fast Breeder Reactors*, International Institute for Applied System Analysis, Luxenburg, Australia, 1977.

3-9.   J. G. Kemeny, *The Accident at Three Mile Island*, Pergamon, New York, 1979.

3-10.   F. C. Williams and D. A. Deese, *Nuclears Non-proliferation: The Spent Final Problem*, Pergamon, New York, 1979.

3-11.   R. Noyes, *Nuclear Waste Cleanup Technology and Opportunities*, Noyes, Porkridge, New Jersey, 1995.

3-12.   National Council Research, *Nuclear Wastes-Research Technologies for Separations and Transmutation*, National Academy Press, Washington DC, 1996.

3-13.   V. Jain, *Environmental Issues and Waste Management Technologies in the Ceramic and Nuclear Industries*, American Ceramic Society, Westerville, 1996.

3-14. International Atomic Energy Agency, *Peaceful Nuclear Explosions III, Applications, Characteristics and Effects*, IAEA, Vienna, 1994.

3-15. G. Friedlander and J. W. Kennedy, *Introduction to Radioactivity*, Wiley, New York, 1949.

3-16. D. Halliday, *Introductory Nuclear Physics*, Wiley New York, 1950.

3-17. Tsoulfanidis, Measurement and Detection of Radiation, Hemisphere, Washington DC, 1983.

3-18. M. Freemantle, *Nuclear Power for the Future*, Chemical Engineering News, 82(7), 31-35, 2004.

PROBLEM SET

1. Compute the mass defect per nucleon for $_{26}Fe^{56}$ if the mass measured by the mass spectrum is 55.847 a.m.u.

2. A radioactive element has a half-life of 5 hours
   a. What is the decay constant?
   b. What fraction will be left after 20 hours?

3. Calculate the energy release in MeV of the following reaction

$$_5B^{10} + _2He^4 \longrightarrow _0n^1 + _7N^{13}$$

   (Note: $_5B^{10}$ = 10.012939 a.m.u.
   $_2\alpha^4$ = 4.002604 a.m.u.
   $_0n^1$ = 1.008665 a.m.u.
   $_7N^{13}$ = 13.00569 a.m.u.)

4. Calculate the radioactivity of ^{16}N in Ci.

THE ATMOSPHERE

*T*he gaseous envelope surrounding us is the earth's atmosphere. It is the working fluid of the earth's heat engine. Most of the radiant energy arriving from the sun is converted into atmospheric heat energy before it is reradiated into space. The fluctuations of the atmospheric system cause weather and climate changes, which form an important part of the earth's history. The **ozone layer**, also part of the atmospheric system, shields the surface of the earth from harsh radiation. In addition to these varied traits, the atmosphere provides an indispensable element of life in the air we breathe. In this chapter, we will address some topics related to the structure and properties of atmosphere that are of an environmental engineering concern, such as the composition and structure of the atmosphere, the greenhouse effect, global warming, natural ice periods, the ozone layer, stratosphere chemistry, and the nature of the thermosphere (ionosphere). In the next chapter, the troposphere, air pollution, control technology, automotive emission, and meteorology will be discussed.

STRUCTURE OF THE ATMOSPHERE

The atmosphere is divided into regions based on the profile of temperature as a function of altitude. These layers are depicted in Figure B-1 (a) and B-1 (b). The lowest layer, which is the region from the surface of Earth to about 12 km, is called the **troposphere**. The troposphere is where the most action is, or at least where most of the action we can observe takes place. Thunder and lightning, rain and rainbows, hail, snow, and spectacular sunsets are all derived from processes occurring in this layer. It is heated from below by solar radiation absorbed at the earth's surface, so its temperature decreases with increasing altitude. This process is referred to as the **lapse rate** (more detail will be given in the latter part of this section). The lapse rate for the troposphere is about $-1°C/100$ meters. This rapid temperature drop leads to cold air masses overlying warmer air masses. The troposphere is unstable and has rapid convection motions. It is also a region of much turbulence due to the global energy flow that results from imbalances in heating and cooling rates between the equator and the poles. These give rise to what is commonly referred to as weather and result in a rapid vertical mixing of gases that enter the troposphere.

A positive lapse rate reflects a region of stability with respect to convection, because the warm air overlies the cool air. The change from a negative to a positive lapse rate is called a **temperature inversion**, and the point at which this change occurs marks a stable boundary between two physically distinct layers of air. The tropopause marks the boundary between the troposphere and the stratosphere. The **stratosphere** is the region from about 12 to 50 km. In this region, the ozone absorbs solar ultraviolet radiation with wavelengths from about 200 to 300 nm leading to a temperature inversion. This positive lapse rate makes the stratosphere very stable in vertical mixing. Usually, the residence times of molecules or particles in this layer are measured on a scale of years.

Because the atmosphere is in the Earth's gravitational field, its density falls with altitude. The mathematical relationship between them will be examined shortly. The troposphere contains 70% of the mass of the atmosphere. Above the stratosphere, extending from 50 to 85 km lies the **mesosphere**, an intermediate region in which the atmosphere grows continuously thinner. Due to the very low pressures, not enough ozone can be formed from molecular oxygen to maintain the temperature inversion. At the top of the mesosphere the temperature is only about $-120°F$, and the pressure is only about 1/1000 of 1 atm.

The **thermosphere**, also known as the **ionosphere**, is the region from about 85 km to the top of the earth's atmosphere. Here, the temperature rises due to the absorption of solar rays in the far ultraviolet region by atmospheric gases, principally oxygen. These far ultraviolet rays, X rays and cosmic rays are sufficiently energetic to ionize molecules and break them into their constituent atoms. An appreciable fraction of the gases in this region exist as atoms or ions.

Barometric Formula

The pressure at any height z in the atmosphere is due to the weight of the air above. Thus, the change of pressure in the vertical direction obeys the relation

$$\left(\frac{dP}{dz}\right) = -\rho_{fluid}\, g$$

[B-1]

where ρ_{fluid} is the density of air. If air is considered to be an ideal gas, the density of the air at any point in the atmosphere can be expressed by

$$\rho_{fluid} = \left(\frac{M}{RT}\right)P = \frac{M}{V}$$

[B-2]

Combining the two preceding equations from above, we obtain the general relation between pressure and temperature at any height z :

$$\left(\frac{dP}{dz}\right) = -\left(\frac{M}{RT}\right)gP$$

[B-3]

If T were constant with height, the equation could be integrated directly to yield

$$P = P_0 \exp\left(\frac{-gMz}{RT}\right)$$

[B-4]

where

M = molecular weight of air (28.97)

P_0 = pressure at ground level

g = gravitational acceleration (9.80 ms^{-2}).

The following illustrates how the atmospheric pressure decreases as the altitude increases (from Equation B-4):

Layers of the Atmosphere	
Altitude (km)	P (atm)
10	0.3
30	0.03
50	0.003

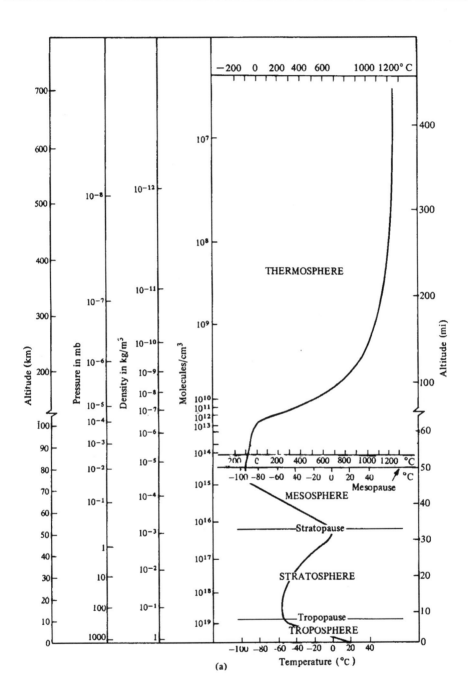

Figure B-1 (a) Vertical distribution of atmospheric properties. (After Miller and Thompson, 1975)

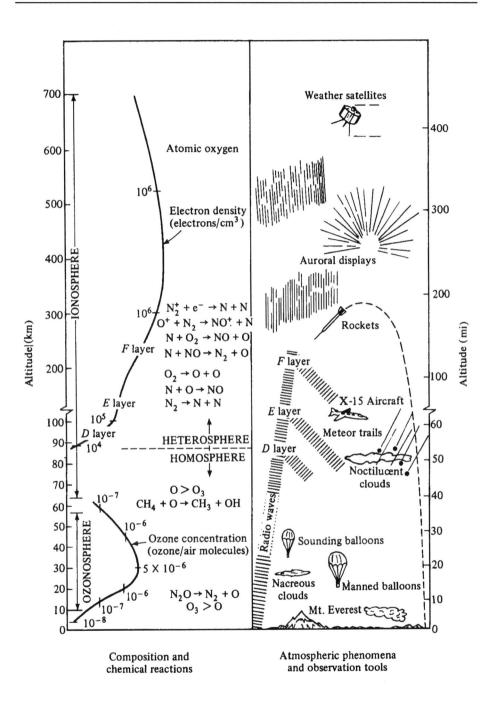

Figure B-1 (b) Vertical distribution of atmospheric phenomena.

Adiabatic Lapse Rate

From basic thermodynamics we have

$$dH = dE + PdV + VdP \qquad \text{[1-10]}$$

We also have the definition

$$C_PdT = dH \qquad \text{[1-28]}$$

the first law,

$$dE = dq - dW = dq - PdV \qquad \text{[1-7,8]}$$

or

$$C_PdT = dq - PdV + PdV + VdP$$

Thus,

$$dq = C_PdT - VdP \qquad \text{[B-5]}$$

where C_P is the specific heat at constant pressure. Assuming it is under **adiabatic conditions**, no heat goes in or out of the system ($dQ = 0$), and we reach

$$C_PdT = VdP \qquad \text{[B-6]}$$

That is,

$$\left(\frac{dT}{dP}\right) = \left(\frac{V}{C_P}\right) \qquad \text{[B-7]}$$

Because

$$\left(\frac{dT}{dz}\right) = \left(\frac{dT}{dP}\right)\left(\frac{dP}{dz}\right) \qquad \text{[B-8]}$$

and from Equation [B-3],

$$\left(-\rho_{\text{fluid}} g\right)\left(\frac{dT}{dP}\right) = -\left(\frac{Mg}{V}\right)\left(\frac{V}{C_P}\right) = -\frac{Mg}{C_P} \qquad \text{[B-9]}$$

The specific heat capacity of dry air is about 7.0 cal/mole K. By substituting C_P into the equation, the adiabatic elapse rate (dT/dz) is determined to be about 1 K per 100 meters. Calculated values for dry adiabatic lapserate (sometimes indicated by F_d) for Venus, Earth, Mars, and Jupiter are respectively 10.7, 9.8, 4.5, and 20.2 K km^{-1}.

GLOBAL ISSUES AND UPPER ATMOSPHERE

4.1 GREENHOUSE EFFECT

The three major constituents of dry air are nitrogen, oxygen, and argon, which account respectively for 78%, 21%, and 0.9% by volume. The remaining constituents of air are present in small amounts that are generally given in parts per million. The most abundant of them is carbon dioxide, forming 0.03% of the total air, while the rest together make up for less than 0.01%. The composition of dry air is remarkably constant over the globe, but the amount of water vapor in the atmosphere varies widely, ranging from 4% by volume to a few ppm. The water vapor pressure is highest at the surface and diminishes very rapidly with increasing height; for example, it is only on the order of a few ppm at the tropopause.

All of the gaseous constituents are transparent to most of the sun's rays. We see that none of them absorbs light significantly in the visible region of the spectrum. The situation is different in the infrared region where ozone, water vapor, and carbon dioxide are the principal absorbers. These gases tend to block a large fraction of the earth's emitted radiation. Therefore, it has often been remarked that the atmosphere is much like the glass wall of a greenhouse, which is transparent to the incoming short-wavelength solar radiation, but which retains (through absorption and re-radiation) the long wavelength infrared rays.

Ozone does not play a large role in the atmospheric greenhouse effect because its infrared absorptions are not very extensive, and its total presence in the atmosphere is not large. However, the absorption of the earth's rays by carbon dioxide and water is of great importance to the climate on our planet. The overall **greenhouse effect** phenomenon, shown in Figure 4-1, heats up the earth's surface so that the light absorbed by carbon dioxide and water can be emitted again. Part of the radiation returns to the surface and raises the surface temperature to be warmer than what it would have been in the absence of an atmosphere. From the earth's radiation balance, it is found that the estimated average temperature of the earth (T_e) should be at 253 K. But actually the surface temperature (T_S) is higher by 13°C (286 K) due to the greenhouse effect. Please see Section 2.3 for the Stefan Boltzman law.

$$S(1-A)\pi R^2 = 4\sigma \pi R^2 T_e^4 \qquad [4\text{-}1]$$

which is

(Solar radiation received by a body) = (Irradiation power by a body)

where

$$T_e = \left[\frac{S(1-A)}{4\sigma}\right]^{\frac{1}{4}} = \text{estimated temperature} \qquad [4\text{-}2]$$

T_s = measured surface temperature

The greenhouse effect coefficient is defined as T_S/T_E. The following table illustrates the coefficient for several bodies:

Body	S-W/m^2	A	T_e(K)	T_s(K)	Coefficient
Earth	1372	.3	255	286	1.13
			(253)		
Venus	2613	.75	232	700	3
			(265)		(2.6)
Mars	583	.15	217	220	1.01
			(230)	(223)	0.96

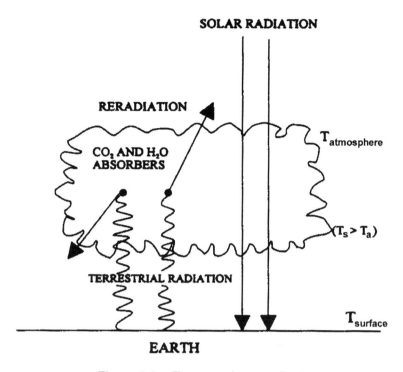

Figure 4-1.   The greenhouse effect.

The coefficient is an indication of whether or not the greenhouse effect is taking place.

Several mathematical models have been proposed to predict the dependence of the surface temperature on the concentration of carbon dioxide (C). A function $E(C, T_S, W)$ was used to explain this relationship, where W is the water vapor pressure. Because the water vapor pressure is a function of T_S, E could be rewritten as $E(C, T_S)$. If the function is exact,

$$\left(\frac{\partial E}{\partial C}\right) dC + \left(\frac{\partial E}{\partial T_s}\right) dT_s = 0 \qquad [4\text{-}3]$$

It is found that $(dT_S) = (3.8°C)dC$, which means that doubling the atmospheric carbon dioxide would raise the global mean surface temperature by 3.8 degrees Celsius.

4.1.1   Carbon Dioxide Cycle

As far as we surface-dwellers are concerned, it is important to know whether the greenhouse effect would be magnified by the increasing amounts of carbon dioxide produced by the rapid burning of fossil fuels. The burning of fossil fuels has a negligible effect on the total oxygen content of the atmosphere, but an appreciable effect has been imposed on the carbon dioxide content. As shown in Figure 4-2, carbon dioxide occupies only a small portion on the carbon balance sheet of the earth. The industrial carbon dioxide production increased exponentially during the last century at a rate of about 3% per year. The concentration of carbon dioxide in the atmosphere at a site in Iceland has been measured continuously since 1958, as shown in Figure 4-3. This process shows that the average carbon dioxide content of the atmosphere has risen more than 5% since 1958 (the current rate is about 1.5 parts per million per year), which is equivalent to 3.18 peta grams of carbon per year. Detailed calculations can be seen as follows:

C (Pg/yr) for Fossil Fuel and Biomass Decomposition

Petroleum	Gas	Coal	Biomass	Σ
2.5	1	2.2	1	6.7

In the United States, carbon consumption is 1.2 Pg/yr., [per capita $\approx$ 14 kg/day (US)]. Past records show that CO_2 increased at a rate of 1.5 ppm/yr. Since the Industrial Revolution, when it was 280 ppm, the CO_2 concentration increases every year, and at present it is 360 ppm.

$$360\, ppm\, of\, CO_2 = \frac{\left(360 m^3\, CO_2\right)}{\left(1\times10^6\, m^3 air\right)} \frac{(mole)}{(22.4\times10^{-3}\, m^3 CO_2)} \frac{(44\, g)}{(mole)} = 0.707 \frac{g}{m^3}\, of\, CO_2$$

$$= \frac{(0.707 g\, CO_2)}{(m^3 air)} \frac{(12\, g\, of\, C)}{(44\, g\, of\, CO_2)} \frac{\left(5.1\times10^{18}\, kg\, air\right)}{\left(1.29\, \frac{kg}{m^3}\, air\right)} = 762\, Pg\, of\, C \quad [4\text{-}4]$$

Therefore,

$$1\, ppm\, of\, CO_2 = \frac{762}{360} = 2.12\, Pg\, of\, C \quad\quad\quad [4\text{-}5]$$

$$\text{and the airborne fraction of } CO_2 = \frac{1.5\, ppm/yr}{\left[(6.7\, Pg/yr)\Big/ (2.12\, Pg/ppm)\right]} \approx 0.5\, .$$

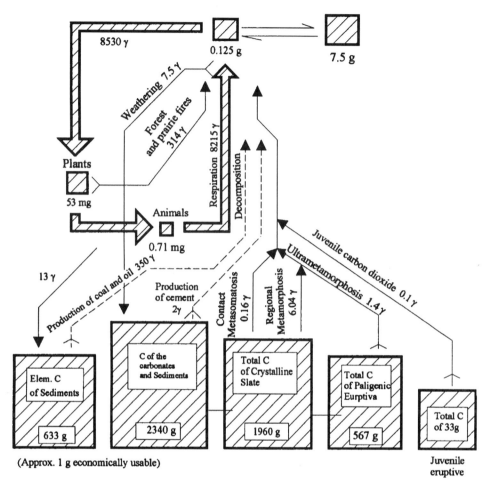

Figure 4-2. The carbon distribution and carbon cycles on Earth (from Dietrich, 1963).

However, the increase is only 50% of the expected rate if all the carbon dioxide emitted by human activity (including the manufacture of cement) during those years were added to the total. Evidently, 50% of the carbon dioxide produced does not stay in the atmosphere. Figure 4-4 shows the earth's carbon dioxide cycle. Much of the carbon dioxide added to the atmosphere is absorbed by the oceans, which contain large amounts of carbon dioxide in the form of bicarbonate ions. Other possible uptakers are the living green plants themselves, the biomass. Experimental studies have shown that plant growth increases in the presence of increased carbon dioxide concentration.

Carbon dioxide levels in the

atmosphere are increasing exponentially.

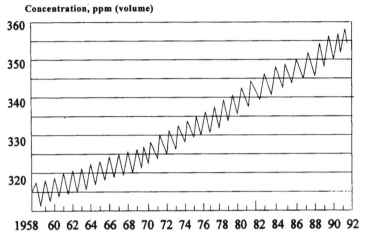

Figure 4-3.   Concentration of carbon dioxide was measured with a continuous recording nondispersive infrared gas analyzer at Mauna Loa Observatory, Hawaii. (Sources: Scripps Institution of Oceanography, National Oceanic & Atmospheric Administration.)

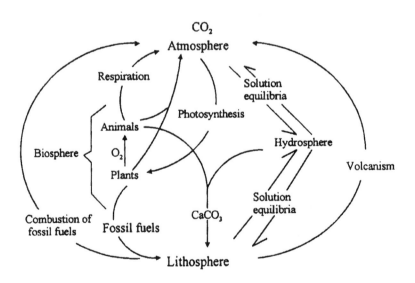

Figure 4-4. The carbon dioxide cycle.

4.1.2 Global Warming and Ice Periods

The mean global temperature has been increasing gradually throughout the first half of this century, a rise that correlates with the increase in carbon dioxide production since the Industrial Revolution. It has been suggested that the greenhouse effect is responsible for this. However, nature also goes through its own cycles such as the ice periods. Table 4-1 gives the geologic time scale. Figure 4-5 illustrates the Quaternary period (0.01 million years before the present) and its subdivision in North America. The average temperature can be traced back over a longer period of time using a variety of geophysical techniques as well as fossil records, as shown in Figure 4-6. One paleothermometer is the ratio of oxygen 18 to oxygen 16 in the Greenland ice core recorded by Vostok for the 160,000 year temperature record, as shown in Figure 4-7. Another example is that the oil yields (in gallon per ton GPT) of the oil shale from the different buried depths corresponding to the geological time scale vary greatly. The higher organic material content of the oil shale would correspond to higher earth temperatures, which, in turn, imply a higher algae growth rate, as shown in Figure 4-8. A summary of the different methods for studying ice periods is as follows:

- Greenland Ice Core

    — Vostok (East Antarctica)
    O^{16}, O^{18} isotope ratio
    160,000 year temperature record

- Tree Ring Research
    (recent years only)

- Ocean Beds — Foraminifera
    amino acids racemization
    using optical activity of carbohydrates

- Green River Oil Shale Core
    since Eocene time (50 M year – now)
    Oil yield (GPT) from kerogen

- Sun Spots
    (maybe Little Ice Age)
    22-yr. cycle and 100-yr. cycle (80 + 180-yr. cycle)

- Milankovitch Cycle (Earth)

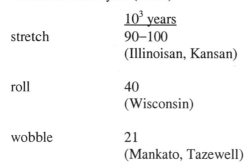

	10^3 years
stretch	90–100
	(Illinoisan, Kansan)
roll	40
	(Wisconsin)
wobble	21
	(Mankato, Tazewell)

The Milankovitch cycle is responsible for most of the ice periods occurring on Earth. As shown in Figure 4-10, it correlates quite well for the past 700,000 yrs. The stretch, roll, and wobble of the Earth are shown in Figure 4-11.

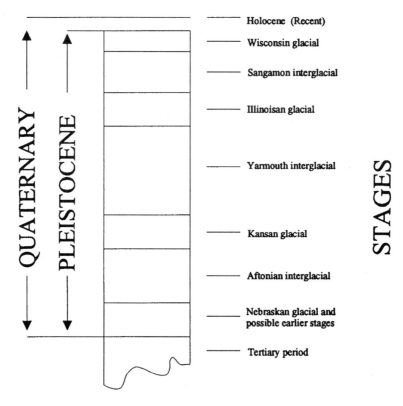

Figure 4-5. The Quaternary period and its subdivisions in North America.

Table 4-1.   Approximate Scale of Geological Time[1]

Era	Period or Epoch	Beginning and End, In 10^4 Years	Approximate Duration In 10^4 Years
Cenozoic[4]	Quarternary		
	Contemporary	0-1	0.8-1.2
	Pleistocene	1-10	1
	Tertiary		69
	Pliocene	1-10	9
	Miocene	10-25	15
	Oligocene	25-40	15
	Eocene	40-60	20
	Paleocene	60-70	10
Mesozoic[4]	Cretaceous	70-140	70
	Jurassic	140-185	45
	Triassic	185-225	40
Paleozoic[4]	Permian	225-270	45
	Carboniferous	270-320	50
	Devonian	320-400	80
	Silurian	400-420	20
	Ordovician	420-480	60
	Cambrian	480-570	90
	Pre-Cambrian IV (Riphean)[2]	570-1200	630
	Pre-Cambrian III (Proterozoic)[3]	1200-1900	700
	Pre-Cambrian II (Archean)	1900-2700	800
	Pre-Cambrian I (Catarchean)	2700-3500	800
	Pregeological era	3500-5000	1500

[1]Based on the geochronological scale of the Commission for Determining the Absolute Age of Geological Formations, published in *Izv. AN SSR*, Geological Series, No. 10, 1960.

[2]Proterozoic II.

[3]Proterozoic I.

[4]total duration of Cenozoic, Mesozoic and Paleozoic is 570.

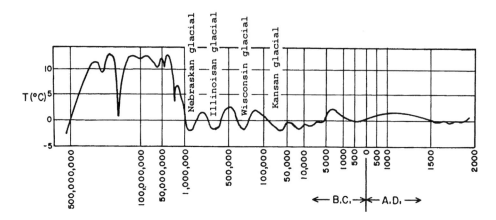

Figure 4-6. Temperature variations in the northern hemisphere (40°N–90°N) as a function of time. (Reprinted, by permission of Wiley, New York, from J. E. Oliver, "Climate and Man's Environment," 1973.)

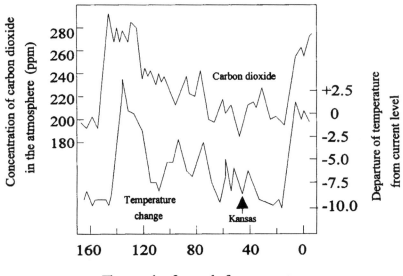

Figure 4-7. CO_2 Concentrations (ppm) and Antarctic temperatures (°C) plotted against age in the Vostok record. Temperatures are referenced to current Vostok surface temperature. (Source: Barnola et al. Reprinted by permission from *Nature*, Vol. 329. 1987, Macmillan Magazines Ltd.)

The assessment of human impacts on the climate must be superimposed on this large natural variation. Figure 4-8 shows an attempt to do this for the carbon dioxide greenhouse effect. In this model, a 3°C per 100% increase in carbon dioxide concentration was added to a projection of the natural temperature fluctuations obtained from an examination of the temperature record over the past several hundred years. If this approach is correct, then the current cooling trend should soon level off and be followed by a rather steep rise in temperature.

Relatively small changes of average temperature could disrupt world food production, and unless there are radical changes in the approach to energy conservation, it seems quite probable that the continued generation of carbon dioxide will be substantial in the future. The implications seem clear enough; carbon dioxide will be an important factor in changing the nature of the Earth's climate.

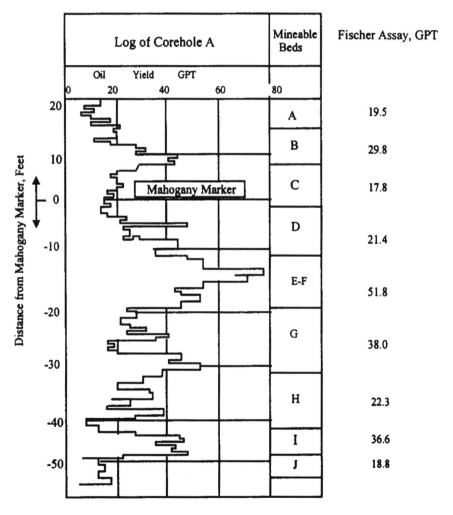

Figure 4-8. Core intervals and minable beds at Anvil Points Mine. Rifle, Colorado.

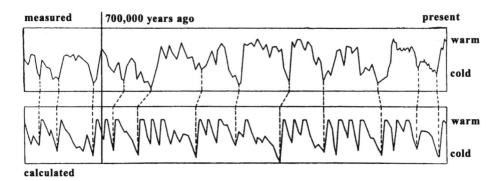

Figure 4-9. Measured and calculated ice periods for 700,000 years ago. (Permission obtained from N. Calder, *The Weather Machine,* Viking Press, 1974)

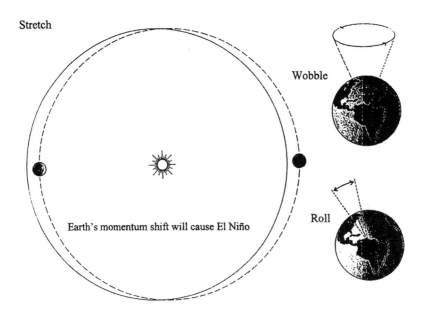

Figure 4-10. Stretch, roll, and wobble of Earth as Milankovitch cycle.

The cultural and economic impact of the Little Ice Age is quite clear. The Golden Period of the Tang Dynasty in China (warm, 700–900 AD) and the Dark Ages in Europe (cold, 1200–1600 AD) can be seen in Figure 4-11. Japan has made a correlation of the GNP index to the short-range temperature change, as shown in Figure 4-12.

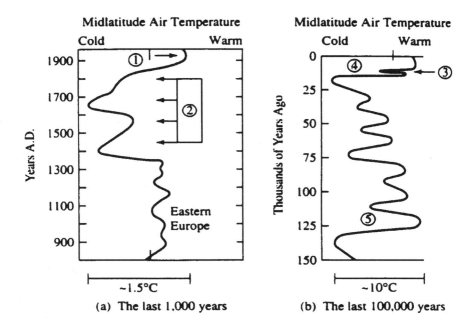

Figure 4-11.   Long-term natural fluctuations in Earth's temperature. (1) Thermal maximum of 1940's, (2) Little Ice Age, (3) Cold Interval, (4) Present interglacial (Holocene), (5) Last previous interglacial (Eemian). [Source: National Research Council (1975). *Understanding Climatic Change, a Program for Action* (Washington, DC: National Academy Press).]

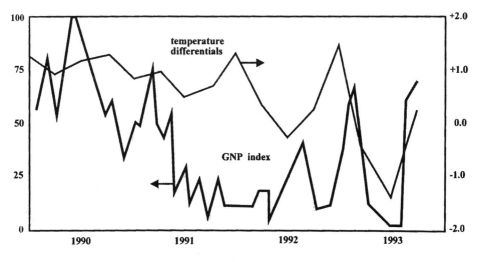

Figure 4-12. Correlation between short-range temperature change and the GNP of Japan.

4.2 OZONE LAYER

Ozone, the triatomic molecule of oxygen (chemical formula O_3), absorbs short-wavelength ultraviolet solar radiation that can be harmful to animal and plant life. Ozone is present in the upper atmosphere in small amounts and prevents most of this radiation from reaching the Earth's surface. If all the ozone in the atmosphere were distributed uniformly over the surface at sea level, it would form a layer only about 3 mm thick. Most of the atmospheric ozone (roughly 95%) is found in the stratosphere. It is this ozone shadow that made life forms as we know them possible on this Earth, and a depletion of the ozone layer is a threat to these same life forms. Ozone is a highly reactive chemical, and the relatively high concentrations of ozone that may occur occasionally during air pollution episodes over parts of North America are harmful to public health and welfare. The background concentration of ozone in the troposphere is about 0.03 ppm. In photochemical smog, such as in Los Angeles, ozone concentration can reach levels as high as 0.5 ppm.

The concentration of stratospheric ozone is determined by the balance between the reactions by which it is created and that by which it is destroyed. Ozone is formed as a consequence of the interaction of UV radiation from the Sun with oxygen to give atomic oxygen (O):

$$O_2 + \text{sunlight } (\lambda < 242 \text{ nm}) \rightarrow 2\,O$$

This reaction is followed by the subsequent reaction of atomic and molecular oxygen.

$$O_3 + O \rightarrow 2\,O_2$$

4.2.1   Beer-Lambert Law and Chapman Layers

Absorption of radiation is governed by the interaction between photons and matter. If radiation of intensity I traverses an absorber of unit area and thickness of dz, the decrease of intensity is given as

$$-dI = Ink_a\,dz \qquad\qquad [4\text{-}6]$$

where n is the atmosphere number density of absorbers that can usually be used as concentration of a given absorber and k_a is a constant with units of area and is referred to as the absorption cross-section. Integration of Equation [4-6] for an incident intensity, I_0, is as follows:

$$I = I_0 \exp\left(-\int_0^z k_a\,dz\right) \qquad\qquad [4\text{-}7]$$

Because n is independent of z,

$$I = I_0 \exp(-nk_a z)$$ [4-8]

This equation is termed Beer-Lambert Law and used in spectroscopy. If it is expressed in transmittance T and the concentration is in molar concentration, then

$$T = \frac{I}{I_0} = \exp(-\varepsilon z)$$ [4-9]

where ε is termed molar absorbance (extinction coefficient) and Figure 4-13 is a spectrum of ozone.

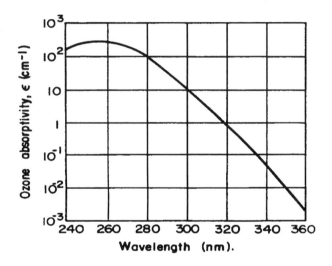

Figure 4-13. Ozone absorptivity as a function of wavelength.

[Example 4-1] If there is a 1% decrease of ozone in the ozone layer, what percentage of the increase in ultraviolet transmittance is for our exposure?

The equivalent thickness of ozone at 0°C and 1 atm is 0.34 cm. on the average.
Differentiate Equation [4-9]:

$$dT = -\varepsilon \exp(-\varepsilon z)dz$$ (a)

or

$$\frac{dT}{T} = -\varepsilon \, dz = -\varepsilon \, z \, \frac{dz}{z} \qquad \text{(b)}$$

after dividing by Equation [4-9]. For the critical regime of sunburn and skin cancer, the wavelength falls between 310 and 290 nm. In Figure 4-14 the product of εz (value of the absorptivity times 0.34) ranges from 1 to 10 for the wavelengths of 310 nm and 290 nm. Therefore, from the above equation (b), a 1% decrease in the ozone layer gives a 1% increase in ultraviolet transmittance at 310 nm, a 3% increase at 300 nm, and a 10% increase at 290 nm. At 260 nm, the transmittance witnesses a 100% increase.

From Equation [B-4], we have

$$P = P_0 \exp\left(\frac{-mgz}{RT}\right)$$

If we express into molecular scale

$$\rho = \frac{MP}{RT} = \frac{mP}{kT}$$

or $R = NR$, then

$$P = P_0 \exp\left(\frac{-mgz}{kT}\right) \qquad [4\text{-}10]$$

in Boltzmann distribution form.

Because the quantity (kT/mg) has the units of length and represents a characteristic length, or

$$P = P_0 \exp\left(\frac{-z}{H_s}\right) \qquad [4\text{-}11]$$

or

$$H_s = kT/mg = \text{scale height}$$

Let atmospheric number density n replace P since n is proportional to P, then

$$n = n_0 \exp\left(\frac{-z}{H_s}\right) \qquad [4\text{-}12]$$

Also, an increase in altitude dz, the atmospheric path that the sun's rays have to traverse, is decreased by $dz \sec \theta$ for a zenith angle θ, or Equation [4-6] becomes

$$dI = \ln k_a \left(dz \sec \theta \right) \qquad\qquad [4\text{-}13]$$

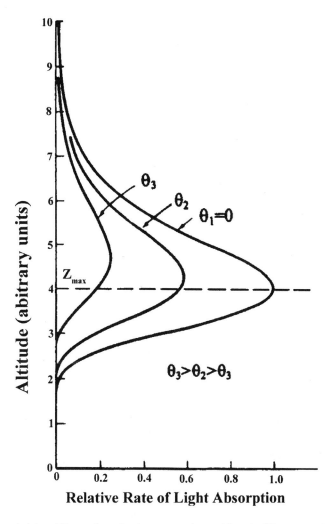

Figure 4-14.   Photochemical energy deposition in Chapman layers.

Combine with Equation [4-12], and we get the following:

$$\frac{dI}{I} = d\left(\ln I\right) = n_o\, k_a \sec\theta \exp\left(\frac{-z}{H_s}\right) dz \qquad [4\text{-}14]$$

On integration and set the boundary condition as

$$I = I_\infty \qquad \text{at } z = \infty$$

then

$$I = I_\infty \exp\left[-n_o\, k_a\, H_s \sec\theta \exp\left(\frac{-z}{H_s}\right)\right] \qquad [4\text{-}15]$$

the rate, P, at which the energy is removed from decrease in intensity per unit path traversed is

$$P = \frac{dI}{dz \sec\theta} = \left(\frac{dI}{dz}\right)\cos\theta$$

$$= I_\infty\, n_o\, k_a \cos\theta \exp\left[-\frac{z}{H_s} - n_o\, k_a\, H_s \sec\theta \exp\left(-\frac{z}{H_s}\right)\right] \qquad [4\text{-}16]$$

This rate is called the Chapman layer function, because the equation will define the shape of the ozone layer. A plot of this function for 3 zenith angles can be found in Figure 4-14. This clearly indicates that the ozone concentrates on 35 km above our atmosphere (refer to Fig. B-1).

4.2.2   Depletion of the Ozone Layer

In the early 1970s, it was realized that human activities result in the addition of certain chlorine, nitrogen, and other catalyst species to the stratosphere, upsetting the balance between production and destruction processes and leading to changes in the total amounts of ozone above the Earth's surface. Over the past two decades, the stratospheric ozone concentration has been declining with the so-called Antarctic ozone hole. Among the chemical agents considered to have the potential to affect stratospheric ozone in various ways are nitrogen oxides (NO_x), chlorofluorocarbons (CFCs), nitrous oxide (N_2O), methyl chloroform (CH_3CCl_3), carbon tetrachloride (CCl_4), methane, and carbon dioxide. A reduction in the stratospheric ozone would lead to an increase in the intensities of UV light reaching the Earth's surface, and the consequences would be harmful.

The oxides of nitrogen are the major chemical family responsible for the ozone decrease in the stratosphere. The addition of significant quantities of NO_x

to the stratosphere—for example, from an increasing concentration of N_2O, detonation of nuclear weapons, or substantial numbers of supersonic aircraft— would result in substantial decreases in the total amount of ozone. Chlorofluo-romethanes are used as aerosol propellants and refrigerants because they are inert. Because of the inertness, they survive in the atmosphere to be transported to the stratosphere, where they are photochemically dissociated to give atomic (Cl). These chlorines, as well as OH and NO, can destroy ozone without a loss of the reactant species. They work as catalysts in chemical reactions.

The ozone destruction can be explained by the **Molina-Rowland scheme**, as follows:

$$CCl_3F + hv \rightarrow CCl_2F + Cl$$

$$Cl + O_3 \rightarrow ClO + O_2$$

$$ClO + O \rightarrow Cl + O_2$$

$$\overline{}$$

$$O_3 + O \rightarrow 2O_2$$

or

$$O(^1D_2) + H_2O \rightarrow 2OH$$

$$OH + O_3 \rightarrow HO_2 + O_2$$

$$HO_2 + O \rightarrow OH + O_2$$

$$\overline{}$$

$$O_3 + O \rightarrow 2O_2$$

Fate of Cl – rainout

$$Cl + HO_2 \rightarrow HCl + O_2$$

$$Cl + CH_4 \rightarrow HCl + CH_3$$

What would be the effect of depletion of the order of 5% of the total ozone? In some estimates, the associated increase in UV radiation reaching the surface of the earth could produce a 10% increase in nonfatal skin cancers. A correlation of melanoma mortality rates and latitudes has been found and is indicated in Figure 4-15.

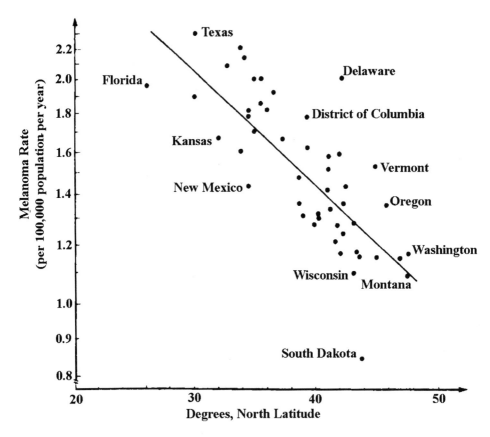

Figure 4-15.   Variation with latitude of human death due to skin melanoma among white males in the United States excluding Alaska and Hawaii. [From Rowland, F.S., Chapter 4.6 in Coyle, J.D., Hill R. R., and Roberts, D.R. (eds) *Light, Chemical Change and Life*, Open University Press, 1982.]

4.2.3   Fate of Halomethanes

Two classes of saturated halocompounds are of importance; the first is chlorofluorocarbons (CFCs), and the second includes both hydrochlorofluorocarbons (HCFCs) and hydrofluorocarbons (HFCs), the latter of which do not contain any chlorine. The CFCs, HCFCs, and HFCs have a nomenclature system, for example, CFC-11, CFC-113 etc. These numbering systems are based on the following:

Number after CFC + 90 = abc (3 digits)

a = No. of C

b = No. of H

c = No. of F

Balance the carbon valence requirement with the number of Cl.

For example,

CFC – 12,	$12 + 90 = 102$	$\therefore CF_2Cl_2$
CFC – 115,	$115 + 90 = 205$	$\therefore CF_3\text{-}CF_2Cl$
CFC – 113,	$113 + 90 = 203$	$\therefore CCl_2F\text{-}CClF_2$
CFC – 11,	$11 + 90 = 101$	$\therefore CFCl_3$

The CFCs have a long atmospheric lifetime; they are also stronger absorbers within the 7 to 13μm atmospheric window. The lifetime of HCFCs is shorter and their breakdowns are relatively quick. They have a modest potential to affect ozone. The same is true for the HFCs; they contain no chlorine atoms to threaten the ozone layer, but they contribute to affect global warming. According to the *Montreal Protocol on Substances That Deplete the Ozone Layer*, all CFCs will be banned by 1996 and all HCFCs by 2030. HFCs will be restricted if other alternatives exist, as shown in Figure 4-16 and Table 4-2.

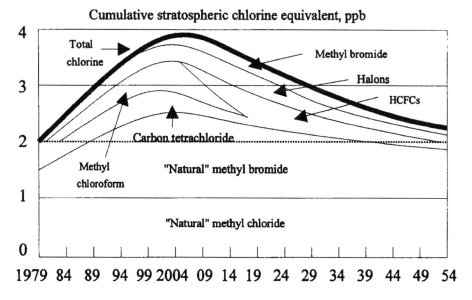

Note: Ozone-depleting effects of bromine atoms in halons and
methyl bromide have been converted to their chlorine equivalents.

Figure 4-16.   Levels of chlorine in the stratosphere will decrease slowly under Montreal protocol. (Source: DuPont)

Table 4-2.   EPA Plans to Phase Out HFC's in Stages

Date [a]	Compounds Affected	Restriction
1994	All HCFCs	Ban on use in aerosols (except medical devices) and plastic foams (except foam insulation).
2003	HCFC-141b (CH_3CCl_2F, ODP^b=0.12)	Ban on production and imports.
2010	HCFC-22 ($CHClF_2$, ODP=0.05) HCFC-142b (CH_3CClF_2, ODP=0.06)	Production and imports frozen at 1989 baseline levels. Ban on use of virgin chemical, unless used as feedstock or for servicing refrigeration or air-conditioning equipment manufactured before Jan.1, 2010.
2015	All other HCFCs	Production and imports frozen at 1989 baseline levels. Ban on use of virgin chemical, unless used as feedstock or for air-conditioning equipment manufactured before Jan.1, 2010.
2020	HCFC-22 HCFC-142b	Ban on production and imports.
2030	All other HCFCs	Ban on production and imports.

Note: HCFCs = Hydrochlorofluorocarbons.

a Effective date is Jan. 1 of the given year.

b ODP = Estimated ozone depletion potential, which depends on chlorine content and atmospheric lifetime.
Potentials are relative to CFC-11, which is assigned a value of 1.0.

Source: *Federal Register.* March 18, 1993. Page 15.014 and Sept. 27, 1993. Page 50.464.

Most of the halomethanes are used in refrigeration. The second largest uses in industry are the blowing agents for foams; others could be dry-cleaning agents. Some recent technological advances include air-conditioning systems that use no refrigerants or compressors—for example, the desiccant system. Actually, this is a combination of evaporative cooling and desiccant drying, as shown in Figure 4-17.

Other types of halomethanes (termed halons) are used in firefighting; they are composed of brominated fluorocarbons, particularly CF_3Br. It seems at present that the fluoroiodocarbon (FIC) CF_3I is a good substitute because its short life will be unlikely to contribute to ozone depletion or global warming.

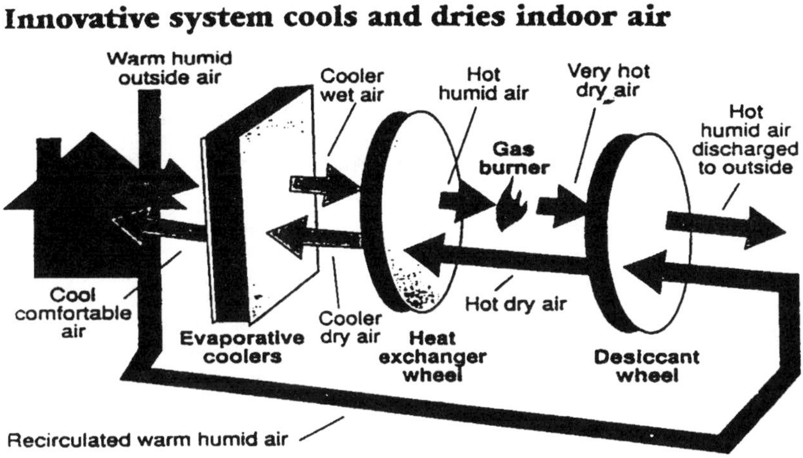

Innovative system cools and dries indoor air

Figure 4-17. Innovative system cools and dries indoor air. (Source: C and EN, Nov. 15, 1993, by P.S. Zurer)

4.3 STRATOSPHERIC CHEMISTRY

Stratospheric chemistry involves a bewildering number of chemical reactions and catalytic cycles. Many key stratospheric reactions involve radicals. Table 4-3 lists some of the reactions and their reaction coefficients. In terms of predicting the effect of a given perturbation on ozone, the primary question is the balance among various species that interact extensively. Current modeling work incorporates as many as 140 reactions with the aid of computers.

In the stratosphere, ozone behaves as a UV shield.

$$O_2 + hv \rightarrow 2O \ (^3P_2 \text{ and } ^1D_2)$$

$$O + O_2 \rightarrow O_3$$

$$O_3 + hv \rightarrow O_2 + O(^1D_2)$$

4.3.1 Photochemistry

By far the most abundant photochemical reaction in the stratosphere is the photolysis of molecular oxygen. The bond dissociation energy of oxygen (5.1 eV) corresponds to a photon of wavelength 243nm. Ultraviolet absorption is weak at 200nm and the excited molecular states can be dissociated into two ground state $^3P_2$ oxygen atoms.

$$O_2 \xrightarrow{\ 200\,nm\ } 2\,O(^3P_2)$$

Below 200nm, excitation leads to dissociation to one ground state $^3P_2$ oxygen atom and one excited $^1D_2$ oxygen atom.

$$O_2 \xrightarrow{<176\,nm} O(^3P_2) + O(^1D_2)$$

As the ultraviolet wavelength continues to decrease, the highest excited states of $^1S_0$ will result.

$$O_2 \xrightarrow{<130\,nm} O(^3P_2) + O(^1S_0)$$

$$O_2 \xrightarrow{<92.3\,nm} 2O(^1S_0)$$

Furthermore, the ionization potential of oxygen molecules is 12.15 eV, which corresponds to a photon wavelength of 102nm. Thus, a more likely reaction below 102nm is photoionization.

$$O_2 \xrightarrow{<102\,nm} O_2^+ + e^-$$

followed by dissociative recombination,

$$O_2^+ + e^- \begin{cases} O(^1S_0) + O(^3P_2) + 64.3\ Kca \\ O(^1S_0) + O(^1D_2) + 19.1\ Kca \end{cases}$$

This reaction plays a role in the photochemical smog cycle. Similar types of reactions occur from the forbidden radiative transitions, as shown in Figure 4-18.

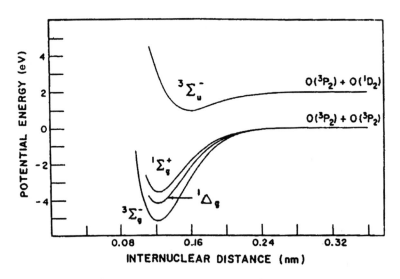

Figure 4-18.   Potential energy curves from O_2. [Curves drawn from data of F. R. Gilmore, *J. Quant. Spectrosc. Radiat. Transfer 5*, 369 (1965)]

Table 4-3. Reactions and Reaction Coefficients in cm-molecule-sec. Units

$O_2 + h\nu$	$\rightarrow$	$O + O$	$\lambda < 2420\text{Å}, k_1 < 10^{-6}$
$O + O_2 + M$	$\rightarrow$	$O_2 + M$	$k_2 = 2.04 \times 10^{-26} \exp(1050/T)$
$O_2 + h\nu$	$\rightarrow$	$O(^1D) + O_2(^1\Delta_g)$	$\lambda \leq 3100\text{Å}, k_{2a} < 5 \times 10^{-6}$
$O_2 + h\nu$	$\rightarrow$	$O + O_2$	$3100\text{Å} \leq \lambda \leq 10,400\text{Å}$
$O(^1D) + M$	$\rightarrow$	$O + M$	$k_4 = 5 \times 10^{-11}$
$O + O_1$	$\rightarrow$	$2O_2$	$k_5 = 1.33 \times 10^{-11} \exp(-2100/T)$
$O + OH$	$\rightarrow$	$H + O_2$	$k_6 = 5 \times 10^{-11}$
$H + O_2$	$\rightarrow$	$OH + O_2$	$k_7 = 2.6 \times 10^{-11}$
$H + O_2 + M$	$\rightarrow$	$HO_2 + M$	$k_8 = 4 \times 10^{-12}$
$HO_2 + O$	$\rightarrow$	$OH + O_2$	$k_9 = 2 \times 10^{-11}$
$OH + O_2$	$\rightarrow$	$HO_2 + O_2$	$k_{10} \leq 10^{-16}$
$OH + CO$	$\rightarrow$	$H + CO_2$	$k_{11} = 10^{-12}$
$HO_2 + NO$	$\rightarrow$	$OH + NO_2$	
$HO_2 + HO_2$	$\rightarrow$	$H_2O_2 + O_2$	$k_{13} = 8 \times 10^{-11} \exp(-1000/T)$
$H_2O_2 + h\nu$	$\rightarrow$	$2OH$	$\lambda < 5650\text{Å}; J_{14} > 5 \times 10^{-4}$
$H_2O + h\nu$	$\rightarrow$	$H + OH$	$\lambda < 2240\text{Å}; J_{15} < 10^{-4}$
$H_2O + O(^1D)$	$\rightarrow$	$2OH$	$k_{14} = 3 \times 10^{-16}$
$H_2O + h\nu$	$\rightarrow$	H_2O	$\lambda = 1.4\mu m$
$H_2O + O$	$\rightarrow$	$2OH$	
$OH + OH$	$\rightarrow$	$H_2O + O$	$k_{19} = K_{13}$
$OH + HO_2$	$\rightarrow$	$H_2O + O_2$	$k_{20} \geq 10^{-11}$
$OH + H_2O_2$	$\rightarrow$	$HO_2 + H_2O$	$k_{21} = 6 \times 10^{-12} \exp(-600/T)$
$NO + O_2$	$\rightarrow$	$NO_2 + O_2$	$k_{22} = 1.33 \times 10^{-12} \exp(-1250/T)$
$NO_2 + O$	$\rightarrow$	$NO + O_2$	$k_{23} = 1.67 \times 10^{-11} \exp(-300/T)$
$NO_2 + O_3$	$\rightarrow$	$NO_3 + O_2$	$k_{24} = 10^{-11} \exp(-3500/T)$
$NO_3 + h\nu$	$\rightarrow$	$NO + O_2$	$J_{21a} = 10^{-2}$ (?)
$NO_2 + h\nu$	$\rightarrow$	$NO_2 + O$	$J_{21b} = 10^{-2}, \lambda < 5710\text{Å}$
$NO_2 + h\nu$	$\rightarrow$	$NO + O$	$J_{26} = 5 \times 10^{-8}, \lambda < 4000\text{Å}$
$NO_2 + NO$	$\rightarrow$	$2NO_2$	$k_{27} = 10^{-11}$
$NO + h\nu$	$\rightarrow$	$N + O$	$J_{28} < 5 \times 10^{-6}$
$N + NO$	$\rightarrow$	$N_2 + O$	$k_{29} = 2 \times 10^{-11}$
$N + O_2$	$\rightarrow$	$NO + O$	$k_{30} = 1.2 \times 10^{-11} \exp(-3525/T)$
$N + O_2$	$\rightarrow$	$NO + O_3$	$k_{31} = 3 \times 10^{-11} \exp(-1200/T)$
$N + OH$	$\rightarrow$	$NO + H$	$k_{32} = 7 \times 10^{-11}$
$N_2O + O(^1D)$	$\rightarrow$	$2NO$	$k_{33a} = 1 \times 10^{-10}$
$N_2O + O(^1D)$	$\rightarrow$	$N_2 + O_2$	$k_{33b} = 1 \times 10^{-10}$
$N_2O + h\nu$	$\rightarrow$	$N_2 + O$	$J_{34} < 5 \times 10^{-7}, \lambda < 3370\text{Å}$

Source: Crutzen, 1972.

$$O_2(^1\Sigma_g^+) \rightarrow O_2(^3\Sigma_g^-) + h\nu \ (762nm)$$

$$O_2(^1\Delta_g) \rightarrow O_2(^3\Sigma_g^-) + h\nu \ (1270nm)$$

These reactions turn out to be the most intense bonds in the atmospheric day glow and which make up the auroral displays. The aurora spectra lines are expressed in the following:

		$\overset{\circ}{A}$
O	$^1S_o \rightarrow {}^1D_2$	5577 (gr)
	$^1D_2 \rightarrow {}^3P_2$	6300
		6363
		6392
N	$^2P_{3/2} \rightarrow {}^4S_{3/2}$ via $^2D_{S/2}$	3466
O_2	$^1\Sigma_q^+ \rightarrow {}^3\Sigma_q^-$	7620
	$^1\Delta_g \rightarrow {}^3\Sigma_q^-$	12700
N_2	$^3\pi \rightarrow {}^3\Sigma$	14700–50300

4.3.2 Term Symbols

The **term symbol** designation can be repeated by $^{2S+1}L_J$, where $S = \Sigma m_s$, $L = \Sigma m_l$, and $J = L + S$ or $L + S - 1$; when $L = 0,1,2,3$, the designation is S,P,D,F. The m_s and m_e are obtained from atomic orbitals. Similarly, an oxygen ion, O^+, or a nitrogen atom can have three states: $^4S_{3/2}$, $^2D_{3/2}$, and $^2P_{3/2}$. For general use, the J value can be eliminated.

These values are used for the sketch below. For O, we have $2s^2$, $2p^4$ (only consider outer orbit). For $2p^4$, we have 4 electrons distributed in 3 p-orbitals, which can be seen as follows:

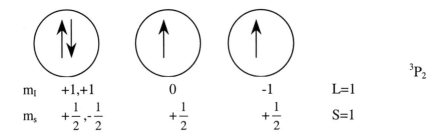

m_l	+1,+1	0	-1	L=1
m_s	$+\frac{1}{2},-\frac{1}{2}$	$+\frac{1}{2}$	$+\frac{1}{2}$	S=1

$^3P_2$

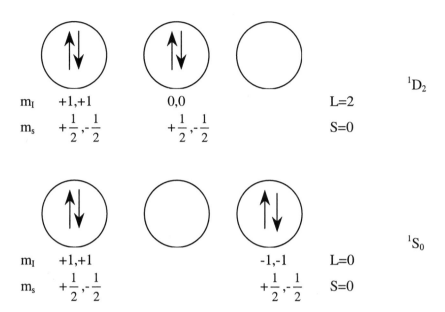

Therefore, the O atom will have the 3 forms, $^3P_2$, $^1D_2$, and $^1S_0$ and stability of those term symbols are as follows:

$$^3P_2 > {}^1D_2 > {}^1S_0$$
$$\leftarrow \quad \text{stability}$$

for example,

$$O_2 \xrightarrow{\ <200\,nm\ } 2O\,(^3P)$$

$$O_2 \xrightarrow{\ <176\,nm\ } O\,(^3P) + O\,(^1D)$$

$$O_2 \xrightarrow{\ <92\,nm\ } 2O\,(^1S)$$

According to above equations for photo desociation of oxygen molecules, the shortest UV will produce the most unstable species.

Alternatively, the term symbol and stability for N(or O$^+$), $2s^2$, $2p^3$ are explained as follows:

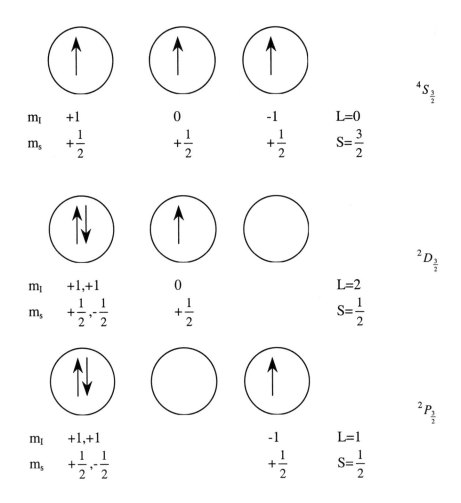

In this fashion, for N,

$$^4S_{3/2} > {}^2D_{3/2} > {}^2P_{3/2}$$

Notice that the term symbol for a diatomic or linear platonic electronic state is $^{2S+1}\Lambda_G$ plus the symmetry, for example, g (gerade) for no change in inversion symmetry, u (ungerade) for the opposite; + for no change in reflection symmetry, − for the opposite. $\Omega = |\Lambda + \Sigma|$, where Λ can have values $0, 1, ..., L$, which are designated respectively by Σ, π, Λ...states, analogous to s, p, d...for the atoms.

For O_2,

$$^3\Sigma_g^- > {}^1\Delta_g > {}^1\Sigma_g^+$$

4.4 THERMOSPHERE AND IONOSPHERE

The purpose of the ionosphere is the production of ions and electrons. The function of the thermosphere is the production of exothermic reactions. Actually, they can be considered as one sphere, as shown in Figure 4-19.

The following is a summary of the chemistry of both thermosphere and ionosphere. Chemical species in these regions are extremely dilute. The names are derived from the properties reflected in these regions.

Production of electron — E-layer

$$O + hv \rightarrow O^+ + e$$
$$O_2 + hv \rightarrow O + O^+ + e$$
$$N_2 + hv \rightarrow N + N^+ + e$$

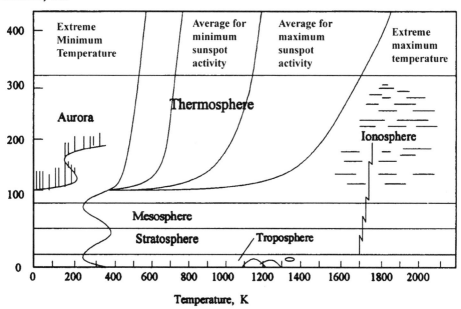

Figure 4-19. Temperature varies more in thermosphere than in low regions of the atmosphere. (Source: C and EN, June 16, 1986 by R.G. Roble)

Production of ions — ionosphere

$$O^+(^4S) + N_2 \rightarrow NO^+ + N$$
$$O^+ + O_2 \rightarrow O_2^+ + O$$
$$NO^+ + e \rightarrow N + O$$
$$O_2^+ + e \rightarrow O + O$$
$$O^+(^2D) + N_2 \rightarrow N_2^+ + O$$
$$N_2^+ + e \rightarrow 2N$$

Chemical heating (exothermic reaction) — thermosphere

$$O_2^+ + e \rightarrow 2O(^3P) + 6.95 \text{ eV}$$
$$\rightarrow O(^3P) + O(^1D) + 4.98 \text{ eV}$$
$$N(^4S) + O_2 \rightarrow O + NO + 1.4 \text{ eV}$$
$$N(^2D) + O_2 \rightarrow O(^1P) + NO + 1.84 \text{ eV}$$
$$O(^1D) + N_2 \rightarrow O(^3P) + N_2 + 1.96 \text{ eV}$$

REFERENCES

4-1.   W. Strauss and S. J. Mainwaring, *Air Pollution*, Edward Arnold Ltd., London, 1984.

4-2.   H. C. Perkins, *Air Pollution*, McGraw-Hill, New York, 1974.

4-3.   J. H. Seinfeld, *Air Pollution*, McGraw-Hill, New York, 1975.

4-4.   Environmental Studies Board, National Research Council," Causes and Effects of Changes in Stratospheric Ozone: Update 1983", National Academy Press, Washington DC, 1984.

4-5.   T. L. Brown, *Energy and the Environment*, Charles E. Merrill, Columbus, OH, 1971.

4-6.   J.H. Seinfeld, *Atmospheric Chemistry and Physics of Air Pollution*, Wiley, New York, 1986.

4-7.   R. M. Baum, "Stratospheric Science Undergoing Change," *C & EN*, September, p. 21 (1982).

4-8.   R. Revelle, "Carbon Dioxide and World Climate," *Scientific Amer.*, *247(2)*, 35 (1982).

4-9.   G. M. Woodwell, "The Carbon Dioxide Question," *Scientific Amer.*, *238(1)*, 34 (1978).

4-10.   A. P. Ingersoll, "The Atmosphere," *Scientific Amer.*, *249(3)*, 162 (1983).

4-11.   D. G. Torr, "The Photochemistry of Atmosphere: Earth and Other Planets and Comets," Academic Press, New York, 1985.

4-12.   R. G. Noble, "Chemistry in the Thermosphere and Ionosphere," *C & EN*, June 16, 23–38 (1986).

4-13.   R.P. Wayne, *Chemistry of Atmospheres*, 2nd ed., Clerendon Press, Oxford, 1991.

4-14.   T.G. Spiro and W.M. Stigliani, *Chemistry of the Environment*, Prentice-Hall, Upper Saddle River, NJ, 1996.

4-15.   A. D. Danilov, *Chemistry of the Ionosphere*, Plenum, New York, 1970.

4-16.   R. W. Bonbel, D. L. Fox, B. D. Turner, and A. C. Stern, *Fundamentals of Air Pollution*, Academic Press, San Diego, 1994.

4-17.   B. M. Smirnov, *Reviews of Plasma Chemistry* Vol. 1, Plenum, New York, 1991.

PROBLEM SET

1. Find the transmittances of ozone for wavelengths 300nm and 280nm at 0°C and 1 atm.

2. Assume that the concentration of suspended particulates in a polluted atmosphere is 170 $\mu g/m^3$. The particulates contain adsorbed sulfate and hydrocarbons comprising 14% and 9% of the weight, respectively. An average person respires 8,500 L of air daily and retains 50% of the particles smaller than 1 μm in diameter in the lungs. How much sulfate and hydrocarbon accumulate in the lungs in one year if 75% of the particulate mass is contained in particles smaller than 1 μm.

TROPOSPHERE: AIR POLLUTION

Air pollution may be defined as any atmospheric condition in which substances are present at concentrations high enough above their normal ambient levels so as to produce a measurable effect on people, animals, vegetation, or materials.

The pollutant types can be generally divided into five categories: carbon monoxide, NO_x, hydrocarbons, SO_x, and particulates. The following table is a summary of their toxicity and residence time.

This chapter is divided into the following sections: particulates, sulfur dioxide, automotive emissions, and tropospheric chemistry. Each section covers the five major air pollutants (SO_x, CO, NO_x, hydrocarbons, and particulates). The nature of their formation, their effects on the environment, and methods of coping with them will also be addressed. Most of the air pollution occurs in the troposphere.

Table 5-1A. Major Air Pollutants in the Atmosphere

Pollutant Types	CO	NO$_x$	HC	SO$_x$	Particulates
Mobile Source* (M ton/yr)	63.8	8.1	16.6	0.8	1.2
Stationary Source*	119	10.0	0.7	24.4	8.9
Tolerance Level ($\mu g/m^3$)	40,000	514	19,300	1,430	375
Relative Toxicity (weighing factor)	1	77.8	2.07	28.0	106.7
Residence Time (yr)	0.1	0.25	2	days to weeks	?

*United States only, both have the same unit.

5.1 PARTICULATES

The first widely recognized form of air pollution was smoke—fine carbon particles arising from the incomplete combustion of fuels, and inorganic ash arising from the noncombustible matter in the fuel. Particulates form a major part of the emissions of air pollutants and come from such diverse sources as cars, steel mills, cement plants, and local dumps. Volcanoes, forest fires, and ocean spray aerosols are among the largest natural sources of world particulate emissions. **Particulates** is a general term for something that exists in the form of minute separate particles, either solid or liquid. Table 5-1 gives the sizes of airborne particulates.

Some necessary definitions are given below:

Aerosol: A dispersion of solid or liquid particles of microscopic size in gaseous media, such as smoke, fog, or mist.

Dusts: Solid particles predominantly larger than colloidal particles, and capable of temporary suspension in air or other gases. They will settle under the influence of gravity.

Fly ashes: Finely divided particles of ash entrained in flue gases arising from the combustion of fuel.

Fumes: The solid particles generated by condensation from the gaseous state and often accompanied by a chemical reaction such as oxidation.

Smoke: Finely divided aerosol particles, which consist mainly of carbon and other combustible material, resulting from incomplete combustion.

Soot: Agglomerations of particles of carbon impregnated with tar, formed by the incomplete combustion of carbonaceous material.

Table 5-1B. Relation between size and properties of particulate matter. (With permission from J.W.Moore, *Environmental Chemistry,* Academic press, 1976, Harcourt Brace Jovanovich Publishers.)

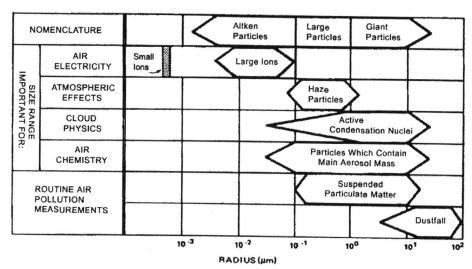

Particle size is given in Table 5-1B in units of **microns**; a micron is one millionth of a meter and is denoted by μm. Atmospheric dust particles have a wide range of sizes, from greater than 100 μm to about 0.001 μm. The comparison of a particle of a human hair to the wavelength of visible light may aid in gaining a feeling for the size of particulates. Health effects are strongly related to particle size. Particles that are less than 1 micron in size are called **submicron particles**. The term "**Aitken particulates**" refers to those particles with radii of less than 0.1 μm. These fine particles scatter light and thereby reduce visibility. They are also the worst causes of lung damage because of their ability to penetrate the innermost passages of the lung, called the **pulmonary region**. This is the region where oxygen is exchanged with carbon dioxide in the blood. There is no effective mechanism for removal of these lodged particles. The lodged particles can cause severe breathing impairment simply by physical blockage and irritation of the delicate lung capillaries. Coal miners' blacklung disease, asbestos workers' pulmonary fibrosis, and city dwellers' emphysema are all associated with the accumulation of such small particles.

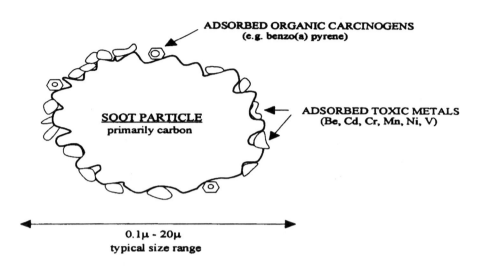

Figure 5-1. Soot particle from the combustion of fossil fuels (Source: Spiro and Stigliani).

Pneumoconiosis is a lung disease caused by prolonged inhalation of metallic or mineral dusts. The fly ashes emitted from coal combustion equipment consist largely of silica (SiO_2), alumina (Al_2O_3), and iron oxide (Fe_2O_3), as shown in Table 5-2. Beryllium oxide (BeO) may come out of a spark plug while driving. Heavy metals, such as mercury, that are often used as fungicides in the paper industry may accompany the ash generated from smoking cigarettes. Some particles are especially dangerous because they may carry toxic chemicals that can interact directly with lung tissue. Soot is finely divided carbon with a loose structure that possesses a large surface area, as shown in Figure 5-1. It is often associated with toxic trace metals such as beryllium, cadmium, chromium, manganese, nickel, and vanadium adsorbed on its surface. Moreover, soot serves as a carrier for toxic organic molecules such as benzo(a)pyrene, which has been implicated as a **carcinogen**, a cancer causing agent. Table 5-3 lists the sources, annual emissions, and health effects of minor air pollutants in the United States.

Particulate matter is removed from the air naturally by **fallout**, frequently referred to as **sedimentation**. Small particles grow to larger ones via collisions and aggregation and gradually settle out. The **settling velocity** in wind-free air is determined by the balance of two forces: frictional forces and gravitational forces. The drag force on a spherical particle of radius r moving at a steady speed u through a fluid of density ρ and viscosity η can be seen in the following equation:

$$F_{\text{friction}} = -6\pi r \eta u$$

Table 5-2. Chemical Analysis of Fly Ash (%)

Chemical Components	Subbituminous Coal	Bituminous Coal
SiO_2	50.50	43.5-57.0
Al_2O_3	33.70	18.0-28.0
Fe_2O_3	6.85	7.9-16.0
CaO	2.80	4.0-10.0
MgO	1.50	1.0-5.5
SO_3	0.75	0.9-3.3
Combustion losses	2.85	(Na_2O+K_2O)

Source: Nowak 1973, Table 1, p. 225.

The gravitational force on the particle is

$$F_{gravitation} = \frac{4}{3}\pi r^3 (\rho_{part} - \rho)g$$

When the particle moves at a steady velocity, it is not accelerating. Therefore, the net force on the particle is zero. According to **Stoke's law**, we can solve for the terminal velocity by equating the two preceding equations:

$$u = \frac{2 gr^2 \left(\rho_{part} - \rho\right)}{9\eta} \qquad [5\text{-}1]$$

The following is an approximation of the terminal velocities for various particle sizes which can be calculated from Equation [5-1].

Table 5-3. Terminal Velocities for Various Particle Sizes

r (μm)	u (cm/sec)
0.1	8×10^{-5}
1	4×10^{-3}
10	3×10^{-1}
100	25
1000	390

The larger the particle size, the faster the terminal velocity is. Customarily, **settling time** is defined as follows:

$$\text{settling time} = (\text{distance traveled})/(\text{terminal velocity}) \qquad [5\text{-}2]$$

For a particle 1 μm in radius, it will take 290 days to travel 1 km. Settling is a relatively minor mechanism (estimated to be 10–20% of the total deposi-

tion rate) for particle removal in the atmosphere. Impaction by wind is a significant mechanism, but the major removal processes are "washout" and "rainout" associated with rainfall. Figure 5-2 illustrates common collection devices
for particulate matter.

Table 5-4. Sources, Annual Emissions, and Health Effects of Minor Air
Pollutants

Substance	Sources[a]	Emissions[a] (tonnes/ year)	Health Effects[b]
Lead	Auto exhaust, industry, solid waste disposal, coal combustion, paint	208,250	Brain damage, behavioral disorders, convulsions, death
Fluorides	Industry, coal combustion	150,400	Mottled teeth, weakening of bone, weight loss, thyroid and kidney injury, death
Vanadium	Coal and petroleum combustion, industry	18,440	Inhibits formation of phospholipids and S-containing amino acids
Manganese	Industry, coal combustion	16,230	Fever, pneumonia
Arsenic	Industry	9,570	Dermatitis, melanosis, perforation of nasal septum, possible carcinogen
Nickel	Coal combustion, industry	6,625	Dermatitis, dizziness, headaches, nausea, carcinogenesis [also $Ni(CO)_4$]
Asbestos	Industry	6,080	Scarring of lungs, lung cancer
Cadmium	Industry	1,962	Gastrointestinal disorder, respiratory tract disturbances, carcinogenic and mutagenic
Mercury	Coal combustion, commercial, industry	777[c]	Tremor, skin eruption, Hallucinations
Beryllium	Coal combustion, industry	156	Lung damage, enlargement of lymph glands, emaciation
Selenium	Ore refining, sulfuric acid manufacture, coal combustion	—	Depression, jaundice, nosebleed, dizziness, headaches

[a] Data from U.S. Environmental Protection Agency, "Air Pollution Emission Factors," AP-42, 2nd ed. USEPA, Research Triangle Park, North Carolina, 1973.

[b] Data from G.L. Waldbott, "Health Effects of Environmental Pollutants." Mosby, St. Louis, Missouri, 1973.

[c] Emissions may be larger than this estimate; see C.E. Billings and W.R. Matson, *Science.* 176, 1232-1233 (1972).

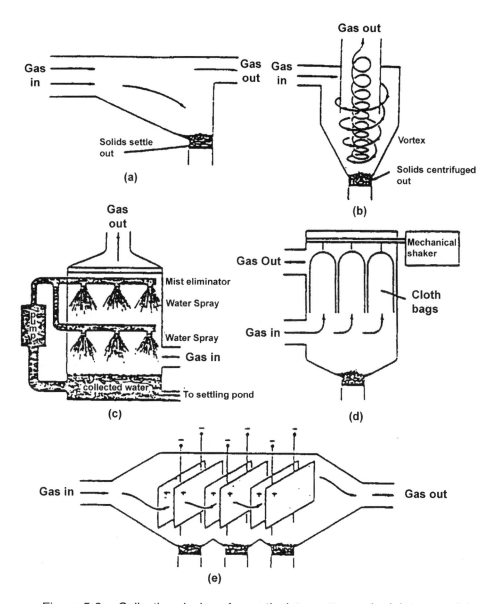

Figure 5-2. Collection devices for particulate matter and minimum particle sizes for which they give efficient removal: (a) settling chamber (d > 50μm); (b) centrifugal separator (d > 1μm); (c) wet scrubber (several other designs also used : d > 0.05 μm; can also collect water-soluble gases); (d) baghouse filter (d > 0.01 μm); and (e) electrostatic precipitator (d > 0.005 μm).

5.2 SULFUR OXIDES

Sulfur can be considered as an impurity in most coals and oils, as shown in Table 5-5. Sulfur dioxide is largely associated with the burning of coal and some crude oils for electricity and heating. This gas has well-known effects on human beings. Most individuals show bronchial response to SO_2 at concentrations of 5 ppm, and more sensitive people at 1 ppm. The odor can be noticed at concentrations that are 10 times lower. However, it is **sulfuric acid aerosol** (which is 10 times more irritating), formed from the oxidation of sulfur dioxide, which causes the most damaging health effects in urban atmospheres. Sulfuric acid aerosol is a **secondary particle** because it is formed by a gas-to-particle conversion process in the atmosphere. In contrast, **primary particles** are emitted directly into the atmosphere without intermediate conversion steps. Sulfur dioxide could be oxidized in the atmosphere to sulfur trioxide, but the conversion is slow in the absence of catalysts. It was found that nitrogen oxide and some metal oxides work as catalysts in this conversion reaction. The reaction can also be greatly accelerated with the aid of the ozone. The ozone donates an oxygen atom to form sulfur trioxide, which, in turn, combines rapidly with water to form sulfuric acid. The sulfuric acid can condense to form aerosol droplets. Besides the adverse health effect, the aerosol is capable of greatly reducing visibility. Figure 5-3 illustrates the conversion process.

Table 5-5.   Elementary and Raw Analyses of Raw Fuels

Sample	C(%)	H(%)	N(%)	S(%)	O(%)	Mineral and Ash(%)
Oil shale (Colorado)	12.4		0.41	0.63		84.6
Oil shale (Alaska)	53.9		0.30	1.50		34.2
Kerogen (Green River)	65.7	9.1	2.13	3.15	8.9	10.2
Lignite (Kineaid)	63.1	5.0	0.64	<0.04	21.5	9.3
Bituminous (Lower Freeport)	77.0	5.7	1.42	1.72	8.1	6.6
Anthracite (Pennsylvania)	87.1	3.8	0.60	<0.30	2.6	7.0
Gilsonite (Tarbon Vein)	84.5	10.0	2.4	0.53	2.6	0.7
Residuum (Mid-continent)	87.7	6.6	0.90	0.75	1.6	1.3
Asphaltene (Baxterville)	84.5	7.4	0.80	5.60	1.7	0.5

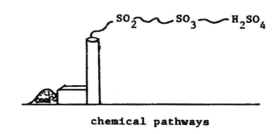

chemical pathways

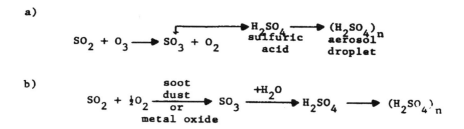

Figure 5-3.   Formation of sulfuric acid aerosol from sulfur dioxide.

5.2.1   Desulfurization

There are many ways of reducing SO_2 concentrations. Burning low-sulfur fuel is one possibility. Coal has a sulfur content of up to 3.4%, but in the Arizona strip mines some coal is found with a sulfur content of only 1%. Also, it is possible to extract the sulfur from the fuel. For oil, this is fairly simple and is currently being done at some refineries.

Sulfur exists as either organic or inorganic forms in coal, as shown in Table 5-6. Pyrite (FeS_2) is the main inorganic sulfur compound. The pyrite can be removed fairly easily by screening powdered coal, which is pulverized for feeding, to large furnaces to improve its combustion characteristics. Also, both pyrite and organic sulfur compounds (such as RSR, RSH, or =CH—S—CH=) can be removed by one of the following reactions:

(1) Displacement

$$=N: + FeS_2 \rightarrow \,=NFe + S_2^{2-}$$

$$R_1S_xR_2 + Nu^- \rightarrow R_1S_xNu + R_2^-$$

$$\text{For example } RSSR + R_3P \rightarrow [\, R_3P^+ - SR, RS^- \,]$$
$$\downarrow$$
$$R_3P = S + RSR \qquad \text{(Nu = nucleophilic agent)}$$

(2) Acid-Base Neutralization

$$4H^+ + FeS_2 + 2e^- \rightarrow 2H_2S + Fe^{2+}$$

$$RSH + OH^- \rightarrow RS^- + H_2O$$

(3) Oxidation

$$[O] + FeS_2 \rightarrow Fe^{2+} + 2S + O^{2-}$$

$$[O] + R_1S_2R_2 \rightarrow R_1SO_3H + R_2SO_3H$$

$$\downarrow H_2O, \Delta$$

$$R_1OH + R_2OH + 2H_2SO_4$$

(4) Reduction

$$FeS_2 + H_2 \rightarrow FeS + H_2S$$

$$FeS_2 + H_2 \rightarrow Fe + H_2S$$

$$R_1S_xR_2 + 4H \rightarrow R_1H + R_2H + H_2S_x$$

(5) Solvent Partition

$$R_1S_xR_2 + X \text{———} \begin{array}{c} R_1 \\ \diagup \\ \diagdown \\ R_2 \end{array} S_x^+X^-$$

(6) Thermal decomposition

$$R_1S_xR_2 \xrightarrow{\Delta} R_1R_2 + S_x$$

$$RCH_2CH_2SH \xrightarrow{\Delta} RCH = CH_2 + H_2S$$

Table 5-6. Sulfur Forms in Selected Bituminous Coal – Worldwide

Region and Country	Location or Mine	Sulfur, Percent w/w[a]			Ratio Pyritic to Organic Sulfur
		Total	Pyritic	Organic	
ASIA					
USSR	Shakhtersky	0.38	0.09	0.29	0.031
China (mainland)	Taitung	1.19	0.87	0.32	2.7
India	Tipong	3.63	1.59	2.04	0.78
Japan	Miike	2.61	0.81	1.80	0.45
Malaysia	Sarawak	5.32	3.97	1.35	2.9
NORTH AMERICA					
U.S.	Eagle No. 2	4.29	2.68	1.61	1.7
Canada	Fernie	0.60	0.03	0.57	0.053
EUROPE					
Germany	—	1.78	0.92	0.76	1.2
United Kingdom	Derbyshire	2.61	1.55	0.87	1.8
Poland	—	0.81	0.30	0.51	0.59
AFRICA					
S. Africa	Transvaal	1.39	0.59	0.70	0.84
AUSTRALIA	Lower Newcastle	0.94	0.15	0.79	0.19
SOUTH AMERICA					
Brazil	Santa Caterina	1.32	0.80	0.53	1.5

a Moisture-free basis, pyrite + sulfate reported as pyrite.

Conjugated bonds between the alpha and sulfur atom tend to stabilize the structure. For example, thiophene and its derivatives such as benzothiophene or dibenzothiophene have sulfur bonds that are quite difficult to cleave.

5.2.2   Removal of Sulfur Dioxide

Because fuel cannot be 100% desulfurized due to technical and economic rea-
sons, the emission of sulfur dioxide from burning fuels is inevitable. To abate
the air pollution, one alternative is to remove the sulfur dioxide from the stack
gases by using chemical scrubbers. This process is often called **flue gas desulfurization (FGD)**. The most common process is the **limestone process**. The
stack gases are passed through a slurry of limestone (calcium carbonate), which
removes sulfur dioxide quite efficiently (and also removes all the HCl which is
co-produced). This reaction, as well as other SO_2 removal processes are summarized here:

Control Technology — Flue Gas Desulfurization (FGD)

- Wet Lime — Limestone Process

$$CaCO_3 \xrightarrow{\Delta} CaO + CO_2 \uparrow$$

$$CaO + SO_2 \rightarrow CaSO_3$$

$$CaO + SO_2 + \frac{1}{2}O_2 \rightarrow CaSO_4$$

$$CaO + H_2O \rightarrow Ca\,(OH)_2$$

- MgO Scrubbing

$$MgO + SO_2 \rightarrow MgSO_3$$

$$MgSO_3 \xrightarrow{\Delta} MgO + SO_2 \uparrow$$

MgO can be easily recycled

- Sodium Citrate Scrubbing

$$NaH_2\,Cit + SO_2 + H_2O \rightarrow NaH(HSO_3 \cdot H_2\,Cit)$$
$$\downarrow H_2O$$
$$S + NaH_2\,Cit + H_2O$$

- Catalytical Oxidation

$$SO_2 \xrightarrow{\;V_2O_5\;} SO_3 \xrightarrow{\;H_2O\;} H_2SO_4$$

A typical limestone is composed of 94% $CaCO_3$ and 1.5% $MgCO_3$; the remainder is inert material. Limestone is cheap and abundant, but the large quantities of calcium sulfate produced pose a major waste disposal problem. A process utilizing sulfate-reducing bacteria to regenerate the limestone has proven technically feasible, but large carbon sources are necessary for bacterial metabolism. Table 5-7 lists all the FGD processes.

5.2.3   Fate of Sulfur Dioxide

The ultimate fate of SO_2 will probably be in the ocean, which can absorb much more than we are now producing. Figure 5-4 illustrates the sulfur cycle. The lifetime of SO_2 in the atmosphere is much shorter than that of CO_2, which is several years. The emissions of SO_2 are 20 times less than those of CO_2, and background air concentrations are a million times less. It is estimated that the lifetime for industrial SO_2 is 3 days, as shown in Table 5-8. Concerns are raised about other species co-produced with SO_2 such as SO, SO_3, COS, CS, etc. They are in very low concentration. Therefore, these co-produced sulfur species are insignificant, as shown in Figure 5-5.

Table 5-7. Summary of SO$_2$ Removal Process

Process and Developer	Description	Chemistry
Sulfite absorption (Wellman-Power Gas)	A solution method for concentrating dilute SO$_2$ via bisulfite formation, crystallization, and thermal regeneration. No reduction or oxidation in the solution step	$SO_2 \text{ (dil.)} + H_2O + Na_2SO_3 \longrightarrow NaHSO_2$ $SO_2 \text{ (conc.)} + H_2O + Na_2SO_3 \xleftarrow{\text{heat}}$
Magnesium Oxide (Chemico/Basic)	Essentially a concentration process using MgO as a collector, followed by regeneration and the production of an SO$_2$ stream.	$SO_2 \xrightarrow{200°-300°F} MgSO_3 \xrightarrow{1400°F} SO_2$ $MgO \xleftarrow{\text{regeneration}}$ contact H$_2$SO$_4$ process
Molten salt (Atomic International)	Dilute SO$_2$ is concentrated by absorption in molten salt as sulfite, and then reduced to sulfide and hence H$_2$S. M stands for metal.	$SO_2 + M_2CO_3 \xrightarrow{800°F} M_2SO_3 + CO_2$ $H_2S + M_2CO_3 \longleftarrow M_2S + H_2O + CO_2$ Claus $\rightarrow$ S + H$_2$O, SO$_2$
Manganese dioxide (Mitsubishi)	SO$_2$ is initially concentrated and oxidized to metal sulfate, followed by regeneration of MnO$_2$ and production of ammonium sulfate.	$SO_2 \longrightarrow MnSO_4 \xrightarrow[\text{air}]{NH_4OH} (NH_4)_2SO_4 + H_2O$ $MnO_2 \xleftarrow{\text{regeneration}}$

Table 5-7 (continued)

Process and Developer	Description	Chemistry
Limestone (TVA, Combustion Engineering, Chemico, others)	Simultaneous reaction of SO_2 with limestone and air oxidation of resulting sulfite to sulfate results in a slag that requires suitable disposal. Reaction may take place inside furnace or in flue-gas scrubber.	$CaCO_3 \xrightarrow{SO_2,\ SO_3,\ air} CaSO_4 + CO_2$
Catalytic (Monsanto)	Accepts hot dilute SO_2 gas stream rather than high-concentration SO_2 for acid-plant feed.	$Air + SO_2 \xrightarrow[V_2O_5]{900°F} SO_3 \xrightarrow{H_2O} H_2SO_4$
Activated Carbon (Westvaco, Hitachi, Chemiebau, others)	All methods depend on absorptive powers of various forms of active carbon to first concentrate and then catalyze oxidation of SO_2 to SO_3 for acid or sulfate production. Fluidized, fixed, and plugged-flow beds have all been employed.	$SO_2 \xrightarrow[\text{active carbon}]{air,\ H_2O} H_2SO_4$
Ammonia scrubbing (Showa Denko)	Adsorption and concentration of SO_2 and air in ammonia solution yields bisulfate and thiosulfate which then forms sulfate, water, and sulfur.	$SO_2 + NH_4OH \xrightarrow{air} HN_4HSO_3 + (NH_4)_2S_2O_3$ $\rightarrow (NH_4)_2SO_4 + H_2O + S$

Table 5-8. Sources, Concentrations, and Major Reactions at Atmospheric Trace Gases

Gas	Anthropogenic Sources	Natural Sources	Background Concentration	Estimated Lifetime	Removal Mechanism
SO_2	combustion of coal and oil	volcanoes	0.002-0.01 ppm	3 days	oxidation to sulfate photochemically or catalytically
H_2S	chemical processes	biological	0.002-0.02 ppm	1 day	oxidation to SO_2
CO	combustion	oxidation of CH_4, oceans	0.12-0.15 ppm	0.1 year	reaction with OH in troposphere and stratosphere, soil removal
NO-NO_2	combustion	Bacterial action in soil	NO: 0.2-2 ppb, NO_2: 0.5-4 ppb	5 days	oxidation to nitrate by photochemical reactions or on aerosol particlces
NH_3	waste treatment	biological decay	6-20 ppb	2 weeks	reaction with SO_2 to form $(NH_4)_2SO_4$, oxidation to nitrate
N_2O	none	biological action in soil	0.25 ppm	4 years	photodissociation in stratosphere
CH_4	combustion, chemical processes	Swamps, paddy fields	1.5 ppm	1.5 years	reaction with OH

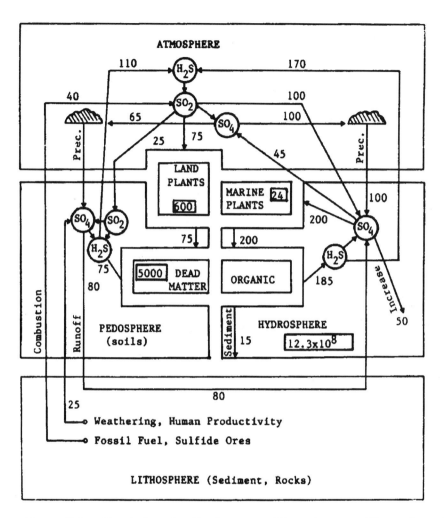

Figure 5-4. Circulation of sulfur in nature. The units are millions of metric tons of sulfur. The enclosed figures are simply amounts; the other figures are amount per year.

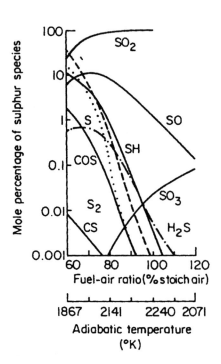

Figure 5-5. Equilibrium distribution of sulfur-containing species in propane-air flames with unburnt gases initially containing 1% SO_2. (After G.M. Johnson, C.J. Mathews, M.Y. Smith and D.V. Williams, *Combust Flame*, 15:211 (1970).)

5.3 AUTOMOTIVE EMISSIONS

The major pollutants from automobile driving is CO among many other pollutants such as particles SO_x and NO_x. All these will cause health problems in the biosphere. For example, the formation of photochemical smog in Los Angeles is related to the pollutant's properties. Two more issues will be addressed here: additives to improve the performance of engines such as oxgenators and anti-knocking agents and modification of the pollutant composition in automobile emissions with a catalytic converter.

5.3.1 Carbon Monoxide

More than 60% of the global anthropogenic carbon monoxide comes from automotive emissions as shown in Table 5-9. Carbon monoxide is an odorless, tasteless, and colorless gas. Upon entering the respiratory system, it combines in the lungs with the hemoglobin in the blood stream to form carboxyhemoglobin, COHb. This process reduces the hemoglobin's capacity to carry oxygen to the body tissues. Both oxygen and carbon monoxide are bound to the iron atoms in the hemoglobin molecules, but the binding of carbon monoxide is 320 times more effective than that of oxygen. Hence, low levels of carbon monoxide can still result in high levels of COHb.

Table 5-9. Estimated Global Anthropogenic Carbon Monoxide Sources in 1970

Source	Emission, Tg
Motor vehicles	222
Other mobile sources	25
Coal combustion	11
Fuel oil combustion	40
Industrial processes	22
Petroleum refining	5
Solid waste disposal	23
Miscellaneous (agricultural burning, etc.)	23
TOTAL	371

At COHb levels of 2–5%, effects such as impairment of time interval discrimination and visual acuity are found in the central nervous system. At levels greater than 5%, there are cardiac and pulmonary functional changes. Figure 5-6 shows the physiological effects of carbon monoxide poisoning.

The principle sources of CO are fossil fuel combustion and the biomass in nature. There is a small fraction of ^{14}CO in the troposphere. The residence time of CO in the troposphere is about 0.1 year. This approximation is based on the following example:

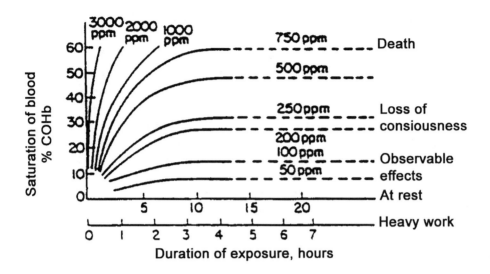

Figure 5-6.   Physiological effects of carbon monoxide poisoning. (Source: Lower Diagram: P. C. Wolfe, *Environ. Sci. Technol.* 5:213 (1971))

[**Example 5-1**]   From fossil fuel combustion rate, estimate the residence time of CO. Also, approximate the production rate of CO from biomass alone.

We will set up the production rate as Equations (a) and (b).

$$\frac{d(CO)}{dt} = P_1 + P_2 - k(CO) \tag{a}$$

P_1 = production rate of CO from biomass
P_2 = production rate of CO from fossil fuel combustion

$$\frac{d(^{14}CO)}{dt} = NP_1 + P_3 - k(^{14}CO) \tag{b}$$

P_3 = production rate of ^{14}CO in troposphere = 290 mole/yr
$N = 1.17 \times 10^{-12}$ = fraction of ^{14}C from biomass

Also, P_2 can be estimated.

We burn 5.7 Pg of C of fossil fuel/yr or (÷ 12)

$$= 475 \text{ T mole of C/yr}$$

Using

$\Phi = 1.4$ (excess of O_2 in atm), Φ here is equivalence ratio at $\Phi = 1.4$

the ratio $CO_2/CO = 1 \times 10^{-1}/115 \times 10^{-3}$ (from Fig. 5-12 by reading off from the graph) or $CO/(CO_2 + CO) \approx 0.015$

Therefore,

$$P_2 = 0.015 \times 475 \text{ T mole} = 7 \text{ T mole/yr}$$

Solving by assuming a state of equilibrium, and setting both Equations (a) and (b) equal to zero,

$$P_1 = 180 \text{ T mole/yr} = 5 \text{ Pg/yr}$$

$$k = 11 \text{ yr}^{-1}, \ 1/k = 0.09 \text{ yr} \approx 0.1 \text{ yr}$$

5.3.2 Nitrogen Oxides

In general, the symbol NO_x indicates a seven oxide mixture of nitrogen, which consists of N_2O, NO, NO_2, NO_3, N_2O_3, N_2O_4, and N_2O_5. Of those seven oxides, the most important two are NO and NO_2. In the United States the mobile sources account for 40% of NO_x, the remainder comes from the combustion of utility industry. The **thermal NO_x** is formed from reactions between nitrogen and oxygen in the air during combustion. NO_x produced in this manner is called the **Zeldovich's model**. Of course, if fuel contains nitrogeneous organics, NO_x can also be produced.

$$N_2 + O = NO + N$$

$$N + O_2 = NO + O$$

$$N + OH = NO + H$$

During combustion the ratio of NO and NO_2 can be a useful indicator. Therefore, the following two reactions are essential:

$$N_2 + O_2 = 2NO$$

$$K_{P_1} = \frac{(P_{NO})^2}{P_{N_2} P_{O_2}} = \frac{(Y_{NO})^2}{Y_{N_2} Y_{O_2}} = 10^{-30} \ (300K) = 7.5 \times 10^{-9} \ (1000K) = 4 \times 10^{-4} \ (2000K)$$

$$NO + \frac{1}{2} O_2 = NO_2$$

$$K_{P_2} = \frac{P_{NO_2}}{P_{NO} (P_{O_2})^{\frac{1}{2}}} = \frac{P_T^{-\frac{1}{2}} Y_{NO_2}}{(Y_{NO})(Y_{O_2})^{\frac{1}{2}}} = 10^6 \ (300K) = 1.2 \times 10^2 \ (1000K) = 3.5 \times 10^{-3} \ (2000K)$$

Here

K_{Pi} = equilibrium constant of the ith reaction.

P_i = partial pressure of component i, atm.

Y_i = mole fraction of component i

P_T = total pressure, atm.

From the two previous equations, it can be seen that

- At flame zone temperature (3000–3600°F), the NO_x concentration can reach 6000–10,000 ppm. The ratio of NO/NO_2 may reach from 500:1–1000:1.

- At flue gas exit temperature (300–600°F), there will be very low NO_x concentration (<1ppm). Therefore, no effort is made for any FGD for the control of NO_x. The ratios of NO/NO_2 at this low temperature is from 1:10,000–1:10.

[**Example 5-2**]   Calculate the NO_x concentration and the NO/NO_2 ratio for a typical flue gas composition of 76% N_2, 4% O_2, 8% CO_2, and 12% H_2O.

We already have K_{P1} at 2000°K, which is close to 3000°F.

$$K_{P_1}(2000K) = 4 \times 10^{-4} = \frac{(P_{NO})^2}{P_{N_2} P_{O_2}}$$

Thus,

$$\left(P_{NO}\right)^2 = 4\times10^{-4}\,(0.76)(0.04)$$

or $\phi_{NO} = 3.49\times10^{-3}$ atm. or eqlm. conc. $= 3490\,\text{ppm}$.

Also,

$$K_{P_2}\left(2000K\right) = 3.5\times10^{-3} = \frac{P_{NO_2}}{P_{NO}\,P_{O_2}^{\frac{1}{2}}}$$

Thus,

$$P_{NO_2} = 3.5\times10^{-3}\,\left(P_{NO}\right)(0.04)^{\frac{1}{2}} = 2.44\times10^{-6}\,\text{atm.}\quad\text{or eqlm. conc.} = 2.4\,\text{ppm}$$

or total NO_x concentration $= 3490 + 2.4 = 3492.4\,\text{ppm}$

Now

$$\frac{P_{NO}}{P_{NO_2}} = \frac{(NO)}{(NO_2)} = 1.42\times10^3$$

which falls in the range of the previous statement.

It was found that production of NO is sensitive to temperature; for example, experiments indicated that below 1600°C, less than 200 ppm of NO are formed, but above 1800°C, several thousands ppm of NO are found. An empirical formula is developed.

$$C_{NO}\,(\text{ppm}) = 5.2\times10^{17}\left(\exp\frac{-72,300}{T}\right)Y_{N_2}\,Y_{O_2}^{\frac{1}{2}}\,t \qquad [5\text{-}3]$$

Here, T is absolute temperature, (time in seconds), and Y_i is the mole fraction. In combustion, usually an equivalence ratio, Φ, is used

$$\Phi = \frac{(A/F)\,\text{actual}}{(A/F)\,\text{stoichiometric}} \qquad [5\text{-}4]$$

which is the quotient of air to fuel ratio; for example, the combustion of gasoline is

$$C_7H_{14} + 10\frac{1}{2}O_2 + 39\,N_2 + \frac{1}{2}Ar \rightarrow 7\,CO_2 + 7\,H_2O + 39\,N_2 + \frac{1}{2}Ar$$

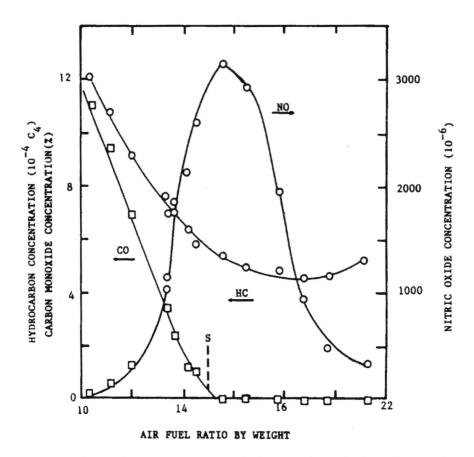

Figure 5-7. The effects of the equivalence ratio on hydrocarbon, carbon monoxide, and nitricoxide exhaust emission. (Source: W.G. Agnew, Research Publication GMR-743 General Motors Corp., 1968)

Here, the stoichiometric ratio is for the combustion of 98g of fuel (C_7H_{14}). It requires 336g of oxygen, 1092g of nitrogen, 20g of argon, and 1478g of air. An approximation for the molar basis is n (1 + 79/21) 29. If 10.5 moles of oxygen is used, then the required air is 1450g (29 is the molecular weight of air). Therefore, the mass ratio of Φ_{mass} = 1450/ 98 = 14.8 and the molar ratio of Φ_{mol} = 1. In Figure 5-7, if Φ < 1, and fuel is rich, then unburned fuel vapors and CO will be emitted. On the other hand, if Φ > 1 and fuel is lean, then NO emission will be increased.

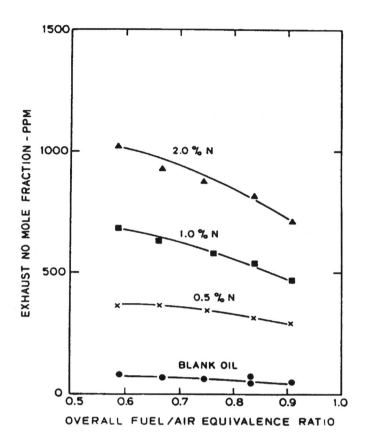

Figure 5-8. Measured exhaust NO mole fractions (dry basis, corrected to stoichiometric) for an oil-fired furnace with various amounts of pyridine (C_5H_5N) added to the fuel.

Another type of NO_x is termed **fuel NO_x**. When a fuel contains organically bound nitrogen, such as coals or fuel oils, the contribution of fuel NO_x to total NO_x production is significant, as shown in Figure 5-8. Most of the United States coals contains % N value ranging from 0.5%–2%. The residual fuels range from 0.1%–0.5%. However, the shale oil may contain up to 6% N. Usually, the C–N bond cracks easier than the N–N bond. This process explains the easy formation of NO. No matter what source the nitrogen comes from, the higher temperature seems to leave a definite overabundance of NO and NO_2 concentration. See Figure 5-9 for fixed Φ and Figure 5-10 for variable Φ. Even the turbine inlet temperature shows this trend (Figure 5-11).

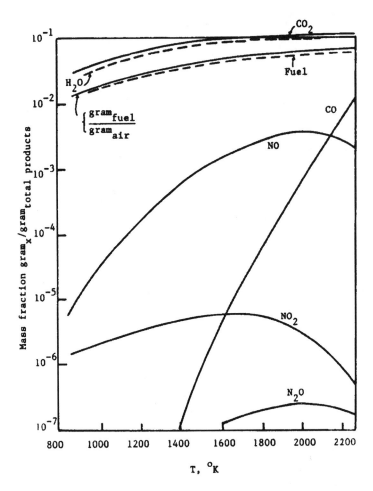

Figure 5-9.   Variation of NO_x and CO production with temperature.

A control strategy is therefore hinged on the decrease of combustion temperature. For example, the **flue gas recirculation** is intended to reroute some of the flue gas to furnace at a lower temperature. The **reduced air preheat** lowers the peak temperature in the flame zone. Also, the **reduced firing rates** have been practiced to reduce the heat released per unit volume. **Water injection** or steam injection can also become effective means for reducing thermal NO_x. Also, flue gas treatments such as the **selective catalytic reduction**, which have been operating in Japan, can convert the NO_x by NH_3 injection.

$$4\,NO + 4\,NH_3 + O_2 \rightarrow 4\,N_2 + 6\,H_2O$$

$$2\,NO_2 + 4\,NH_3 + O_2 \rightarrow 3\,N_2 + 6\,H_2O$$

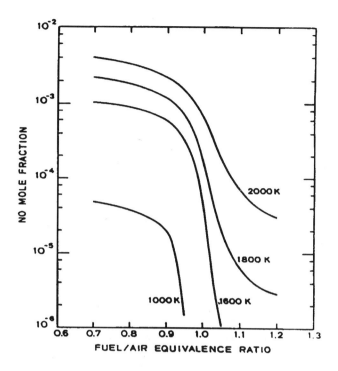

Figure 5-10. Equilibrium NO mole fractions for premixed methane-air combustion at pressure of 1 atm.

Also, the wet absorption method has been used. For example, at high pH,

$$NO + MnO_4^- + 2OH^- \rightarrow NO_2^- + MnO_4^{2-} + H_2O$$

and at low or neutral pH,

$$NO + MnO_4^- \rightarrow NO_3^- + MnO_2$$

Finally, the adiabatic flame temperature and equilibrium product distribution for the constant adiabatic combustion of n-octane is shown in Figure 5-12. The molar ratio of different products can be evaluated from different values of equivalence ratio.

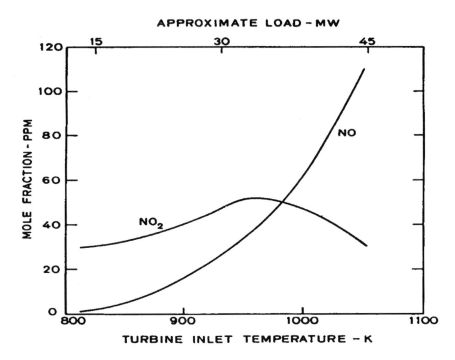

Figure 5-11. NO and NO₂ exhaust mole fractions for a gas turbine operating on natural gas for various loads. (From Ref. 21. Reprinted with permission from Gordon and Breach Science Publishers S.A. G.M. Johnson and M.Y. Smith, *Combustion Sci. Technol.* 19:67 (1978))

5.3.3 Photochemical Smog

Photochemical smog was first recognized as a problem in Los Angeles in 1943, and has since been detected in many cities of the world. It refers to the complex mixture of products formed from the interaction of sunlight with two major components of automobile exhaust, nitric oxide (NO_x), and hydrocarbons.

Furthermore, stable meteorological conditions such as an inversion layer favor the creation of smog. The presence of smog, which is highly oxidizing in nature, is known by its effects—cracking of stressed rubber, eye and throat irritation, unpleasant odor, plant damage, and decreased visibility.

The chemistry of photochemical smog is complex; a flow chart for the chemistry of smog formation is given in the figure on the next page. Basically, it involves a catalysis of the oxidation of hydrocarbons in air to form products that are themselves reactive and irritating to biological tissue. The catalysis requires sunlight and the key chemical ingredient, nitrogen dioxide. At the beginning of the smog formation reaction, however, there is only nitric oxide (NO). It is the combination of hydrocarbon with nitric oxide that forms nitrogen dioxide. The process is set in motion by a reaction with the ozone to produce hydrocarbon free radicals, highly reactive molecules with unpaired electrons. Because of the unpaired electron, the free radical combines readily with molecular oxygen to form a reactive oxygen species that can easily transfer an oxygen atom to nitric oxide, which in turn forms more nitrogen dioxide. This process regenerates the free radicals and works as a chain reaction. The free radicals eventually combine with oxygen and nitrogen dioxide to form the oxidation products (eye and lung irritants), and gradually the hydrocarbon supply is depleted. The most potent eye and lung irritants in smog are peroxyacetylnitrates (PAN).

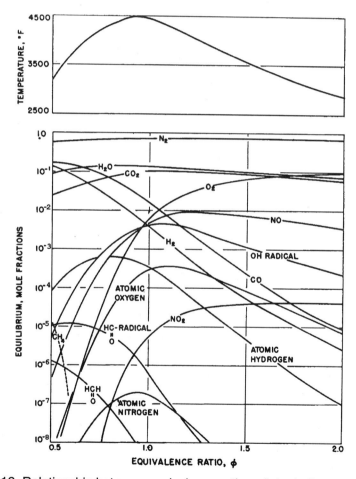

Figure 5-12. Relationship between equivalence ratio, adiabatic flame temperature, and equilibrium product distribution for the constant volume adiabatic combustion of n-octane. Initial temperature 77°F and initial pressure 10 atm.
(Source: R.J. Stefferson et. al. *Engineering Bull.* Purdue University, Engineering Extensive Sr., No. 122, 1966)

Los Angeles smog is comparable to London fog as an industrial event. London fog requires smoke (combustion), fog (location), and coal volatile (industry); Los Angeles smog requires NO_x (combustion), sunlight (location), and hydrocarbons (industry—mostly transportation). The schematics are indicated here:

Industrial Events

London Fog

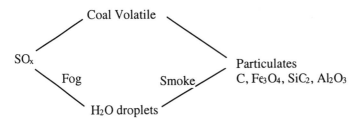

Los Angeles Smog

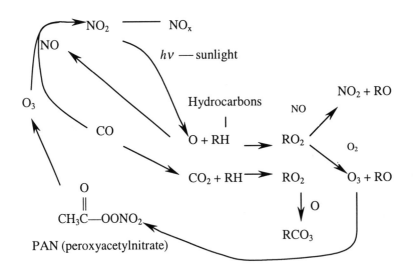

5.3.4   Antiknocking Agents

The **combustion reaction** in an internal combustion engine involves a series of radical reactions. Without suitable control, the air/fuel mixture tends to ignite spontaneously as it is heated during the compression stroke in the cylinder. This sets off a small explosion preceding the desired one, which is induced by the spark when the fuel is maximally compressed. This preignition, or knocking, decreases the efficiency of the engine and also sets up harmful stresses that increase engine wear. To correct this, an antiknocking agent is often added to control the radical reaction.

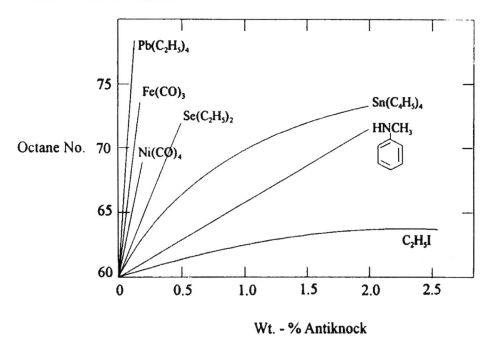

Figure 5-13.   Relative effects of several metallo-organic compounds as antiknock agents. [From Fig. 4, Livingston, *Ind. Eng. Chem.*, 43:663 (1951)]

Figure 5-13 gives the relative effects of several antiknocking agents. It shows tetraethyl lead to be the most effective. Even with a low dosage required, it results in a higher motor octane number. Due to the concern with inorganic lead acting as an agent to cause a variety of effects on human health, including liver and kidney damage, gastrointestinal damage, mental health effects in children, and abnormalities in fertility and pregnancy, the use of $Pb(C_2H_5)_4$ as an antiknocking agent is diminishing.

An alternative to using tetraethyl lead is to change the composition of gasoline so that it is less likely to generate the free radicals during preignition. Aromatic hydrocarbons such as BTX (benzene, toluene, and xylene) do not easily break down into radicals and could be added to upgrade the quality of gasoline, but they are more costly. Some other antiknocking agents such as MMT (methylcyclopentadienyl manganese tricarbonyl) have been chosen, but they are less effective and have a great potential to generate carbonyl radical, a very toxic by-product.

Following the phase out of lead alkyl additives, the oxygenated gasoline program went into effect in 1992 requiring gasoline to contain 2.7% by weight of oxygen. To maintain required oxygen content, corresponding imports of oxygenators have to be blends. The most common one is methyl t-butyl ether (MTBE). MTBE is used as an oxygenater for reformulated gasoline. In many gasoline blends the MTBE content can reach 11–15% because gasoline needs an oxygen content of 2.0–2.7%; thus MTBE is not an additive. The production was 7.1 M tonnes in 1993 and was anticipated to reach 15.6 M tonnes in 1998. MTBE is a suspected human carcinogen. During 1996, a jet ski event was held in Anaheim, California, and 15 M gal. of water from Orange County could not be recycled due to high contamination levels. Three drinking water wells in Santa Monica were found to contain up to 600 ppb of MTBE in 1996. Furthermore, many other ethers, such as dipropyl ether (DIPE), ethyl t-butyl ether (ETBE) and t-anyl methyl ether (TAME) are also used for some gasoline blends to substitute the use of MTBE. The fate of MTBE is that it can be oxidized into t-butyl formate (air) or hydrolized into t-butyl alcohol (water).

Antiknocking Agents

Tetraethyl Lead

$$Pb(CH_3CH_2)_4 \rightarrow Pb + 4CH_3CH_2$$

MMT (Methyl Cyclopentadienyl Manganese Tricarbonyl)

$$CH_3 \quad \boxed{\bigcirc} \quad Mn(CO)_3 \quad \longrightarrow \quad Mn + CH_3C_5H_5 + 3\,CO$$

MTBE (Methyl Ter-Butyl Ether)

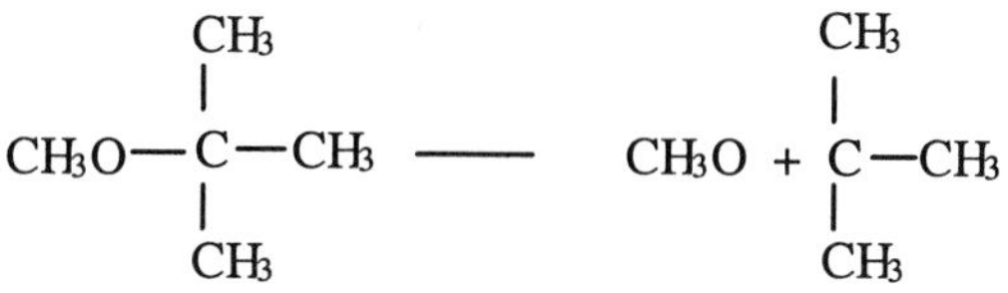

$$CH_3O-\underset{\underset{CH_3}{|}}{\overset{\overset{CH_3}{|}}{C}}-CH_3 \quad \longrightarrow \quad CH_3O + \underset{\underset{CH_3}{|}}{\overset{\overset{CH_3}{|}}{C}}-CH_3$$

There is a considerable amount of additives introduced when using transportation fuel, as shown here:

Additives

Stabilizer
Prevention of gum formation (e.g., Ketimines)

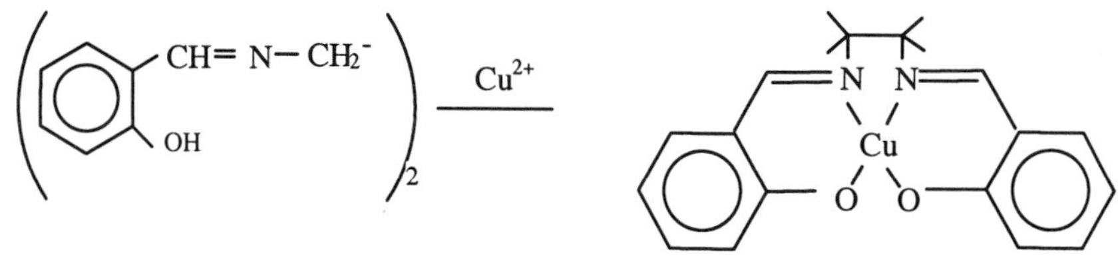

Pour Point Depressor

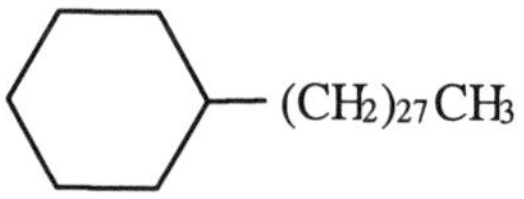

Antioxidant

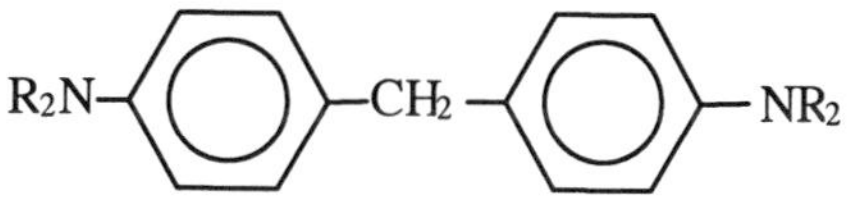

High Pressure Additive

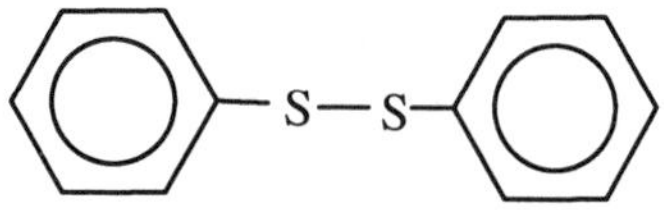

Corrosion Inhibitors

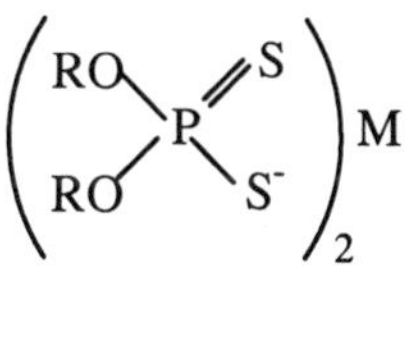

5.3.5   Catalytic Converters

As previously mentioned, it is obvious that automotive emissions cause air pollution problems. Stringent standards have been set for the concentrations of nitrogen oxides as well as hydrocarbons and carbon monoxide under the U.S. Federal Clean Air Act. The conversion efficiency must satisfy all three pollutants with a narrow range of equivalence ratio, as shown in Figure 5-14. As known in the science of combustion, there are conflicting requirements for getting rid of different pollutants — that is, more complete combustion is necessary to lower the concentrations of HC and CO, while lower temperatures are required to decrease the concentration of NO. A **catalytic converter** is a solution that eliminates pollutants from the exhaust gases before they are vented to the atmosphere. The overall chemistry of catalytic conversion is rather complicated. HC and CO are oxidized to form carbon dioxide and water. However, the oxidation of NO would produce nitrogen dioxide, which is itself a pollutant. Instead, the nitric oxide must be reduced to molecular nitrogen, which is stable. Thus, two catalytic converters are required — one for reduction and the other for oxidation, as shown in Figure 5-15. Finely divided platinum is used in both converters. Figure 5-16 shows some of the reactions involved.

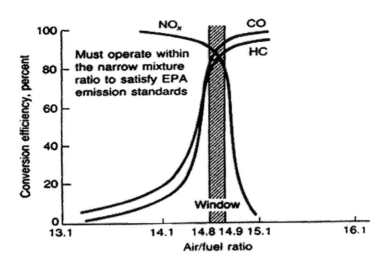

Figure 5-14.   Effect of the air/ fuel ratio on conversion efficiencies of a three-way catalyst. (Source: Niepoth, G.W., Gumbeton, J.J., and Haefner, D.R., *Closed Loop Carburetor Emission Control Systems*, 71st annual meeting, Air Pollution Control Association, June 1978. Used by permission of General Motors Corporation).

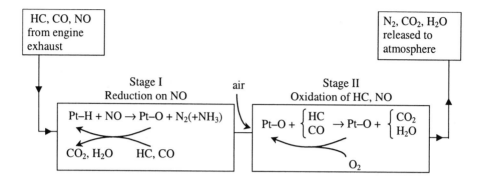

Figure 5-15. Catalytic converter for treating auto emissions.

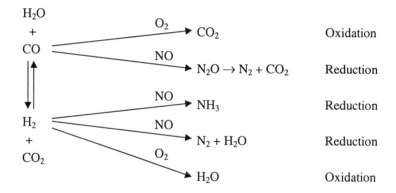

Figure 5-16. Reactions in catalytic converters.

A fundamental requirement of the converter is that the gasoline is un-leaded. If leaded gasoline were burned, the lead compounds would be expelled as small solid particles of lead sulfate and oxide. These compounds would deposit on the microscopic pores of the converter and cover the catalytic metals, thereby reducing the capability of the emission conversion.

5.4 TROPOSPHERIC CHEMISTRY

The troposphere contains the air we breathe; from it falls the water we drink. Its chemistry consists mainly of the oxidation of reduced molecules released by human and nonhuman activities on the Earth's surface. Apart from CO_2, N_2, and N_2O, which are unreactive, the major emissions are SO_2, H_2S, CO, NO and

NO$_2$, NH$_3$, and hydrocarbons. It has been estimated that the amount of pollutants produced by the human race is 1 ton/yr per capita, which is about 5–6 pounds/day. The residence time for various atmospheric trace gases was determined based on the rate of release and background concentration (listed in Table 5-8). It ranges from 1 day to a few years. Due to the low reactivity of molecular oxygen, the rate of oxidation of the reduced gases depends on the concentration of the catalysts in the air. The major catalysts appear to be oxygen and hydroxyl radicals. The hydroxyl radicals are especially important because they control the rate at which many trace gases are oxidized and removed from the atmosphere. In the proceeding text, there are some reactions of hydroxyl and other radicals in the troposphere. Also, the reactions with CO, CH$_4$, and NO are summarized. For details, see Figure 5-17.

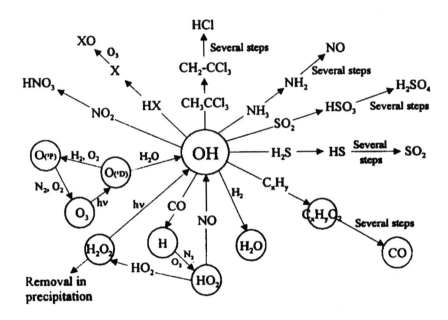

Figure 5-17. Photochemistry of OH radical controls trace gas concentration.

Hydroxyl Radical in Troposphere

$$O_3 + hv \,(1200 > \lambda > 315 \text{ nm}) \rightarrow O(^3P) + O_2$$

$$O(^3P) + O_2 + M \rightarrow O_3 + M \qquad (M = N_2)$$

$$O_3 + hv \,(\lambda < 315 \text{ nm}) \rightarrow O(^1D) + O_2$$

$$O(^1D) + H_2O \rightarrow 2OH$$

Reaction with CO + CH$_4$

$$CO + OH \rightarrow CO_2 + H$$

$$CH_4 + OH \rightarrow CH_3 + H_2O$$

$$H + O_2 \rightarrow HO_2 \quad \text{(hydroperoxyl)}$$

$$CH_3 + O_2 \rightarrow CH_3O_2 \quad \text{(methylperoxyl)}$$

$$HO_2 + NO \rightarrow NO_2 + OH$$

$$HO_2 + O_3 \rightarrow 2O_2 + OH \quad \text{(Regenerate)}$$

$$HO_2 + OH \rightarrow H_2O + O_2 \quad \text{(terminate)}$$

$$HO_2 + HO_2 \rightarrow H_2O_2 + O_2 \quad \text{(rain out)}$$

Reaction with NO Supersonic Transport (SST)

$$HO_2 + NO \rightarrow NO_2 + OH$$

$$NO_2 + h\nu \rightarrow NO + O$$

$$NO + OH \rightarrow HNO_2$$

$$NO_2 + OH \rightarrow HNO_3 \quad \text{(rain out)}$$

In all cases except SO$_2$ and CO, the natural emissions of these trace gases greatly exceed the contribution from human activities, and in all cases the removal rates are fast enough that background concentrations are far less than the levels that might affect human health. However, these rates are not fast enough to cope with the high local concentrations that result from urban congestion and lead to serious air pollution.

We have already emphasized that photochemical smog is based on a free radical mechanism. However, free radical chemistry and inhibitors are also key elements of pollutant elimination.

Diethylhydroxyamine (DEHA), an inhibitor, is a clear, volatile liquid with only a mild odor. So far it has proven to be nontoxic and nonmutagenic. Some tests should be conducted on it by dispersing it into the atmosphere. The following is a summary of its functions as a smog inhibitor and a flame retardant:

- Inhibition of Photochemical Smog

  DEHA (diethyl hydroxylamine)

  $(C_2H_5)_2NOH + OH \rightarrow (C_2H_5)_2NO + H_2O$

- Flame Retardants

  Tris [tris (1,2,3—dibromipropyl)phosphate]

  $O=P—(O—CHBr—CH_2Br)_3$

  ↓

  $PO_4 + 3CHBr—CH_2Br$

  ↓

  $Br + CH = CH$

A radical inhibitor is a stable radical. Asphaltene in asphalt is an example of such; it will inhibit even the radiation from cosmic rays. Thus, asphalt roof shingles have certain advantages. Other radiation shields or aging retardants are as follows:

Radical Inhibitor

- Vitamin C

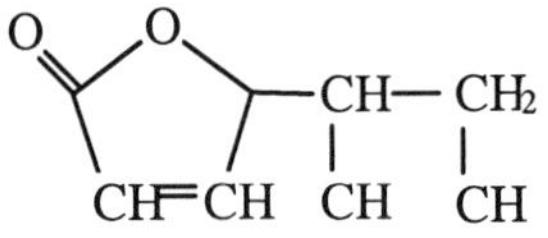

- BHT

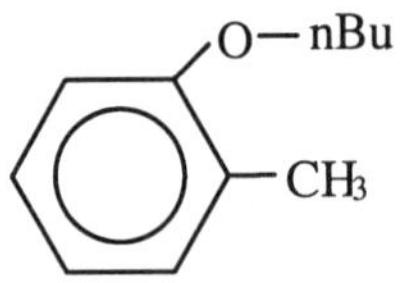

- MEA

  $NH_2CH_2CH_2SH$

- WR 2721

$$NH_2 - (CH_2)_3 - NH(CH_2)_3 - S - \overset{\overset{\displaystyle O}{\|}}{\underset{\underset{\displaystyle OH}{|}}{P}} - OH$$

- AET

$$H_2N-(CH_2)_3-S-C\overset{\displaystyle NH_2}{\underset{\displaystyle NH}{<}}$$

- Cystein

$$HOOC-\underset{\underset{\displaystyle NH_2}{|}}{CH}-CH_2-SH$$

(Aging retardant)

5.5 METEOROLOGY INFLUENCED BY POLLUTION

Even though the total input and output of radiant energy to and from the Earth is essentially in balance, it is not in balance at every point on the Earth. The amount of energy reaching the Earth's surface depends, in part, on the nature of the surface (sea vs. land, for example) and the degree of cloudiness, as well as on the latitude of the location. The uneven distribution of energy resulting from latitudinal variation in insulation and from differences in absorptivity of the Earth's surface leads to the large scale air motions of the earth. In particular, the tendency to transport energy from the tropics to polar regions, thereby redistributing energy inequalities on the Earth, is the overall factor governing the general circulation of the atmosphere.

To predict the general pattern of macroscale air circulation on the Earth, we must consider both the tendency for thermal circulation and the influence of **Coriolis forces**, as shown in Figure 5-18. Meteorology is of an environmental engineer's concern, because it is both related to the weather and pollution problems in the troposphere. Transport of the pollutants depends heavily on the meteorological conditions. The pattern of general circulation shown in Figure 5-18 does not represent the actual state of atmospheric circulation on a given day. The irregularities of land masses and their surface temperatures tend to disrupt the smooth global circulation patterns. Another influence that tends to break up zonal patterns is the Coriolis force. The boundary layers and the unevenness of land surfaces will be discussed in Chapter 6.

Air that converges at low levels toward regions of low pressure must also execute a circular motion because of Coriolis forces. Low pressure systems are usually associated with warm fronts, giving steady rain and drizzle, and cold fronts, giving heavy local rain. The effect of friction at the surface is to direct the winds at low levels in part toward the region of low pressure, producing an inward spiraling motion.

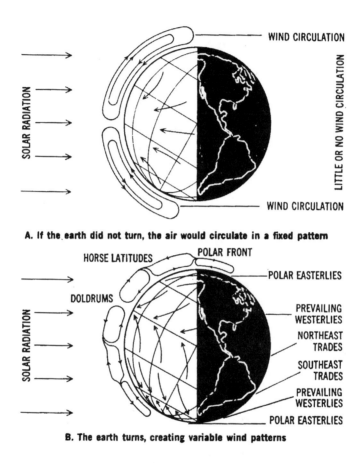

A. If the earth did not turn, the air would circulate in a fixed pattern

B. The earth turns, creating variable wind patterns

Figure 5-18.   Global wind patterns. (Source: American Lung Association)

This vortex-like motion is called a **cyclone**. The center of a cyclone is usually a rising column of warm air. Similarly, a low-level diverging flow from a high pressure region will spiral outward. No wave fronts and no wind and dispersion, which cause serious air pollution episodes, are associated with high pressure systems. Such a region is called an **anticyclone**.

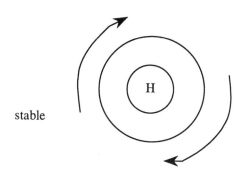

stable

In the Northern Hemisphere, the motion of a cyclone is counterclockwise and that of an anticyclone, clockwise.

5.5.1 Dispersion of Pollutants

Turbulent diffusion has two approaches: **Eulerian** and **Lagrangian**.

For the Eulerian approach, the species are relative to a fixed coordination system; for the Lagrangian approach, concentration changes are relative to moving fluid. The following are descriptions for both approaches:

Eulerian Approach

For $i = 1, 2, ..., N$ species, then

$$\frac{\partial C_i}{\partial t} + \frac{\partial}{\partial x_j} U_j C_i = D_i \frac{\partial^2 C_i}{\partial x_j \partial x_j} + R_i (C_1 ... C_N, T) + S_i (x, t) \qquad [5\text{-}5]$$

C_i = concentration of ith species

U_j = jth component of fluid velocity

D_i = molecular diffusivity of species i in fluid

R_i = rate of generation of species i by chemical reaction

S_i = rate of addition of species i at location of $\overline{x}$ = (x_1, x_2, x_3) and time t

Lagrangian Approach

For a single particle at location and time $(\overline{x'}, t')$, considering subsequent motion at a later time t, the probability is

$$\psi(\overline{x}, t) d\overline{x} \qquad [5\text{-}6]$$

at time t, the element is

$$x_1 \rightarrow x_1 + d\,x_1$$

$$x_2 \rightarrow x_2 + d\,x_2 \qquad\qquad [5\text{-}7]$$

$$x_3 \rightarrow x_3 + d\,x_3$$

Probability density function is

$$\int_{-\infty}^{\infty}\int_{-\infty}^{\infty}\int_{-\infty}^{\infty} \psi(\bar{x},t)\,d\bar{x} = 1$$

$$\psi(\bar{x},t) = \int_{-\infty}^{\infty}\int_{-\infty}^{\infty}\int_{-\infty}^{\infty} Q(\bar{x},t|\bar{x'},t')\psi(\bar{x'},t')d\bar{x'} \qquad\qquad [5\text{-}8]$$

We will now discuss how to determine the atmospheric concentration of pollutants emitted from single sources. We are usually concerned with horizontal dispersion (y-direction) and vertical dispersion (z-direction). **Horizontal dispersion** depends on the wind direction and speed. Meteorologists commonly use the **wind rose** to present wind direction and wind speed data. Figure 5-19 illustrates pollution rises in SO_2 concentration of 250 mg/m^3 in a chemical plant.

Vertical dispersion depends on temperature profiles in the atmosphere. Figure 5-20 illustrates five possible **temperature profiles**. (1) **Adiabatic lapse rate**: T decreases with height such that any vertical movement imparted to an air parcel will result in the parcel maintaining the same T or density as the surrounding air (Neutral stable). (2) **Superadiabatic**: A rising air will be warmer than its environment so it becomes more buoyant (Unstable). (3) **Subadiabatic**: A rising air parcel is cooler than its surroundings so it becomes less buoyant and returns (Stable). (4) **Isothermal**: Temperature constant with height. (Stable). (5) **Inversion**: Temperature increases with height (Extremely stable).

The atmospheric stability can determine the plane shapes out of a stack, as shown in Figure 5-21. In the figure the broken lines represent the adiabatic lapse rate and the solid lines depict the prevailing conditions. At the theoretical adiabatic lapse rate, the atmosphere is uniform, and coning occurs. Under unstable conditions, looping is observed. The dispersion is greater, but associated fluctuations will cause segments of the plume to reach ground level close to the stack.

Under stable conditions, the plume will have little tendency to disperse — a condition known as fanning.

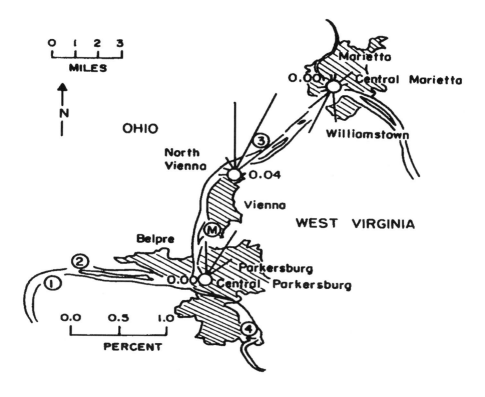

Figure 5-19. Pollution roses, with SO_2 concentrations greater than $250\mu g/m^3$. The major suspected sources are the four chemical plants, but the data indicate that Plant 3 is the primary culprit.

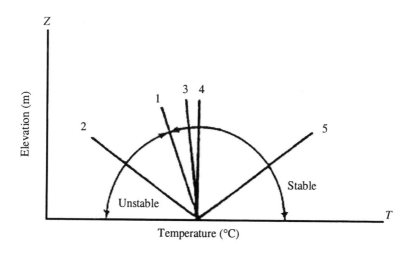

Figure 5-20. Temperature profiles.

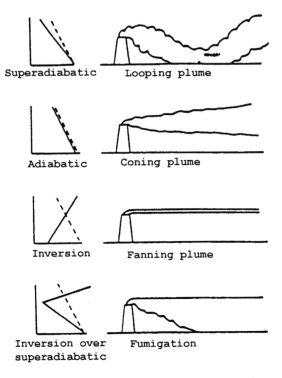

Figure 5-21.   Plume shapes and atmospheric stability.

5.5.2   Gaussian Model

There are several models available for predicting the downwind concentrations of a single source. Attempts have been made to correlate the dispersion coefficients with atmospheric conditions. Table 5-10 notes the atmospheric conditions and Figure 5-22 gives the dispersion coefficients. The following example explains how to use these figures and the Gaussian model. By applying the **Gaussian function**, the solution for the plume concentration takes the form where

$$C(x, y, z, H) = \frac{Q}{2\pi\, \sigma_y \sigma_z U} \exp\left[-\frac{1}{2}\left(\frac{y}{\sigma_y}\right)^2\right]\left[\exp\left(-\frac{1}{2}\left(\frac{z-H}{\sigma_z}\right)^2\right)\right] + \exp\left[-\frac{1}{2}\left(\frac{z+H}{\sigma_z^2}\right)^2\right]$$

[5-9]

C = concentration in g/m^3

Q = source strength in g/sec

U = average wind speed in m/sec

σ_y, σ_z = dispersion coefficients in y and z direction in meters

H = effective height of source emission in meters

Table 5-10.   Atmospheric Stability Key for Figure 5-22

Surface Wind Speed	Day Incoming Solar Radiation (Sunshine)			Night Thinly Overcast	
(at 10 m), m/sec	Strong	Moderate	Slight	4/8 Low Cloud	3/8 Cloud
2	A	A-B	B	—	—
2-3	A-B	B	C	E	F
3-5	B	B-C	C	D	E
5-6	C	C-D	D	D	
6	C	D	D	D	D

[Example 5-3]   For a sunny summer afternoon, estimate the ground level concentration 200 meters downward from a stack if the wind speed is 4m/sec. The height of the stack is 20 m and the efficient concentration (Q) out of it is 0.01 kg/sec.

From Table 5-10 it is found that type B is most fitted to the conditions (sunny afternoon and U = 4m/sec). Then from Figure 5-22, σ_y = 36 and σ_z = 20 at x = 200 m.

At plane center line $y = 0$.

At ground level $z = 0$.

Thus, C =

$$\frac{0.01}{2(3.14)(4)(100)(40)} \exp\left[-\frac{1}{2}\left(\frac{0}{100}\right)^2\right]\left[\exp\left(-\frac{1}{2}\left(\frac{0-20}{40}\right)^2\right) + \exp\left(-\frac{1}{2}\left(\frac{0+20}{40}\right)^2\right)\right]$$

$$= 9.95 \times 10^{-8}(1)(e^{-1/8} + e^{-1/8})$$

$$= 1.76 \times 10^{-7} \text{ kg/m}^3 = 176 \text{ }\mu\text{g/m}^3$$

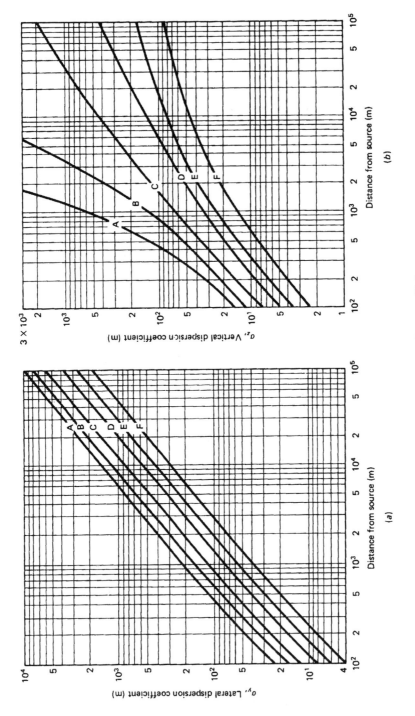

Figure 5-22.　Dispersion coefficients.

REFERENCES

5-1. W. L. Chameides and D. D. Davis, "Chemistry in the Troposphere," C & E N, p. 39 (Oct. 1982).

5-2. J. H. Seinfeld, *Air Pollution*, McGraw-Hill, New York, 1975.

5-3. H. C. Perkino, *Air Pollution*, McGraw-Hill, New York, 1974.

5-4. W. Strauss and S. J. Mainwaring, *Air Pollution*, Edward Arnold, London, 1984.

5-5. T. G. Spiro and W. M. Stigliani, *Environmental Issues in Chemical Perspective*, State University of New York Press, Albany, New York, 1980.

5-6. H. J. Sanders, "Flame Retardants," C & E N, p. 22 (Apr. 24, 1978).

5-7. J. B. Edwards, *Combustion*, Ann Arbor Science, Ann Arbor, Michigan, 1974.

5-8. I. Glassman, *Combustion*, Academic, New York, 1977.

5-9. D. J. Patterson and N. A. Henein, *Emissions from Combustion Engines and Their Control*, Ann Arbor Science, Ann Arbor, Michigan, 1973.

5-10. A. W. Demmler, "Automotive Catalysis," Auto. Eng. 85, 29-32 (1977).

5-11. G. Baumbach, *Air Quality Control: Formation and Sources, Dispersion, Characteristics and Impact of Air Pollutants*, Springer-Verlag, Berlin, 1996.

5-12. R. M. Whitcomb, *Non-Lead Antiknock Agents for Motor Fuels*, Noyes, Porkridge, New Jersey, 1975.

5-13. N. Calder, *The Weather Machine*, Viking Press, New York, 1974.

PROBLEM SET

1. Using the figures of combustion product distribution versus equivalence ratio (Figure 5-12), and given a 10 gallon gasoline (isooctane) consumption in an automobile without emission control, approximate the following:
   a. The volume ratios of CO and CO2 at $\Phi = 1$, 0.7 and 1.3
   b. The weight in lbs. of the CO and CO2 produced at $\Phi = 1$.

2. Estimate the concentration of SO_2 downwind of a 1,000 MW power
    plant burning 10,000 tons of 1% sulfur coal per day. The stack height is
    250 meters. The wind speed has been measured on a clear day as 5
    m/sec at the top of a 10 m tower. Find the pollutant concentration at a
    distance of 1 and 5 km away if the meteorologists estimate the hourly
    mean value of diffusion coefficients in y and z directions to be:

x(m)	y(m)	z(m)
1	140	125
5	540	500

THE HYDROSPHERE

*I*n addition to sunlight and air, water is an indispensable element for most plant and animal life. A casual observation of the world map would suggest that the supply of water is endless, since it covers over 80% of the earth's surface. Unfortunately, we cannot use it directly; over 95% is in the salty oceans, 2% is tied up in the polar ice caps, and most of the remainder is beneath the Earth's surface. Therefore, only a small fraction of the water in the world is available for human use. This section will focus on hydrology, the chemical properties of water, the chemistry of runoff, hardness and acid rain, groundwater movement, and contaminated aquifers. We especially emphasize the chemical aspects of underground water contamination and the chemical restoration of contaminated aquifers. The following three chapters will describe the hydrosphere in greater depth. The major topics of the upcoming chapters will be pollution in natural water and the chemical treatment of wastewater.

Water is not present merely in seas and lakes. Table C-1 gives a picture of the total hydrosphere. All water is locked into a recycling process called the

hydrologic cycle, shown in Figure C-1. One third of the solar energy absorbed by the earth's surface goes into the hydrologic cycle. Solar energy, especially in the tropics, evaporates water from the ocean surface, filling the air mass above with large quantities of water vapor. When these warm, moist, maritime masses conflict with cool, dry air over large land areas, some of the water vapors precipitate out. Seventy percent of the annual precipitation evaporates or is transpired by plants, while the remaining 30% goes into the stream flow. About a quarter of this stream flow is used for various purposes that can be broadly divided into irrigation, municipal water supplies, industrial uses, and electric utilities. Some of the water is evaporated during use, but most of it eventually finds its way into the ocean. Figure C-2 illustrates the water cycle.

One definition for **runoff**, also called **rainfall excess** in the hydrological cycle shown previously, is the rainfall that is not absorbed by soil. Depending on local surface and soil conditions,

$$R = cP \qquad\qquad\qquad\text{[C-1]}$$

and the storage is $(1 - c)P$. From the mass balance,

$$R = P - I - I_A - E - T \qquad\qquad\qquad\text{[C-2]}$$

Table C-1. The Total Hydrosphere

Atmosphere	$\begin{cases}\text{Rain}\\\text{Hail, sleet and snow}\\\text{Water vapor}\end{cases}$	
Biosphere	$\begin{cases}\text{Body fluids (external)}\\\text{Cell liquid (internal)}\\\text{Biopolymer hydration (bound)}\end{cases}$	
Lithosphere	$\begin{cases}\text{Groundwater}\\\text{Juvenile water}\\\text{Water on hydration}\end{cases}$	
Hydrosphere	Terrestrial	$\begin{cases}\text{Springs and swamps}\\\text{Ponds, lakes, ice and snow cover}\\\text{Streams, river, glaciers}\\\text{Estuaries}\end{cases}$
	Marine	$\begin{cases}\text{Seas}\\\text{Oceans}\\\text{Interstitial water in marine sediments}\end{cases}$

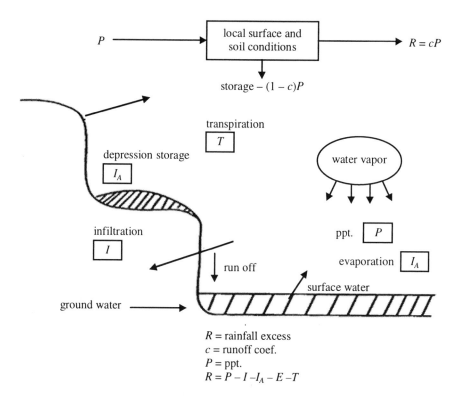

Figure C-1. This figure depicts the hydrologic cycle.

where R is rainfall excess, P is **precipitation**, I is **infiltration**, I_A is **depression storage**, E is **evaporation**, T is **transpiration**, and c is the **runoff coefficient**. Usually rainfall excess is a function of time.

It should be emphasized that for water usage

$$\text{Withdrawals} = \text{consumption} + \text{returns}$$

For example, with regards to irrigation in the United States, freshwater usage consists of 50% consumption and 50% return; however, for thermal electric power, there is 2% consumption and 98% return. The U.S.G.S., giving examples of withdrawals that serve a variety of ends, concludes that there are nearly 2000 gal of water withdrawn per person per day. Out of the 2000 gal. of water, the domestic use accounts for only 100 gal of personal use in the home, while the bulk is industrial (e.g., 60% or 1200 gal—it takes 100,000 gal to make one automobile) and agricultural (e.g., 35% or 700 gal for irrigation).

In the lithosphere-hydrosphere, the immense water in the oceans may be a result of the squeezing of the material of the planet (lithosphere). Estimation has been made that during earth's geological history, hot springs on continents and the ocean floor release 6.6×10^{16} g of water per year, which will be 2×10^{26} g of water. A case of lithosphere-hydrosphere-biosphere interaction can be exemplified by the metals industry. Primary metals industry utilizes nearly 1 trillion gal. of process water per year of which iron and steel-making require 80% while the copper and aluminum industries consume the remainder.

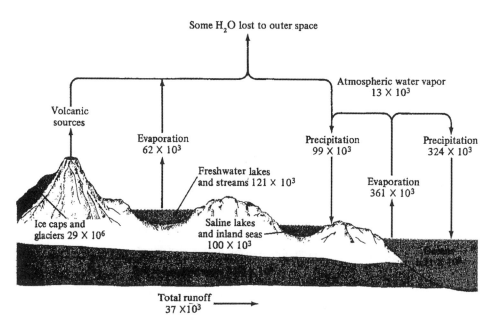

Figure C-2. The water cycle. Volumes of water in each reservoir are given in km^3. Flows from one reservoir to another are in km^3/year. Data from L. Hodges. "Environmental Pollution," pp. 126-127. Holt, New York, 1973. Also from J. Harte, "Consider a Spherical Cone, a Course in Environmental Problem Solving", William Kauffman, Los Altos, CA, 1985.

WATER PROPERTIES AND GROUNDWATER

6.1 CHEMISTRY OF RUNOFF

Direct runoff or overland flow (also called the Hortonian overland flow) is the runoff of precipitation without infiltration. This is important in understanding the water quality as unaltered by discharge, bad land, or urban runoffs with disturbances of the soil structure. A cross-section view is illustrated in Figure 6-1.

The mixing models have been developed to express this type of rapid flow generation. The samples may be two-component systems, as in the following:

$$C_t Q_t = C_g Q_g + C_s Q_s \qquad [6\text{-}1]$$

and

$$Q_t = Q_g + Q_s \qquad [6\text{-}2]$$

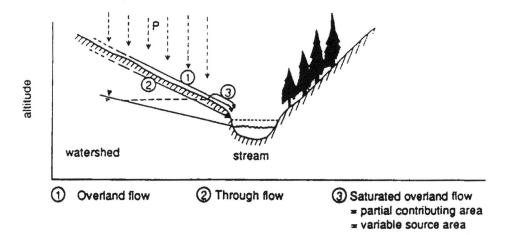

Figure 6-1. Conceptual models of peak runoff generation.

where C is the concentration of a given solute in mm/L, Q is the discharge in L/s, and the subscripts g, s, and t indicate respectively groundwater, surface (direct rapid), and total runoff components. From Equations [6-1] and [6-2],

$$\frac{Q_s}{Q_g} = \frac{C_g - C_t}{C_t - C_s} \qquad [6\text{-}3]$$

or

$$C_t = \frac{C_g + C_s \left(\dfrac{Q_s}{Q_g}\right)}{1 + \dfrac{Q_s}{Q_g}} \qquad [6\text{-}4]$$

Assuming,

$$\frac{Q_s}{Q_g} = \beta \, Q_t \qquad [6\text{-}5]$$

Here, Q_t is actually the stream discharge. If we define the difference between groundwater and rapid runoff concentration as

$$C_d = C_g - C_s \qquad [6\text{-}6]$$

then

$$C_t = \left(\frac{C_d}{1 + \beta Q_t} \right) + C_s \qquad [6\text{-}7]$$

Johnson has plotted the **concentration in stream discharge** for the Hubbard Brook catchments against $1/(1 + \beta Q_t)$ to obtain a slope of $C_d = C_g - C_s$ and an intercept of C_s. The plots are expressed in Figure 6-2. The decreasing current actions with discharges such as Na^+, Mg^{2+}, and Ca^{2+} can be explained as dilution of the high current action in deep soil water as a result of the weathering of silicate rock. The increase in aluminum concentration is due to the high concentration of acid in the topsoil.

The seasonal fluctuation of NO_3^- and K^+ is due to biological activities of uptaking.

Other models are summarized here:

$$C_t = aQ_t^{-1/n}$$

$$C_t = aQ_t^{-1/n} + C_s$$

$$C_t = b - a \log Q_t$$

$$C_t = a \exp(-bQ_t + 1/n)$$

$$C_t = b - aQ_t^{1/n}$$

$$C_t = (C_g - C_s)/(1 + aQ_t^{1/n}) + C_s$$

$$C_t = a(Q_t - Q_b)^{-1/n} + C_s$$

$$C_t = C_g/(1 + aQ_t)^{1/n} \qquad [6\text{-}8]$$

where the subscripts t, b, s, and g are, respectively total runoff, base flow (before peak flow), surface runoff, and groundwater; a, b, and n are arbitrary parameters used for fitting.

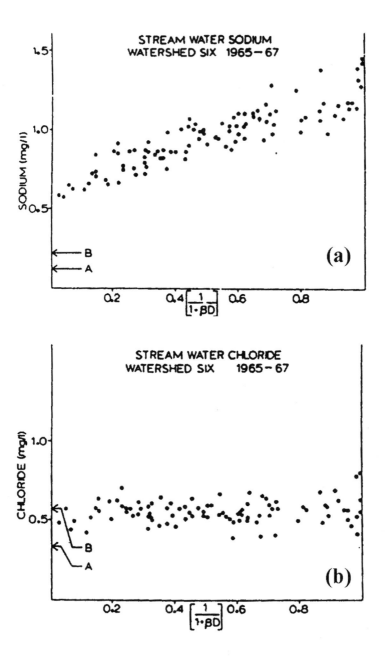

Figure 6-2. Variation of concentrations in Hubbard Brook streamflow with the reciprocal of discharge (note that D stands for total discharge = Q_t). Reprinted with permission from Johnson et. al., 1969. Copyright by the American Geophysical Union.

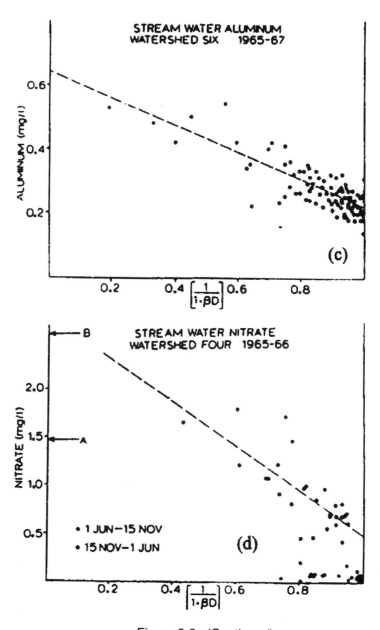

Figure 6-2. (Continued)

6.2 CHEMICAL PROPERTIES OF NATURAL WATERS

Most water molecules have a molecular weight of 18. In the water molecule, both hydrogen atoms are located on the same side of the oxygen atom, and their bonds with the oxygen atom are 105° apart. The hydrogen atoms carry a positive charge while the oxygen atom is negatively charged. Because of this distribution of charge, H_2O is a strongly dipolar molecule. The water molecule dipoles attract each other and form aggregates through bonds that are known as hydrogen bonds, as shown in Figure 6-3. It is thought that these aggregates in water at room temperature can reach sizes of up to 100 H_2O molecules.

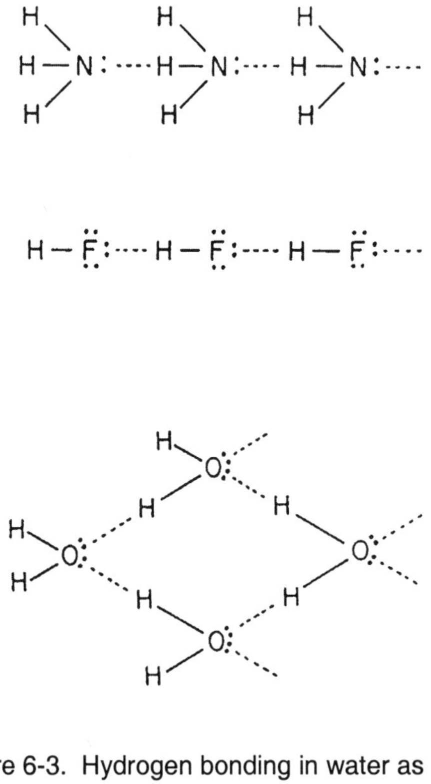

Figure 6-3. Hydrogen bonding in water as compared to ammonia and hydrogen fluoride.

Water is by far the most abundant liquid on the surface of the earth, and it is essential to all known forms of life. Its properties in the liquid state are, however, highly anomalous. Hydrogen bonding in water is responsible for many of its unusual properties. Water is a dihydride of oxygen; dihydrides of other elements in the same family of the periodic table as oxygen are H_2S, H_2Se, and H_2Te. These heavier molecules are all gases at atmospheric pressure and room temperature (25°C). Water is a liquid that becomes a gas only when the temperature is increased to 100°C and above. This can also be seen from a comparison of melting and boiling points for the hydrides of the elements in the first row of the periodic table, as shown here:

Melting Point and Boiling Point of First Row Elements

	CH_4	NH_3	H_2O	HF
Melting point (°C)	-182	-33	0	-83
Boiling point (°C)	-164	-78	100	20

Not only does water melt and boil at a higher temperature than any of its neighboring hydrides, but it also has a larger range of temperature over which it is a liquid. One characteristic that makes the planet Earth uniquely suitable for the evolution of life is that its surface temperature over most of its extent lies within the liquid range of water.

Water is far denser than its related species at any given temperature; the maximum density occurs at 4°C. That means it expands upon freezing, whereas almost all other substances contract upon solidification. The reason for this is that ice is an open-structured substance that is less dense than the liquid water from which it forms. This property has far-reaching ramifications. If solid H_2O were denser than liquid H_2O, ice would form at the bottom of natural bodies of water rather than at the top, and lakes would freeze from the bottom upward. Consequently, life in its present form in aquatic systems would not exist, because natural bodies of water would freeze solidly whenever the temperature fell below the freezing point of water.

The polarity of water is an important factor in determining its solvent properties. The minerals that make up the earth's crust are largely inorganic solids in which positively and negatively charged ions exist in a lattice structure held together by very strong electrostatic forces. These forces must be overcome upon dissolution. Water greatly assists this process by reducing the magnitude of these forces by a factor of about 80 (relative to air). This is the value of its dielectric constant (D). The large value of the dielectric constant stems from the large value of the dipole moment resulting from the partial charges in the hydrogen and oxygen atoms. Water, with this dipolar character, has the power to surround a positively charged ion with the negatively charged part of its molecule (or the reverse), thereby isolating the ions from their surrounding ions. The ion, surrounded by water molecules, can then leave the crystal ions and move out into solution.

Besides these characteristics, water possesses a high surface tension (70 dyne/cm), which is important in surface phenomena, droplet formation in the atmosphere, and many physiological processes including transport through biomembranes, a large heat capacity (75.5 KJ/mole K), a high heat of fusion (44.1 KJ/mole), and a large thermal conductivity that provides an important heat transfer mechanism in stagnant systems such as cells. The large latent heat produces a stable liquid state. Water itself is quite inert; the dissociation is small. Water is also highly transparent, which thickens the biologically productive euphatic zone. Because it is such a large body, water often serves as a sink for many substances. For example, DDT can be found in ice caps.

6.3 ACIDITY AND HARDNESS

Here, we will spend some time discussing hardness and some environmental aspects related to acid rain.

6.3.1 Hardness

Calcium, magnesium, and other divalent cations in water can combine with organic radicals in soaps to form undesirable precipitates; for example,

$$2 \ C_{17}H_{35}COO^- + Ca^{++} = Ca(C_{17}H_{35}COO)_2$$

Hard waters, or waters containing these divalent ions, also produce scales in hot-water pipes, heaters, boilers, and other units in which the temperature of the water is increased materially. The hardness of waters varies considerably from place to place. In general, surface waters are softer than groundwaters. The hardness of water reflects the nature of the geological formations with which it has been in contact.

A summary of chemical properties is shown in Table 6-0.

The hardness of water is derived largely from contact with soil and rock formations. Rain water as it falls upon the earth is incapable of dissolving the tremendous amounts of solids found in many natural waters. The ability to dissolve is gained in the soil when carbon dioxide is released by bacterial action. The soil water then becomes highly charged with carbon dioxide, which exists in equilibrium with carbonic acid. Under the low pH conditions that develop, basic materials, particularly limestone formations, are dissolved.

$$H_2CO_3 + CaCO_3 = Ca^{++} + 2HCO_3^-$$

Table 6-0. Summary of Chemical Properties

Property	Comparison with Normal Liquids	Significance
State	Liquid rather than gas, like H_2S, H_2Se, and H_2Te	Provides life media
Heat capacity	Very high	Moderates environmental temperatures, good heat transport medium
Latent Heat of Fusion	Very high	Moderating effect, tends to stabilize liquid state
Latent Heat of Vaporization	Very high	Moderation effect, important in atmospheric physics and in precipitation-evaporation balance
Density	Anomalous maximum of 4°C (for pure water)	Freezing from the surface and controls temperature distribution and circulation in bodies of water
Surface Tension	Very high	Important in surface phenomena, droplet formation in the atmosphere, and many physiological processes including transport through biomembranes
Dielectric Constant	Very high	Good solvent
Hydration	Very extensive	Good solvent and mobilizer of environmental pollutants, alters the biochemistry of solutes
Dissociation	Very small	Provides a neutral medium but with some availability of both H^+ and OH^- ions
Transparency	High	Thickens biologically productive euphatic zone
Heat Conduction	Very high	Can provide an important heat transfer mechanism in stagnant systems such as cells

Because limestone is not pure carbonate but includes impurities such as sulfates, chlorides, and silicates, these materials become exposed to the solvent action of the water as the carbonates are dissolved, and they pass into the solution too.

Calcium and magnesium cause by far the greatest portion of the hardness occurring in natural waters. Usually, hardness is classified into carbonate and noncarbonate hardness with respect to anions associated with the metallic ions. The part of the total hardness that is chemically equivalent to the bicarbonate plus carbonate alkalinities present in a water is called **carbonate hardness** (formerly called temporary hardness because through prolonged boiling it can be caused to precipitate). The amount of hardness in excess of the carbonate hardness is called **noncarbonate hardness** (formerly called permanent hardness because it cannot be removed or precipitated by boiling).   Noncarbonate hardness cations are associated with sulfate, chloride, and nitrate anions.

Lime $Ca(OH)_2$ and soda ash Na_2CO_3 are two convenient and economical materials used in industry to remove excess hardness.

$$Ca(OH)_2 + Ca^{++} + 2\ HCO_3^- = 2CaCO_3(s) + 2H_2O$$

Ammonia can be used to remove scales in the bathtub due to hard water.

$$Ca^{++} + 2HCO_3^- + 2NH_3 = CaCO_3 + 2NH_4^+ + CO_3^-$$

In industry, ion exchange resins or zeolites are also used to exchange sodium ions with divalent ions to produce soft water.

In addition to surfactants, detergent formulation contains other agents called **builders**, which are intended to eliminate precipitation by the positive ions. They can do this by tying the ions up, either in a soluble form or in a precipitate that can settle easily. Chemicals that tie up the ions in a soluble form are called **chelating agents**. Sodium tripolyphosphate (STP) is a commonly used chelating agent because it is cheap and has the advantage of rapidly breaking down in the environment to sodium phosphate. Phosphate ions can also serve as nutrients for plants and buffers in aqueous systems. Nevertheless, consumption of phosphate in detergents has decreased by 67% over the past 10 years in the United States because the natural bodies of water are being overfertilized as a result of using STP.

Sodium nitrilotriacetate (NTA), $N(C_2H_2O_2)_3Na_3$, is another chelating agent that has considerable promise; but there is concern that, in addition to binding to calcium, NTA could bind to heavy metals and mobilize these toxic elements in the environment. NTA is also suspected to be related to some birth defects. Sodium silicates and sodium borate are two alternatives. A famous household softener, Borax, contains $Na_2B_4O_7 \bullet 10H_2O$.   Incidentally, detergents are sometimes called **soft detergents** if they are biodegradable.

6.3.2   Acid Rain

The acidity of the rain and snow (**acid rain**) falling on widespread areas of the world has been rising during the past three decades or so. This may be a result of the combustion of tremendous quantities of fossil fuels. The United States itself annually discharges approximately 50 million metric tons of sulfur and nitrogen oxides into the atmosphere. Through a series of complex chemical reactions, these pollutants can be converted to acids (see Table 6-1).

Table 6-1. Sulfuric and Nitric Acids are Major Sources of Acidity in Precipitation

Substance	Concentration In Precipitation (mg per liter)	Contribution to Free Acidity[a] (microequivalents per liter)	Contribution to Total Acidity[a] (microequivalents per liter)
H_2CO_3	0.62[b]	0	20
NH_4^+	0.92	0	51
Al, dissolved	0.05[c]	0	5
Fe, dissolved	0.04[c]	0	2
Mn, dissolved	0.005[c]	0	0.1
Total organic acids	0.34	2.4	4.7
HNO_3	4.40	39	39
H_2SO_4	5.10	57	57
Total[d] :		98	179

a at pH 4.01

b equilibrium concentration

c average value for several dates

d Data from a sample of rain collected at Ithaca, NY, on Oct. 23, 1975.

The Hubbard Brooks Experimental Forest in the White Mountains of New Hampshire provides the longest known record for pH of precipitation in the United States (see Figure 6-4). It was found that in Norway, where the acidity of rain and snow is the highest, the acidity of lake water is likewise high (see Figure 6-5).

Pure water is known to be in equilibrium with atmospheric carbon dioxide and results in a pH of 5.6 due to the release of hydrogen ions from carbonic acid, H_2CO_3.

$$CO_2 + H_2O = H_2CO_3$$

$$H_2CO_3 = HCO_3^- + H^+$$

$$HCO_3^- = CO_3^{2-} + H^+$$

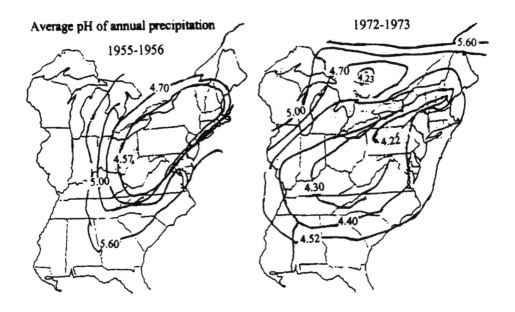

Figure 6-4. Acidity of precipitation has increased markedly in the Eastern United States (Sources: C.V. Cogbill and G.E. Likens; C.V. Cogbill, Thomas Burton, Patrick Brezonik, and Gray Henderson)

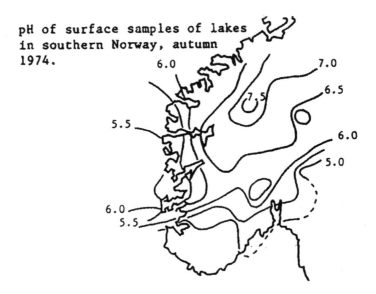

Figure 6-5. pH of surface samples of lakes in southern Norway, autumn 1974.

Potentially strong acid will yield a pH of 4 and below. Sulfur dioxide can be acidized in the atmosphere (ca. 1-5% hr^{-1}). It has two absorption bands for troposphere wavelength: one at ca. 384 nm, giving rise to triplet state SO_2, and the other at ca. 294 nm, causing formation at a higher energy excited singlet state.

$$SO_2 + h\nu = \bullet SO_2$$

$$\bullet SO_2 + O_2 = SO_3 + O$$

$$SO_2 + O = SO_3$$

$$SO_3 + H_2O = H_2SO_4$$

The reaction in cloud-droplets appears to yield the bisulfite and sulfite.

$$SO_2(g) + H_2O = SO_2(aq)$$

$$SO_2(aq) + H_2O = H_3O^+ + HSO_3^-$$

$$HSO_3^- + H_2O = H_3O^+ + SO_3^{2-}$$

The sulfite can be oxidized by atmospheric oxygen, and this process is catalyzed by metal ions such as Cu^{2+} and Fe^{3+}.

$$2SO_3^{2-} + O_2 = 2SO_4^{2-}$$

Other oxidizing agents are also present, for example, O_3 and H_2O_2.

$$HSO_3^- + O_3 = HSO_4^- + O_2$$

$$SO_3^{2-} + H_2O_2 = SO_4^{2-} + H_2O$$

Oxidation of NO_2 to nitric acid is quite fast (ca. 10% h^{-1}). Reactions with an OH radical or with ozone are as follows:

$$NO_2 + OH = HNO_3$$

$$NO_2 + O_3 = NO_3 + O_2$$

$$NO_3 + NO_2 = N_2O_5$$

$$N_2O_5 + H_2O = 2HNO_3$$

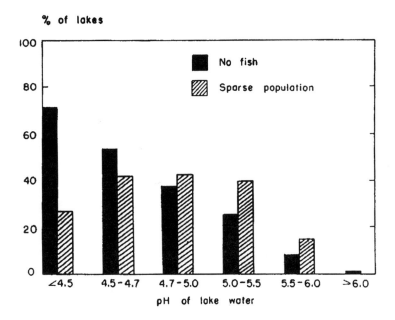

Figure 6-6. Fish population declines as the acidity of lake water increases.

HCl can be directly released from coal combustion ions. Through incineration, all these chemical species have been modified by NH_3 in the atmosphere.

Hundreds of lakes in North America and Scandinavia have become so acidic that they can no longer support fish life. Figure 6-6 illustrates the relationship between the fish population and the acidity of lake water. Besides the aquatic effects of acid precipitation, the yield from agricultural crops can be reduced as a result of both the direct effects of acids on foliage and the indirect effects resulting from the leaching of minerals from soils. The productivity of forests may be affected in a similar manner. The leachate of heavy metals (e.g., Al) to lakes or rivers is one of the causes of fish being unable to survive. In this case, the presence of Al^{+3} in water is toxic to fish due to the reaction of minerals therein at a low pH level.

6.4 GROUNDWATER

Groundwater is water that has percolated downward from ground surface through the soil pores. Formations of soil and rock (porous media) that have become saturated with water are known as groundwater reservoirs or aquifers. Water is withdrawn from aquifers by wells. One of the giant aquifers in the United States is called Ogallalo aquifer which underlies parts of 8 states. In some parts the rate of water use is 2 to 3 order of magnitude greater than rate of water discharge. In some parts of this aquifer, contamination has been found of salts that have been used in agriculture production.

6.4.1 Aquifer

Subsurface water occurs in two different zones: the upper zone has both air and water filling the cracks and pores between particles of soil and rock, whereas the lower zone is filled with water only. As indicated in Figure 6-7, the upper zone is termed the **unsaturated zone (vadose zone)**, which contains vadose water, while the lower zone is called the **saturated zone**, which contains groundwater. The vadose water is not available for use; on the contrary, the groundwater supplies most of our drinking water. Between the two layers, there is a transition region called the capillary fringe, and just above the groundwater is the water table. Usually, an unconfined aquifer is only restricted in the bottom by the presence of a confining bed or layer. If the aquifer is restricted in both the top and bottom with the confining layers, then it is termed a **confined aquifer**, as shown in Figure 6-8. The confining layers can be referred to as **aquitards** or **aquicludes**. Sometimes water can also be located in the unsaturated zone; this is trapped water and it is usually referred to as a **perched water table**. The water level in an artesian well, which is leveled to the recharge area, is termed the **piezometric surface** or the **potentiometric surface**.

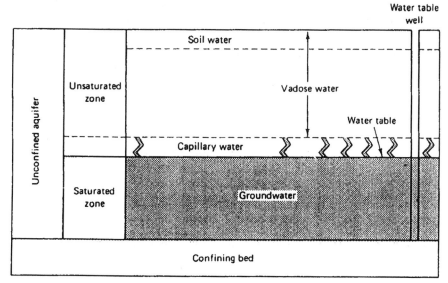

Figure 6-7. An unconfined aquifer is made up of saturated and unsaturated zones. A well penetrating the saturated zone would have water at the level of water table.

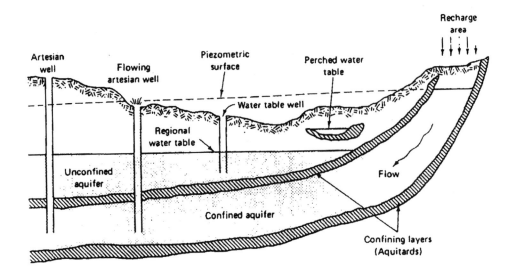

Figure 6-8. A confined aquifer and artesian wells.

Porosity is defined as the ratio of the volume of voids to the total volume of the aquifer rock. The specific yield or the specific porosity is a true measure of the amount of water that can be drained out from the aquifer rock. In an unconfined aquifer, the **specific yield** is the volume of water available per unit area per unit decline in water table. This value is equivalent to the storage coefficient of a confined reservoir.

The **hydraulic gradient** for an unconfined aquifer is defined as the slope of the water table or the piezometric surface. In a microscopic sense, it is the change in head divided by the change in horizontal distance, or

$$dh/dL \qquad\qquad [6\text{-}9]$$

If one knows the difference in the heads of the two wells, one can predict the direction of flow.

6.4.2 Darcy's Law and Groundwater Movement

Darcy's law states that the flow rate is proportional to the cross sectional area times the hydraulic gradient:

$$Q = KA\frac{dh}{dL} \qquad\qquad [6\text{-}10]$$

where

Q = the flow or the gross discharge and

$$Q = qA \quad (L^3/T) \tag{6-11A}$$

and

q = Darcy velocity (L^3/L^2T)
A = cross-sectional area (L^2)
K = hydraulic conductivity (L/T)

Sometimes,

$$dh/dL = (\Phi_1 - \Phi_2)/L = \text{hydraulic gradient} \tag{6-11B}$$

and

Φ = hydraulic potential (L)
L = length (L).

The hydraulic potential (Φ) can be expressed as

$$\Phi = z + P/\rho g \tag{6-12}$$

(see Bernoulli equation in Equation [6-65] in Section 6.4.3)

where

z = depth (L)
P = hydrostatic pressure $(ML^{-1}T^{-2})$
ρ = density of fluid (M^3/L)

Thus,

$$q = K(P_1 - P_2)/\rho g L \quad \text{for } z_1 = z_2 \tag{6-13}$$

Often Darcy's law is used to measure the **permeability** of the medium (either soil or rock). In this case, the equation is written as

$$q = \frac{k}{\eta} \frac{dP}{L} \tag{6-14}$$

where

k = intrinsic permeability of the medium (L^2)
η = viscosity of the fluid $(ML^{-1}T^{-2})$
v = kinetic viscosity of the fluid (L^2/T)

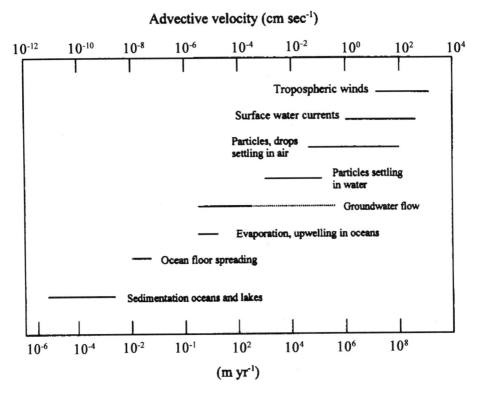

Figure 6-9. Characteristic velocities associated with major transport processes near the Earth's surface.

Thus,

$$K = \frac{\rho g L q}{(P_1 - P_2)} = \frac{\rho g k}{\eta} = \frac{kg}{v} \qquad [6\text{-}15]$$

For example, the viscosity of water at 20°C is 1 cp (0.01 g/cm sec)

$$g = 980 \text{ cm/sec}^2$$

$$k/K = 10^{-5} \text{ (cm sec)} \qquad [6\text{-}16]$$

A conversion between hydraulic conductivity (K) and permeability (k) can be found in Figure 6-10.

In fluid mechanics, a well-known equation for tube flow, called the **Poiseuille equation,** is

$$u = \frac{1}{8} \frac{d\rho}{\eta dz} a^2 \qquad [6\text{-}17]$$

where a is the tube radius, z is the streamwise direction, and m is the **hydraulic radius,** which is

$$m = \frac{\text{cross-sectional area of tube}}{\text{wetted perimeter of tube}} \quad [6\text{-}18]$$

For cylindrical tube of radius $a/m = \pi a^2 / 2\pi a = a/2$. Assuming $k_0 = 2$, the generalized Poiseuille equation becomes

$$u = \frac{m^2}{k_0 \eta} \frac{\Delta P}{L_e} \quad [6\text{-}19]$$

In Equation [6-19], because $\Delta P/ L_e = dp/ dz$, $\Delta P/ L_e$ can be expressed as pressure gradient. In a real aquifer for porous media, the flow is random and the m value can be estimated as

$$m = \frac{\text{void volume}}{\text{wetted surface of porous medium}} = \frac{\phi V}{V(1-\phi)S_o} \quad [6\text{-}20]$$

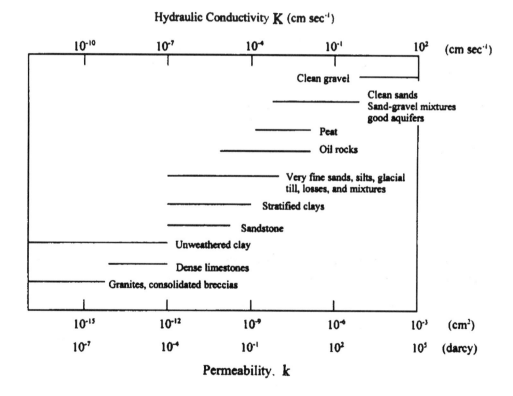

Figure 6-10. The coefficients of hydraulic conductivity and permeability of sediments and rocks.

Here, ϕ is **porosity** (ratio of voids/total volume), V is volume of porous medium, and S_0 is specific surface of the porous medium.

$$S_o = \frac{\text{porous medium surface area}}{\text{unit volume of porous medium solids}} \qquad [6\text{-}21]$$

Furthermore, Darcy velocity assumes the media have 100% porosity. The actual velocity is often higher due to the solid portion in the medium. Thus the actual velocity, u (cm/ sec), is

$$u = \frac{q}{\phi} \qquad [6\text{-}22]$$

Substituting this equation and the equation for the definition of m [Eq. 6-18] into the generalized Poiseuille equation [Eq. 6-19], then

$$q = \frac{1}{k_o \eta} \frac{\phi^3}{S_0^2 (1-\phi)^2} \frac{\Delta P}{L_e} \qquad [6\text{-}23]$$

L_e here is the length of one of the average twisted flow tubes in aquifer. We would prefer to consider overall thickness of the sections of porous medium. Thus, we add a constant,

$$K' = \frac{L_e}{L} k_o \qquad [6\text{-}24]$$

where L is the length of porous medium section and K' is Kozeny constant. After substitution,

$$q = \frac{\phi^3}{K'\eta S_0^2 (1-\phi)^2} \frac{\Delta P}{L} \qquad [6\text{-}25]$$

This is called the **Carman-Kozeny equation**. Finally, if we compare this equation with Darcy's law equation, then we have an equation for the permeability,

$$k = \frac{\phi^3}{K' S_0^2 (1-\phi)^2} \qquad [6\text{-}26]$$

Thus, the Poiseuille equation and Darcy's equation are connected through the definition of permeability.

Figure 6-9 gives the characteristic velocities associated with major transport processes near the earth's surface. The advective velocity of groundwater flow ranges from 10^{-6} to 10^{-1} cm/sec. Figure 6-10 illustrates some of the coefficients of the hydraulic conductivity and permeability of sediments and rock. It can easily be found that the ratio of the permeability to the hydraulic conductivity is about 10^{-5} cm/sec as previously mentioned. One unit often used by petroleum engineers is termed darcey, and

$$1 \text{ darcey} = 10^{-8} \text{ cm}^2 = 10^{-12} \text{ m}^2 = 1 \text{ } \mu\text{m}^2$$

or

$$1 \text{ darcey} = 10^{-3} \text{ cm/s if expressed by K.}$$

A much lower permeability unit is adopted by food engineers and is called a **barrier**, where

$$1 \text{ barrier} = 10^{-10} \text{ cm/s}$$

if expressed by K. This unit is used for gas transport.

The gas permeability is one trillionth of a darcey. Figure 6-10A is an illustration of oxygen permeability of selective polymer membranes.

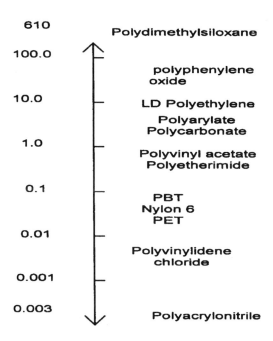

Figure 6-10A. Oxygen permeabilities of selected commercial polymers. (Scale is in the range of barriers, 1 barrier = 10^{-10} cm/s).

Transmissibility (*T*) is also a convenient parameter to use. The detailed definitions of hydraulic conductivity (*K*) and *T* are illustrated in Figure 6-11. The mathematical relationship between *K* and *T* is

$$T = KD \qquad\qquad [6\text{-}27]$$

where *D* is the saturated thickness of the aquifer.

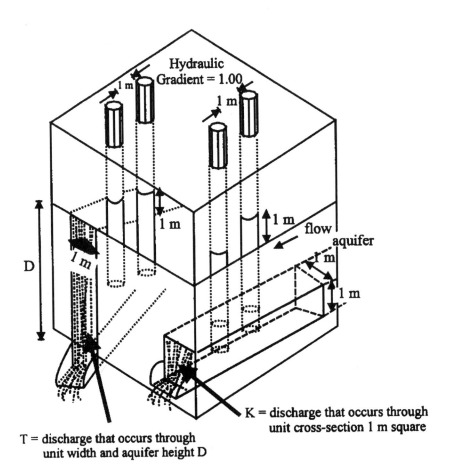

Figure 6-11. Illustration of definition of permeability (K) and transmissibility (T). (Source: Johnson Division. UOP. *Ground Water and Wells*. Reprinted with permission.)

[Example 6-1] For an artesian aquifer in the coastal region of Florida, the vertical drop is 10 m over a distance of 100 Km. Find the residence time of the water in this 100 Km long aquifer where the average porosity of the medium is 30%.

The hydraulic gradient = $\Delta\Phi/L = 10\text{m}/100\text{Km} = 10^{-4}$
For a good aquifer, from Figure 6-5, $K = 10^{-1}\text{cm/sec}$

Then $q = K\Delta\Phi/L = (10^{-1} \text{ cm sec}^{-1})(3 \times 10^7 \text{ sec yr}^{-1})(10^{-4}) = 300 \text{ cm yr}^{-1} =$ 3m yr^{-1}
$u = 3\text{m yr}^{-1}/30\%$
The residence time = $10^5\text{m}/10\text{m yr}^{-1} = 10{,}000$ yr.

Permeability depends greatly on:
- the grain size of the sediment, r,
- the sediment porosity (ratio of voids/ total volume), ϕ, and
- the packing arrangement of particles.

Many models have been proposed to correlate permeability with these parameters.   Table 6-2 lists some of them, and Figure 6-12 shows the dependence of permeability k on sediment ϕ. Generally, the higher the porosity, the higher the permeability.
When a well is pumped, the water table (unconfined aquifer) or the piezometric surface (confined aquifer) forms a cone of depression in the vicinity of the well. Assuming the drawdown is small in comparison to the depth of the aquifer, and the flow to the well is horizontal and radial, then

$$Q = KA\frac{dh}{dr} = K2\pi rh\frac{dh}{dr} \qquad [6\text{-}28]$$

Because the cross-sectional area is cylindrical around the well, then

$$\int_r^{r_1} Q\frac{dr}{r} = 2\pi K \int_h^{h_1} hdh$$

or

$$Q\ln\frac{r_1}{r} = \pi K\left(h_1^2 - h^2\right)$$

or

$$Q = \frac{\pi K \left(h_1^2 - h^2 \right)}{\ln\left(\dfrac{r_1}{r} \right)}$$ [6-29]

If $h = h_w$, which is the drawdown, then the drawdown can be calculated if the pumping rate, Q, is known.

Table 6-2. Coefficient of Hydraulic Permeability k (cm^2 or Approximately 10^8 Darcy) as a Function of Sediment Porosity ϕ (Fraction) and Particle Radius r (cm)

Equation	Explanatory notes
1. $2.47 \times 10^{-19} r^2$	Krumbein and Monk (1942), sands, $0.005 < r < 0.1$cm.
2. $k = \text{constant} \times \phi^5$	Terzaghi (1925) in Rieke and Chilingarian (1974, p. 148), $0.2 < \phi < 0.8$.
3. $k \propto \phi^9 r^2$	Sands, from data of Bear and Weyl (1973); proportionality factor depends on ϕ $0.25 < \phi < 0.4$.
4. $k = 10^{-9} \left(\dfrac{\phi}{1-\phi} \right)^7$	Marine clayey sediments (Bryant et al., 1974).
5. $k = 7.25 \times 10^{-11} \left(\dfrac{\phi}{1-\phi} \right)^7$	Aragonitic sediment, $0.2 < \phi < 0.7$, from data of Robertson (1967). Original data also obey log $K = 17.0 + 13.6$.
6. $k = \dfrac{\phi^3}{(1-\phi)^2} \times \dfrac{c_k}{S_s^2}$	The Carmen-Kozeny equation, where c_k is the Kozeny constant and S_s is the surface area of the pore space per unit volume of solid (Bear, 1972, p. 166). For beds made of spherical and cylindrical particles, see Equations 7, 8, and 9.
7. $k = \dfrac{\phi^3}{(1-\phi)^2} \times \dfrac{r^2}{45}$	Spherical particles, $c_k = 1/5$ and $S_s = 3/r$ (cm^{-1}) in Equation 6.
8. $k = \dfrac{-2\ln(1-\phi) + 4(1-\phi) - (1-\phi)^2 - 3}{1-\phi} \times \dfrac{r^2}{8}$	Circular cylinders, flow parallel to cylinder axis. R is cylinder radius. The Kozeny constant c_k given by Happel and Brenner (1973, p. 393). For straight circular cylinders, $S_s = 2/r$ (cm^{-1}) in Equation 6.
9. $k = \dfrac{-\ln(1-\phi) - \dfrac{1-(1-\phi)^2}{1+(1-\phi)^2}}{1-\phi} \times \dfrac{r^2}{8}$	Circular cylinders, flow perpendicular to cylinder axis, parameters as in Equations 8 and 6. For a random network of cylinders, Happel and Brenner (1973) recommend a weighted sum of k from Equations 8 and 9, in proportions of 2/3 and 1/3.

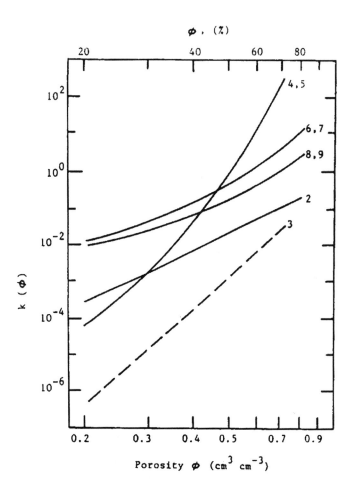

Figure 6-12. Dependence of the permeability k on sediment porosity ϕ, according to the models listed in Table 6-2.

6.4.3 Stream Function and Turbulent Flow

Some basic concepts of hydraulics and hydrodynamics are introduced here. A **path line** is the trace made by a single particle over a period of time in the flow of a liquid. The path line shows the direction of the particle's velocity. **Streamlines** show the mean direction of a number of particles at the instant of time. A series of curves tangent to the means of the velocity vectors are streamlines. Path lines and streamlines are identical in the steady flow of a liquid in which there are no fluctuating velocity components. Such flow is either that of an ideal functionless fluid or of one so viscous that no eddies are formed. This is a **laminar** type of flow. In contrast, another type of flow is a **turbulent** flow in which the path lines and streamlines are not coincident. The path lines are irregular while the streamlines are everywhere tangent to the local mean temporal velocity. When a dye or a tracer is frequently injected into the flow to trace the motion of fluid particles, it is termed a **streak line** or a **filament line**.

A **stream function** Ψ, based on the continuity prime cycle, is a mathematic expression of the **flow field**. In a two-dimensional case, let $\Psi(x,y)$ be the streamline near the origin then $\Psi + d\Psi$ represents the second streamline. In Figure 6-13, $d\Psi$ will be the flow carried out between the two streamlines. For continuity, referring to the triangular fluid element, an incompressible fluid is

$$d\Psi = -vdx + udy \qquad [6\text{-}30]$$

Also, for $d\Psi$,

$$d\Psi = \frac{\partial \Psi}{\partial x}dx + \frac{\partial \Psi}{\partial y}dy \qquad [6\text{-}31]$$

From these two equations, we note

$$u = \frac{\partial \Psi}{\partial y}$$

$$v = -\frac{\partial \Psi}{\partial x} \qquad [6\text{-}32]$$

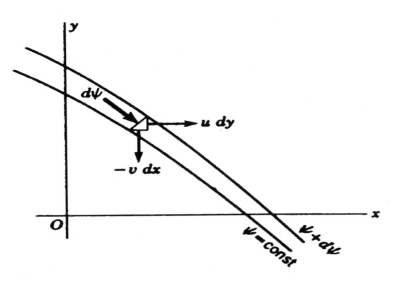

Figure 6-13.   Stream function.

Here, u and v are velocity components at any point of a two-dimensional flow field. The equation of continuity may be expressed in terms of Ψ:

$$\frac{\partial u}{\partial x} + \frac{\partial v}{\partial z} = 0 \qquad [6\text{-}33]$$

then

$$\frac{\partial}{\partial x}\frac{\partial \Psi}{\partial y} - \frac{\partial}{\partial z}\frac{\partial \Psi}{\partial x} = 0 \qquad [6\text{-}34]$$

or

$$\frac{\partial^2 \Psi}{\partial x \partial y} = \frac{\partial^2 \Psi}{\partial z \partial x} \qquad [6\text{-}35]$$

There is also a velocity potential, Φ, for the velocity vector

$$\nabla \Phi = u \qquad [6\text{-}36]$$

or

$$u_i + v_j + w_k = \frac{\partial \Phi}{\partial x}i + \frac{\partial \Phi}{\partial y}j + \frac{\partial \Phi}{\partial z}k$$

in two-dimensional space,

$$u = \frac{\partial \Phi}{\partial x}$$

$$v = \frac{\partial \Phi}{\partial y} \qquad [6\text{-}37]$$

For substituting the continuity equation into the preceding equation, we get the following:

$$\frac{\partial^2 \Phi}{\partial x^2} + \frac{\partial^2 \Phi}{\partial z^2} = 0 \qquad [6\text{-}38]$$

Also, there is a **vorticity** defined as the curl of velocity.

$$w = \nabla \times u \qquad [6\text{-}39]$$

The solution of the preceding equation is by a determinant.

$$\nabla \times u = \begin{vmatrix} i & j & k \\ \frac{\partial}{\partial x} & \frac{\partial}{\partial y} & \frac{\partial}{\partial z} \\ u & v & w \end{vmatrix} \qquad [6\text{-}40]$$

The solution for a two-dimension case is

$$d\omega = \left(\frac{\partial v}{\partial x} - \frac{\partial u}{\partial y} \right) dx dy$$

[6-41]

Substituting the velocity potential into the vorticity equation,

$$\omega = \frac{\partial v}{\partial x} - \frac{\partial u}{\partial y} = \frac{\partial}{\partial x} \left(-\frac{\partial \Phi}{\partial y} \right) - \frac{\partial}{\partial y} \left(-\frac{\partial \Phi}{\partial x} \right)$$

$$= -\frac{\partial^2 \Phi}{\partial x \partial y} + \frac{\partial^2 \Phi}{\partial y \partial x} = 0$$

[6-42]

There is no vorticity, and flow is irrotational. This type of flow is termed **potential flow**.

The **streamline function** is

$$d\Psi = \frac{\partial \Psi}{\partial x} dx + \frac{\partial \Psi}{\partial y} dy$$

[6-43]

and the velocity potential is

$$d\Phi = \frac{\partial \Phi}{\partial x} dx + \frac{\partial \Phi}{\partial y} dy$$

[6-44]

Using velocity components u and v,

$$d\Psi = -v dx + u dy$$

[6-45]

$$d\Phi = -u dx - v dy$$

[6-46]

Along a streamline Ψ = constant and $d\Psi = 0$

$$\frac{dy}{dx} = \frac{v}{u}$$

[6-47]

and along an **equipotential line**, shown in Figure 6-14, Φ = constant and $d\Phi$ = 0.

$$\frac{dy}{dx} = -\frac{u}{v}$$

[6-48]

The streamlines and equipotential lines are orthogonal, which means they are perpendicular to each other. In this manner, Φ and Ψ are required to form an orthogonal network forming a **flow net** provided irrotationality (condition for Φ) and continuity (condition for Ψ) are satisfied.

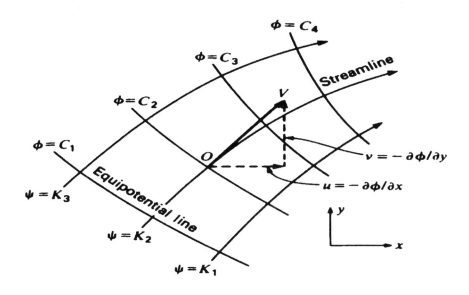

Figure 6-14.   Flow net.

[**Example 6-2**]   A flow is defined as $u = 2x$ and $v = -2y$. Find the stream function and potential function for this flow, and plot the flow net.

Check for continuity.

$$\frac{\partial u}{\partial x} + \frac{\partial v}{\partial y} = 2 - 2 = 0$$

It is now possible to have a stream function:

$$d\Psi = -vdx + udz = 2ydx + 2xdy$$

$$\Psi = 2xz + c_1$$

Check for irrotationality.

$$\frac{\partial v}{\partial x} - \frac{\partial u}{\partial y} = 0 - 0 = 0$$

Therefore, a potential function exists:

$$d\Phi = -udx - vdy = -2xdx + 2ydy$$

$$\Phi = -\left(x^2 - y^2\right) + c_2$$

Given the numerical values of Ψ and Φ, curves can be drawn as depicted by Figure 6-15. (Only one quadrant is shown.)

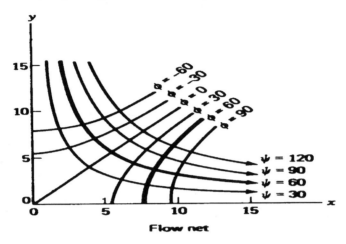

Figure 6-15. Graph for the Example 6-2. (Only one quadrant is shown here).

We can now summarize some relations as follows:

$$\nabla \cdot u = 0 \qquad\qquad [6\text{-}49]$$

The preceding equation represents the **continuity** equation,

$$\nabla \times u = \omega = 0 \qquad\qquad [6\text{-}50]$$

which represents no vorticity, and its flow is **irrotational**,

$$\nabla \Phi = u \qquad\qquad [6\text{-}51]$$

which represents the **velocity potential**. It can be seen from the preceding equations that

$$\nabla \cdot u = \nabla \cdot \nabla \Phi = \nabla^2 \Phi = 0 \qquad\qquad [6\text{-}52]$$

which represents the **Laplace equation**. Now, if we take the steady state of the Euler equation,

$$u \cdot \nabla u = -\frac{1}{P} \nabla P + g \qquad\qquad [6\text{-}53]$$

Using a scalar field of g

$$g = \nabla G$$

and

$$G = -gZ \qquad [6\text{-}54]$$

Thus,

$$u \cdot \nabla u = -\frac{1}{\rho} \nabla P + \nabla G \qquad [6\text{-}55]$$

From vector algebra identity,

$$u \cdot \nabla u = \nabla(\frac{1}{2} u \cdot u) - u \times (\nabla \times u) \qquad [6\text{-}56]$$

Therefore, from the right-hand side,

$$\nabla(\frac{1}{2} u \cdot u) - u \times \omega = -\frac{1}{\rho} \nabla P + \nabla G \qquad [6\text{-}57]$$

When evaluating along a streamline in the flow, the pressure gradient term can be expressed as

$$\frac{1}{\rho} \nabla p = \nabla \left(\int \frac{dP}{\rho} \right) \qquad [6\text{-}58]$$

or

$$\nabla \left(\int \frac{dP}{\rho} + \frac{1}{2} u \cdot u - G \right) = u \times \omega \qquad [6\text{-}59]$$

To form a dot product of both sides

$$u \cdot \nabla \left(\int \frac{dP}{\rho} + \frac{1}{2} u \cdot u - G \right) = u \cdot (u \times \omega) \qquad [6\text{-}60]$$

Therefore,

$$u \cdot \nabla \left(\int \frac{dP}{\rho} + \frac{1}{2} u \cdot u - G \right) = 0 \qquad [6\text{-}61]$$

and it follows that

$$\int \frac{dP}{\rho} + \frac{1}{2} u \cdot u - G = \text{constant along a streamline} \qquad [6\text{-}62]$$

or

$$\frac{P}{\rho} + \frac{U^2}{2} + gZ = \text{constant along a streamline} \qquad [6\text{-}63]$$

This equation is the **Bernoulli equation** for the irrotational flow where P, U, and Z are the variables.   Alternatively the Bernoulli equation can be expressed as

$$\frac{P}{\delta} + \frac{U^2}{2g} + Z = \text{constant} \qquad [6\text{-}64]$$

because $\delta = \rho g$

Another way to express the Bernoulli equation is

$$H = \frac{P}{\gamma} + Z + \frac{U^2}{2g} \qquad [6\text{-}65]$$

P/γ is **pressure head**, representing the energy per unit weight stored in the fluid by virtue of the pressure; Z is **elevation head**, representing the potential energy per unit weight of fluid, and $U^2/2g$ is the **velocity head**, representing the kinetic energy per unit weight of fluid. The sum of the three terms, which all have the dimension of (L), is termed the **total head**, H. For a frictionless incompressible fluid with no machine between 1 and 2,

$$H_1 = H_2 \qquad [6\text{-}66]$$

For a real fluid

$$H_1 = H_2 + h_L \qquad [6\text{-}67]$$

where h_L is called the **head loss**. If there is a machine between 1 and 2, then

$$H_1 + h_m = H_2 + h_L \qquad [6\text{-}68]$$

where h_m is the energy head portion, contributed by a machine. If the machine is a pump, then $h_m = h_p$; if the machine is a turbine then $h_m = -h_t$, and the energy head has to be extracted.

Flow is not limited to a laminar flow, induct, or conduit and even along a flat plate. If there is a viscous sublayer, which will transfer a negative momentum to the overlying fluid, a nonuniformity in the x-direction will be developed, and eventually the flow will transform into a turbulent.   This transformation from a laminar boundary layer, through a transition zone, to a turbulent boundary layer is depicted in Figure 6-16.   Clark analyzed the turbulence problem and we will follow his presentation for the boundary layer. He begins with the x-component of Navier-Stokes equation of a two-dimensional flow.

$$u\frac{\partial u}{\partial x} + v\frac{\partial u}{\partial y} = -\frac{\partial p}{\partial x} + v\left(\frac{\partial^2 u}{\partial x^2} + \frac{\partial^2 u}{\partial y^2}\right) \qquad [6\text{-}69]$$

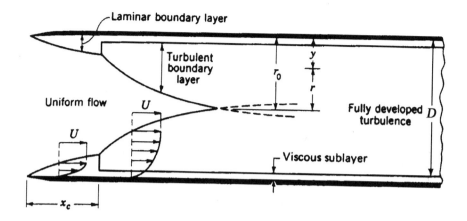

Figure 6-16. Development of boundary layer in a pipe.

If the **boundary layer thickness**, δ, is thin, then the outside edge of the boundary layer, Δy, is small. And, accordingly, due to Bernoulli equation, $\Delta p = 0$; thus, one can drop the first term on the right-hand side. Let us proceed with the component of the viscous terms. First, one wants to estimate the magnitude of individual relatives, u, when compared with the overall velocity, U. From Boolean algebra,

$$u \subset (U)$$ [6-70]

which is along the x-direction. Along the y-direction,

$$\frac{\partial u}{\partial y} \subset \left(\frac{U}{\delta} \right)$$ [6-71]

or the estimation of

$$\frac{\partial^2 u}{\partial y^2} \subset \left(\frac{U}{\delta^2} \right)$$ [6-72]

Also, for the x-direction,

$$\frac{\partial u}{\partial x} \subset \left(\frac{U}{x} \right)$$

$$\frac{\partial^2 u}{\partial x^2} \subset \left(\frac{U}{x^2} \right)$$

Therefore,

$$\frac{\dfrac{\partial^2 u}{\partial y^2}}{\dfrac{\partial^2 u}{\partial x^2}} \subset \frac{\dfrac{U}{\delta^2}}{\dfrac{U}{x^2}} \subset \left(\frac{x}{\delta}\right)^2 \qquad [6\text{-}73]$$

because $x \gg \delta$, for the majority of the boundary layers. Thus, the term $\dfrac{\partial^2 u}{\partial x^2}$ of the Navier-Stokes equation (Eq. 6-69) can be dropped. Hence, it becomes

$$u\frac{\partial u}{\partial x} + v\frac{\partial u}{\partial y} = v\frac{\partial^2 u}{\partial y^2} \qquad [6\text{-}74]$$

Also, the two-dimensional continuity equation,

$$\frac{\partial u}{\partial x} + \frac{\partial v}{\partial y} = 0 \qquad [6\text{-}75]$$

which is the boundary-layer equation and the boundary conditions are

$$u = v = 0 \;\; at \;\; y = 0$$
$$u = U \;\; at \;\; y = \infty$$

Using the stream function, Equation [6-30], to substitute to Equation [6-74] we obtain

$$\frac{\partial \Psi}{\partial \Psi}\frac{\partial^2 \Psi}{\partial x \partial y} - \frac{\partial \Psi}{\partial x}\frac{\partial^2 \Psi}{\partial y^2} = v\frac{\partial^3 \Psi}{\partial y^3} \qquad [6\text{-}76]$$

A solution of Equation [6-76] can be obtained by transformation of variables.

$$\Psi(x, y) = (v\, U\, x)^{\frac{1}{2}}\, g\,(\eta) \qquad [6\text{-}77]$$

and η, an independent variable, is

$$\eta = \frac{y}{2\left(\dfrac{vx}{U}\right)^{\frac{1}{2}}} \qquad [6\text{-}78]$$

Now the following transforming relations (Eqs. 6-79 to 6-83) can be successively derived:

$$u = \frac{\partial \Psi}{\partial y} = \frac{U}{2}g'(\eta) = \frac{U}{2}g' \qquad [6\text{-}79]$$

$$-v = \frac{\partial \Psi}{\partial x} = \frac{1}{2}\left(\frac{vU}{x}\right)^{\frac{1}{2}}(g - \eta g')$$ [6-80]

$$\frac{\partial^2 \Psi}{\partial x \partial y} = -\frac{U\eta}{4x}g''$$ [6-81]

$$\frac{\partial^2 \Psi}{\partial y^2} = \frac{U}{4}\left(\frac{U}{vx}\right)^{\frac{1}{2}}g''$$ [6-82]

$$\frac{\partial \Psi^3}{\partial y^3} = \frac{U^2}{8vx}g'''$$ [6-83]

Substituting the preceding relations into the differential equation, Equation [6-76],

$$g''' + gg'' = 0$$ [6-84]

This differential equation, Equation [6-84], can be solved with

$$g = g' = 0 \ at \ \eta = 0$$
$$g' = 2 \quad at \ \eta = \infty$$

The numerical solutions are listed in Table 6-3. From the table $u < U$, initial $u \to U$, and $u/U = 1$. Now we can find boundary thickness layer $y = \delta$ and from Equation [6-78] we can derive the following:

$$\frac{y}{x} = \frac{2\eta}{\left(\frac{Ux}{v}\right)^{\frac{1}{2}}} = \frac{2\eta}{(Re_x)^{\frac{1}{2}}} = \frac{\delta}{x}$$ [6-85]

Here, Re_x is **boundary layer Reynold number**:

$$Re_x = \frac{Ux}{v}$$ [6-86]

We can calculate that as the boundary layer thickness decreases, the boundary layer Reynold number increases.

Another approach to turbulence flow is from the consideration of stress tension, τ_{ij}. The **shear stress** acting normal to the y-axis and in planes parallel to the x-axis is

$$\tau_{yx} = \mu\left(\frac{\partial U_y}{\partial x} + \frac{\partial U_x}{\partial y}\right)$$ [6-87]

Table 6-3.   Values of Absolute Roughness *e* for New Pipes

	Feet	Millimeters
Drawn tubing, brass, lead, glass, centrifugally spun cement, bituminous lining, transite	0.000005	0.0015
Commercial steel or wrought iron	0.00015	0.046
Welded-steel pipe	0.00015	0.046
Asphalt-dipped cast iron	0.0004	0.12
Galvanized iron	0.0005	0.15
Cast iron, average	0.00085	0.25
Wood stave	0.0006 to 0.003	0.18 to 0.9
Concrete	0.001 to 0.01	0.3 to 3
Riveted steel	0.003 to 0.03	0.9 to 9

Note: $\dfrac{e}{D} = \dfrac{e\ in\ feet}{D\ in\ feet} = \dfrac{e\ in\ mm}{D\ in\ mm} = 10^{-1} \times \dfrac{e\ in\ mm}{D\ in\ cm}$

Because there is velocity in the y-direction,

$$\tau_{yx}\Big|_{y=0} = \mu \frac{\partial u_x}{\partial y}\Big|_{y=0} \qquad [6\text{-}88]$$

From Equation [6-79],

$$\frac{\partial u}{\partial y} = \frac{U}{4}\left(\frac{U}{vx}\right)^{\frac{1}{2}} g'' \qquad [6\text{-}89]$$

or

$$\tau_{yx}\Big|_{y=0} = \frac{\mu U}{4}\left(\frac{U}{vx}\right)^{\frac{1}{2}} g''\Big|_{y=0} = \frac{\mu U}{4}\left(\frac{U}{vx}\right)^{\frac{1}{2}} g''(0)$$

Now $g''(0)$ can be obtained from Table 6-3 as $g''(0) = 1.32824$ or

$$\tau_{yx}\Big|_{y=0} = 0.322\mu U\left(\frac{U}{vx}\right)^{\frac{1}{2}}$$

$$= 0.332\rho\ U^2\left(\frac{v}{Ux}\right)^{\frac{1}{2}}$$

$$= 0.332\rho\ U^2\ Re_x^{-\frac{1}{2}} \qquad [6\text{-}90]$$

The **skin-friction drag coefficient** for the flat plate is

$$f_f = \frac{\tau_{yx}|_{y=0}}{\frac{1}{2}\rho U^2} = 0.664 Re_x^{-\frac{1}{2}} \qquad [6\text{-}91]$$

This is also referred to as the **Fanning friction factor** or **pipe-function factor** if flow is in cylindrical conduits. The shear stress at the boundary as expressed in Equation [6-89] is a function of U^2 and, hence, it is possible to adopt a **shear velocity**, $U*$

$$U* = \left(\frac{\tau_0}{\rho}\right)^{\frac{1}{2}} \qquad [6\text{-}92]$$

The shear velocity is related to the eddy diffusity of momentum v_t by

$$v_t = \kappa U * Z \qquad [6\text{-}93]$$

Here, κ is the von Kámán constant. Large eddies and swirls are responsible for disturbed flow that typifies turbulent flow. Roughness in a pipe is usually defined by a **roughness height**, e, which can be illustrated as in Figure 6-17. If the boundary thickness is lower than e, then the pipe is rough; on the other hand, if the boundary thickness is much higher than e, then it is a smooth pipe. A **friction factor**, f, is generally used. In practice.

$$f = 4 f_f \qquad [6\text{-}94]$$

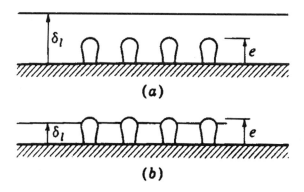

(a)

(b)

Figure 6-17. Turbulent flow near boundary. (a) Relatively low R, $\delta_l > e$. If $\delta_l > 6e$, the pipe behaves as a smooth pipe. (b) Relatively high R, $\delta_l < e$. If $\delta_l < 0.3e$, the pipe behaves as a wholly rough pipe.

For the three regions in relation to roughness,

- Smooth-pipe flow, $\delta > 6e$

$$f^{-1/2} = 2 \log Re\, f^{1/2} - 0.8$$

- Transitional flow, $6e > \delta > 0.3e$

$$f^{-1/2} = -2 \log \left(\frac{e/D}{3.7} + \frac{2.51}{Re}\, f^{-1/2} \right)$$

- Rough-pipe flow, $\delta < 0.3e$

$$f^{-1/2} = 2 \log \frac{D}{e} + 1.14 \qquad\qquad [6\text{-}95]$$

(D is diameter of the pipe)

A **Moody diagram**, as Figure 6-18, is useful for the estimation of e, as is in Table 6-4. The shear velocity can also separate the three regions.

- Hydraulically smooth regime

$$0 \le \frac{eu_*}{v} \le 5$$

- Transition regime

$$5 \le \frac{eu_*}{v} \le 70$$

- Completely rough regime

$$\frac{eu_*}{v} \ge 70 \qquad\qquad [6\text{-}96]$$

Furthermore, the boundary-layer Reynold number can also be useful.

- Laminar

$$Re_x < 2 \times 10^5$$

- Transitional

$$2 \times 10^5 < Re_x < 3 \times 10^6$$

- Turbulent

$$Re_x > 3 \times 10^6 \qquad\qquad [6\text{-}97]$$

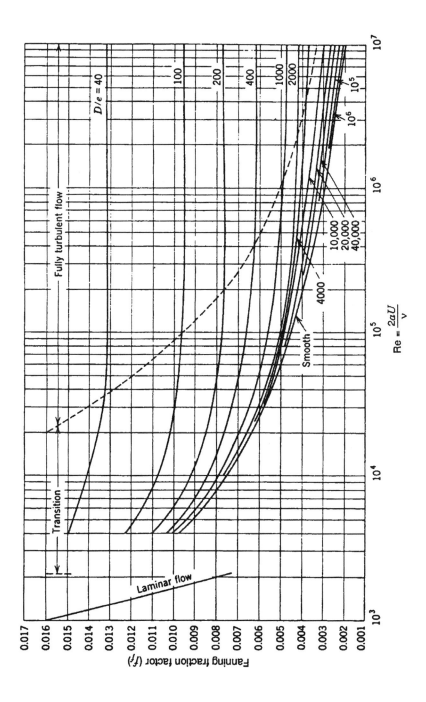

Figure 6-18. Friction factor for pipes (Moody diagram).

Table 6-4. Solution of Laminar Boundary Layer Problem

$\eta = \dfrac{y}{2}\sqrt{\dfrac{U}{vx}}$	f'	$\dfrac{u}{U}$	f''
0	0	0	1.32824
0.2	0.2655	0.1328	1.3260
0.4	0.5294	0.2647	1.3096
0.6	0.7876	0.3938	1.2664
0.8	1.0336	0.5168	1.1867
1.0	1.2596	0.6298	1.9670
1.2	1.4580	0.7290	0.9124
1.4	1.6230	0.8115	0.7360
1.6	1.7522	0.8761	0.5565
1.8	1.8466	0.9233	0.3924
2.0	1.9110	0.9555	0.2570
2.2	1.9518	0.9759	0.1558
2.4	1.9756	0.9878	0.0875
2.6	1.9885	0.9943	0.0454
2.8	1.9950	0.9962	0.0217
3.0	1.9980	0.9990	0.0096
3.2	1.9992	0.9996	0.0039
3.4	1.9998	0.9999	0.0015
3.6	1.9999	1.0000	0.0005
3.8	2.0000	1.0000	0.0002
4.0	2.0000	1.0000	0.0000
5.0	2.0000	1.0000	0.0000

Source: Reproduced from Welty, Wicks, and Wilson, *Fundamentals of Momentum, Heat, and Mass Transport*, p. 173, 1976. Reprinted by permission of John Wiley & Sons, Inc.

Another way for differentiating the three regions is shown in Figure 6-19. In this case,

$$U^+ = \frac{U_z}{U_*}$$

$$Y^+ = \frac{YU_*}{v}$$

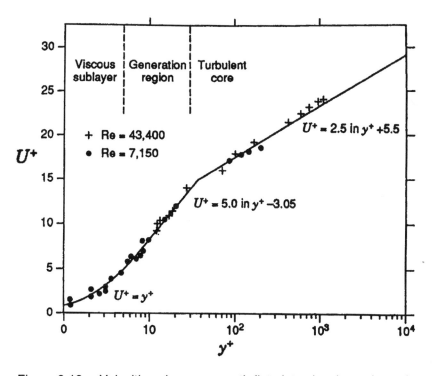

Figure 6-19. Velocities above a smooth flat plate showing universal velocity laws. (Source: Reproduced from Brodkey and Hershey, *Transport Phenomena: A Unified Approach,* McGraw-Hill Co.)

- Viscous sublayer

$$U^+ = Y^+ \text{ for } Y^+ \leq 5$$

- Generation region

$$U^+ = 5.0 \ln Y^+ - 3.05 \text{ for } 5 < Y^+ < 30$$

- Turbulent region

$$U^+ = \kappa^{-1} \ln Y^+ + 5.5 \text{ for } Y^+ > 30 \qquad [6\text{-}98]$$

These **logarithmic laws** also can be modified for application to a number of environmental problems, such as rivers, streams, ocean bottom surfaces, and various Earth surfaces. We define a roughness length, Z_0, for the surfaces, as shown in Table 6-5.

$$Z_0 = \frac{e}{30} \qquad [6\text{-}99]$$

The velocity distribution in turbulent regions can be given by a **universal logarithmic law.**

$$U^+ = \frac{1}{\kappa} \ln \frac{y}{z_0} \text{ for } y \geq z_0 \qquad [6\text{-}100]$$

Real laminar and turbulent flows always evolve boundary layers. The universal logarithmic law will be suitable for the data obtained, regardless of whether by pipes or flat plates. Furthermore both velocity and distance are dimensionless. For low Reynolds number flows, environmental problems such as drag on a place or transport through porous media will have applications.

Table 6-5. Roughness Lengths for Various Surfaces

Surface	z_0 (m)
Very smooth (e.g., ice)	10^{-5}
Snow	10^{-3}
Smooth sea	10^{-3}
Level desert	10^{-3}
Lawn	10^{-2}
Uncut grass	0.05
Fully grown root crops	0.1
Tree covered	1
Low-density residential	2
Central business district	5-10

Source: Seinfeld, Atmospheric Chemistry and Physics of Air Pollution, p. 495, 1986. Reprinted with permission of John Wiley & Sons, Inc.

6.4.4   Hydrodynamic Control for Contaminant Plume

When a toxic substance enters the groundwater, a complex situation exists. Even taking the simple case, the toxicant can be in the suspended particulate or in the dissolved form. As shown in Figure 6-20, both can also be in the sediments, either as sediment particulates or in the form of entering interstitial water. Often the organic liquid will create multiphase flow problems. The **nonaqueous phase liquid** (NAPL) can be separated into two different classes: one is light in density, the **light nonaqueous phase liquid** (LNAPL) such as petroleum hydrocarbons; the other is heavy, called the **dense nonaqueous phase liquid** (DNAPL) such as chlorine-containing hydrocarbons.

The spreading of a tracer in a two-dimensional uniform flow filled in an isotopic sand will tend to give plumes that are elliptical in shape, because longitudinal dispersion is stronger than the transverse dispersion, as shown in

Figure 6-21. In an actual field experiment, when a chloride tracer (Cl⁻) is co-injected with carbon tetrachloride (CTET) and perchloroethylene (PCE) into an aquifer, there are separations into different zones after 21 months, as shown in Figure 6-22.

This retardation of the transport of most organic compounds in aquifers is actually due to their solubility in water, which can be predicted from the partition coefficients for octanol-water, K_{ow}, as shown in Figure 6-23 and Table 6-6. Generally, if the fraction of organic carbon in water is known (such as fulvic acids, humic acids, or humin), the distribution coefficient for the sediments can be evaluated.

$$K_d' = K_{oc} f_{oc} \qquad [6\text{-}101]$$

where K_d' is the distribution coefficient, and

$$K_d' = s/C \qquad [6\text{-}102]$$

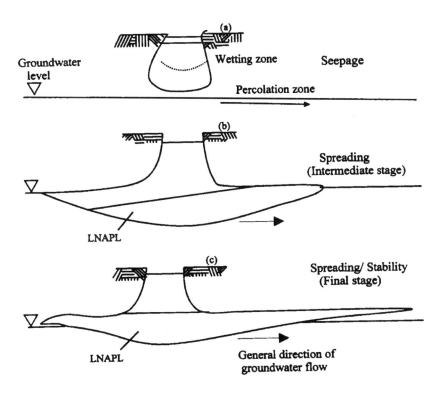

Figure 6-20. Stages of subsurface hydrocarbon migration (as approved in Proc. Nat. Water Well Assn. 2nd Canadian/American Conf. on Hydrogeology, Bauff, Alberta, 1985, pp. 31-35).

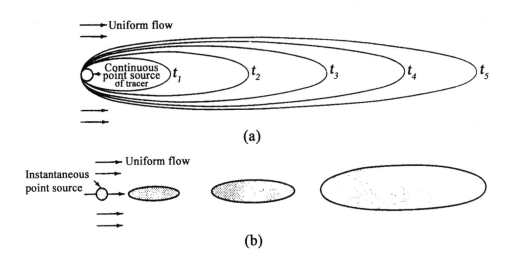

Figure 6-21. Spreading of a tracer in two-dimensional uniform flow field in
an isotropic sand: (a) continuous tracer feed with step function initial
condition; (b) instantaneous point source. (Source: Freeze/Cherry,
Groundwater, 1979, p. 394. Reprinted by permission of Prentice Hall, Inc.,
Englewood Cliffs, New Jersey.)

where s is adsorbed concentration in mg/kg dry soil or ppm and C is the
concentration in water in mg/L. A scaled expression of $K_d{}'$ is

$$K_d = \frac{q}{C} = \frac{\rho_b s}{\phi C} = \frac{\rho_b}{\phi} K_d' \qquad\qquad [6\text{-}103]$$

where ρ_b is the bulk density of the sediments, ϕ is the porosity of the sediment
here, and q is the adsorbed concentration expressed in mg/L of the pore water.
The conversion factor is

$$\frac{\rho_b}{\phi} = 6 \ \text{kg/L} \qquad\qquad [6\text{-}104]$$

This **corrected distribution coefficient**, K_d is an indication of the
movement of a given compound in aquifer. Usually a retardation factor, R, is
expressed as

$$K_d = 6K_d'$$

and

$$R = 1 + K_d \qquad\qquad [6\text{-}105]$$

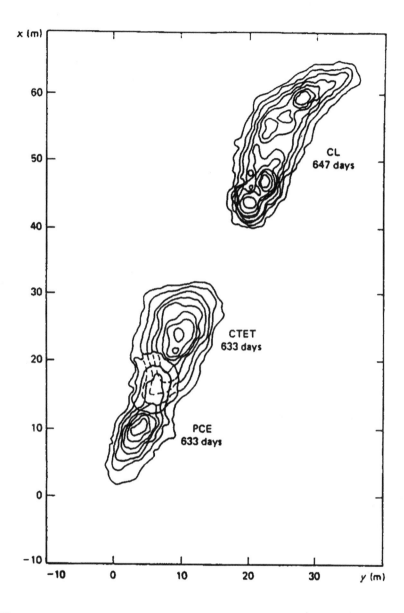

Figure 6-22.   Plume separation for chloride (CL), carbon tetrachloride (CTET), and tetrachloroethylene (PCE) 21 months after injection into an actual aquifer. (Source: Roberts, Goltz, and MacKay, 1986. "A Natural Grade Experiment on Solute Transport in a Sand Aquifer, 3, Retardation Estimates and Mass Balances for Organic Solutes." *Water Resources Research* 22(13):2047-2058, The American Geophysical Union.)

Table 6-6 .   (a) Partition Coefficients for Octonol-Water (k_{ow}) and Organic Carbon-Water (K_{oc}).Estimation of K_{oc} from K_{ow} (After Karlckhoff S. W., Chemosphere, 10, p. 833-846, 1981)

Compound	Log K_{ow}	Log K_{oc}	Compound	Log K_{ow}	Log K_{oc}
Hydrocarbons and chlorinated hydrocarbons			*Carbamates*		
3-methyl cholanthrene	6.42	6.09	Carbaryl	2.81	2.36
Dibenz(a,h) anthracene	6.50	6.22	Carbofuran	2.07	1.46
7,12-dimethylbenz(a) anthracene	5.98	5.35	Chloropropham	3.06	2.77
Tetracene	5.90	5.81	*Organophosphates*		
9-methylanthracene	5.07	4.71			
Pyrene	5.18	4.83	Malathion	2.89	3.25
Phenanthrene	4.57	4.08	Parathion	3.81	3.68
Anthracene	4.54	4.20	Methylparathion	3.32	3.71
Naphthalene	3.36	2.94	Chlorpyrifos	3.31	4.13
Benzene	2.11	1.78			
1,2-dichloroethane	1.45	1.51	*Phenyl ureas*		
1,1,2,2-tetracholroethane	2.39	1.90	Diuron	1.97	2.60
1,1,1-trichloroethane	2.47	2.25	Fenuron	1.00	1.43
Tetrachloroethylene	2.53	2.56	Linuron	2.19	2.91
γ HCH (lindane)	3.72	3.30	Monolinuron	1.60	2.30
α HCH	3.81	3.30	Monuron	1.46	2.00
β HCH	3.80	3.30	Fluometuron	4.34	2.24
1,2-dichlorobenzene	3.39	2.54	*Miscellaneous Compounds*		
pp'DDT	6.19	5.38			
Methoxychlor	5.08	4.90	13Hdibenzo(a,i) carbazole	6.40	6.02

Table 6-6 . (a) continued

22', 44', 66' PCB	6.34	6.08
22', 44', 55' PCB	6.72	5.62
Chloro-s-triazines		
Atrazine	2.33	2.33
Propazine	2.94	2.56
Simazine	2.16	2.13
Trietazine	3.35	2.74
Ipazine	3.94	3.22
Cyanazine	2.24	2.26
2,2'biquinoline	4.31	4.02
Dibenzothiophene	4.38	4.05
Acetophenone	1.59	1.54
Terbacil	1.89	1.71
Bromacil	2.02	1.86

(b) Estimation of K_{oc} from K_{ow} by the Expression $\log K_{oc} = a \log K_{ow} + b$ (After Schwartzenbach R. P. and J. Westall, *Environ. Sci. Technol. 15*, p. 1360-1367, 1981)

Regression Coefficient		Correlation Coefficient	Number of Compounds	Type of chemical
a	b			
0.544	1.337	0.74	45	Agricultural chemicals
1.00	-0.21	1.00	10	Polycylic aromatic hydrocarbons
0.937	-0.006	0.95	19	Triazines, nitroanilines
1.029	-0.18	0.91	13	Herbicides, insecticides
1.00	-0.317	0.98	13	Heterocylic aromatic compounds
0.72	0.49	0.95	13	Chlorinated hydrocarbons and alkylbenzenes
0.52	0.64	0.84	30	Substituted phenyl ureas and alkyl-N-phenyl carbamates

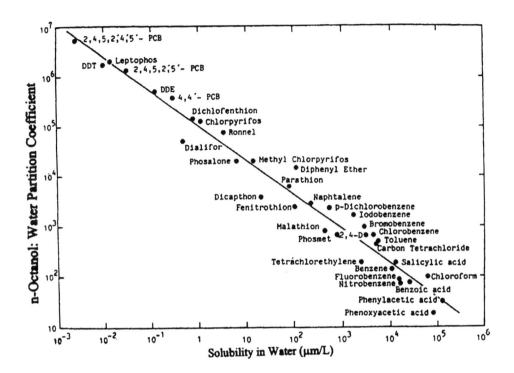

Figure 6-23. Relationship between solubility of a number of organic compounds and their octonol/water distribution (or partition) coefficient. (Reprinted with permission from Chiou et al., 1977. American Chemical Society.)

[**Example 6-3**] Calculate the retardation of lindane and of PCB in groundwater flow in a sediment containing 0.3% organic matter.

In Table 6-6, the K_{oc} of lindane = $10^{3.3}$

or

$$\log K_d' = 3.3 + \log 0.003 = 0.8$$

$$K_d' = 6.3 \text{ mL/g}$$

$$K_d = (6.3)(6) \cong 38$$

Similarly, for PCB,

$$\log K_d' = 5.6 + \log 0.003 = 3.1$$

$$K_d' = 1200 \text{ mL/g and } K_d = 7200$$

Thus, lindane moves 40 times slower, and PCB moves 7,200 times slower than water velocity.

The manipulation of hydraulic gradients for the control and removal of a groundwater plume is termed **hydrodynamic control**. A method to protect a production well (for drinking water use) from the approaching of a moving plume is to place a row of injection wells between the plume and the production well, as shown in Figure 6-24. The extraction well is used to remove the contaminants in the plume (which usually have to go through a treatment facility), and then use the purified water to reinject through the rows of injection wells.

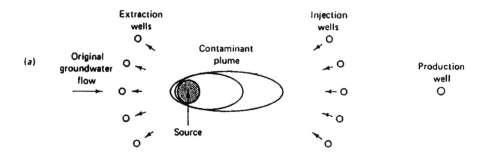

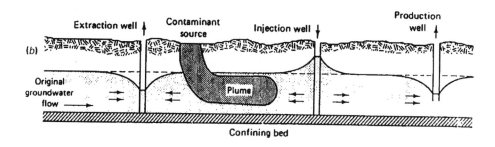

Figure 6-24. Hydrodynamic control with injection wells and extraction wells to protect a production well. (a) plan view; (b) cross-section. (After Masters)

Extraction wells should be placed closer to the head of the plume within the capture zone. The **capture zone** is the region within which all the flow lines converge on the extraction well (flow net). As shown in Figure 6-25, the envelope surrounding the capture zone can be calculated out as

$$y = \pm \frac{Q}{2Bq} - \frac{Q}{2\pi Bq} \tan^{-1}\left(\frac{y}{x}\right) \qquad [6\text{-}106]$$

where B is aquifer thickness and

$$\tan \phi = \frac{y}{x}$$

Hence, for $0 \le \phi \le 2\pi$,

$$y = \frac{Q}{2Bq}\left(1 - \frac{\phi}{\pi}\right) \qquad [6\text{-}107]$$

For $x \to \infty$, $\phi = 0$, $y = Q/2Bq$; thus, the maximum total width of the capture zone = $2(Q/2Bq) = Q/Bq$. For $\phi = \pi/2$, $x = 0$, $y = Q/4Bq$; thus, the width of the capture zone to the y-axis = $Q/2Bq$.

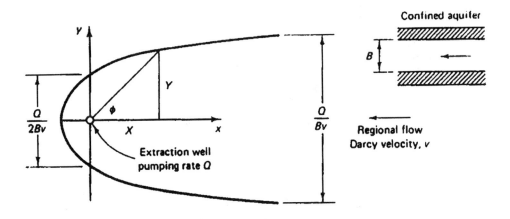

Figure 6-25. Capture-zone-type curve for a single extraction well located at the origin in an aquifer with regional flow velocity v and thickness B, and pumping at the rate Q. (Based on Javandel and Tsang, 1986.)

6.4.5   Chemical Restoration of Contaminated Aquifers

The conventional method for containment of contaminant plumes is to construct
a wall of impermeable material surrounding the plume from the surface to the
aquitard. The slurring cutoff wall or grout curtain is constructed with backfilled
material consisting of a mixture of soil and bentonite.   Such material is
relatively impermeable but cannot serve as a foolproof guarantee for
impermeability by toxic substances. For all the geotechnical fabric material,
Yen has used a biopolymer of the polyester type, poly β-hydroxybutyrate (PHB)
and poly β-hydroxyvelerate (PHV) derived from common bacteria such as
Alcaligenes eutrophus, which can be grown under anaerobic conditions in soil.
According to laboratory experiments, the relative permeability has reduced
down to a million fold. In this manner the source of pollution can be cut off by
the biobarrier technique. Furthermore, zones that are relatively contaminated
can be also isolated from the more concentrated ones if bioremediation is
applied to the aquifer. This is termed **zonal bioremediation**.

Besides pump-and-treat, **flushing** is considered a good method for aquifer
cleanup. The method is based on the fact that the sediment can adsorb the
molecules from the solution.

$$q = K_d C \qquad [6\text{-}108]$$

where

q = sorbed concentration (mass/L pore water)

K_d = distribution coefficient

C = solute concentration (mass/L pore water)

The overall change in concentration due to transport, dispersion, and
sorption can be written as the following expression:

$$\left(\frac{\partial C}{\partial t}\right)_x = -u\left(\frac{\partial C}{\partial x}\right)_t + D_L\left(\frac{\partial^2 C}{\partial x^2}\right)_t - \left(\frac{\partial q}{\partial t}\right)_x \qquad [6\text{-}109]$$

Neglecting the term of dispersion of Equation [6-109],

$$\left(\frac{\partial C}{\partial t}\right)_x = -u\left(\frac{\partial C}{\partial x}\right)_t - \left(\frac{\partial q}{\partial t}\right)_x \qquad [6\text{-}110]$$

The second term of Equation [6-110] can be expressed as

$$\left(\frac{\partial q}{\partial t}\right)_x = \left(\frac{dq}{dC}\right)\left(\frac{\partial C}{\partial t}\right)_x \qquad [6\text{-}111]$$

After substitution into Equation [6-110], which can be written

$$\left(1+\frac{dq}{dC}\right)\left(\frac{\partial C}{\partial t}\right)_x = -u\left(\frac{\partial C}{\partial x}\right)_t \qquad [6-112]$$

Now, for $C(x,t) = $ constant, we have by definition

$$dC(x,t) = 0 = \frac{\partial C}{\partial x}dx + \frac{\partial C}{\partial t}dt$$

or

$$\left(\frac{\partial C}{\partial t}\right)_x = -\left(\frac{\partial C}{\partial x}\right)_t\left(\frac{\partial x}{\partial t}\right)_C \qquad [6-113]$$

Substitute Equation [6-113] to Equation [6-112], then

$$u = \left(\frac{\partial x}{\partial t}\right)_c\left(1+\frac{\partial q}{\partial C}\right) \qquad [6-114]$$

Then allow

$$u_{c_i} = \left(\frac{\partial x}{\partial t}\right)_C \qquad [6-115]$$

as the concentration of i's species, and then we obtain from Equation [6-114]

$$u_{c_i} = \frac{u}{1+\left(\dfrac{dq}{dC}\right)} \qquad [6-116]$$

The pore-water flow velocity is inversely related to the pore volume of fixed length L from an aquifer to the flushed, or

$$\frac{u}{u_{c_i}} = \frac{V_{c_i}}{V_0} \qquad [6-117]$$

where

V_0 = pore volume of column with length L (m^3)

V_{c_i} = pore volume required to flush the column for C_i to arrive at L

Therefore, using Equation [6-116] we obtain

$$\frac{dq}{dC} = \frac{V_{C_i}}{V_0} - 1 \equiv V_{c_i}^* \qquad [6-118]$$

There is a simple relationship between the slope of adsorption isotherm at C_i and the number of pore volumes that must flush the column to arrive at that concentration. Thus, $V_{c_i}^*$ is termed the **flushing factor**, as shown in Figure 6-26.

[Example 6-4] Concentration of trichloroethylene (TCE) in groundwater is 10 mg/L in an aquifer. How many pore volumes need to be flushed in order for the TCE concentration to be below 1 mg/L? Assume that an approximate sorption isotherm is $q = 6 + 6 \ln C$.

Differentiation of the isotherm yields

$$\frac{dq}{dC} = \frac{6}{C_{TCE}} = V^*$$

and

$$V^*_{(c=1)} = 6$$

Thus, in order to meet the condition $V_c/V_0 = 1 + 6 = 7$, the pore volume of 7 times of the original has to be pumped out from the area to lower TCE. Also, $K_d = \dfrac{q}{C} \approx 1.4$ will give a much lower estimate. (See Equation 6-103)

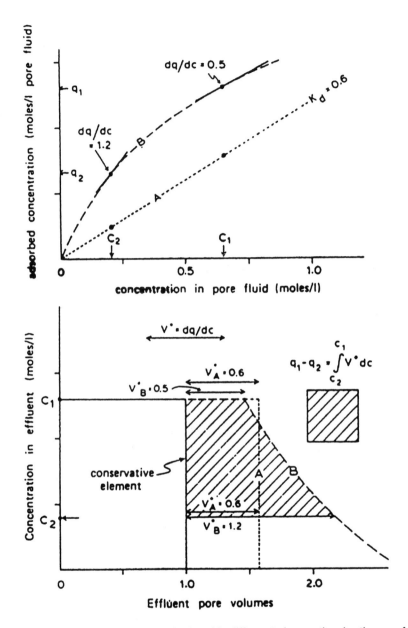

Figure 6-26. Elution of chemicals with different desorption isotherms from a column: substance A with linear desorption isotherm, B with convex desorption isotherm (left), Isotherm effects on breakthrough curves. Desorption isotherm can be obtained by graphical integration as shown for B (right).

REFERENCES

6-1. T. G. Spiro, and W. M. Stigliani, *Environmental Issues in Chemical Perspective*, State University of New York Press, Albany, New York 1980.

6-2. V. L. Snoeyink, and D. Jenkins, *Water Chemistry*, Wiley, New York, 1980.

6-3. C. N. Sawyer, and P. L. McCarty, *Chemistry for Environmental Engineering*, McGraw-Hill, New York, 1978.

6-4. W. Stumm, and J. J. Morgan, *Aquatic Chemistry*, Wiley, New York, 1981.

6-5. G. E. Likens, "Acid Precipitation," C & EN, 22, 29 (1976).

6-6. L. R. Ember, "Acid Pollutants: Hitchhiker Ride the Wind," C & EN, 20, Sept. 14 (1981).

6-7. U.S. Environmental Protection Agency, "Acid Rain," EPA-600/8-79-028, October (1979).

6-8. A. Lerman, *Geochemical Process: Water and Sediment Environments*, Wiley, New York, 1979.

6-9. G. Gambolati and G. Verri, *Advanced Methods for Groundwater Pollution Control*, Springer-Verlag, Berlin, 1995.

6-10. R. L. Daugherty, J. B. Franzini, and E. J. Finnemore, *Fluid Mechanics with Engineering Applications*, 8th ed., McGraw-Hill, New York, 1985.

6-11. M. M. Clark, *Transport Modeling for Environmental Engineers and Scientists*, Wiley, New York, 1996.

6-12. G. M. Fair, J. C. Geyer, and D. A. Okun, *Water and Wastemaker Engineering*, Vol. 1, *Water Supply and Wastewater Removal*, Wiley, New York, 1966.

6-13. G. M. Masters, *Introduction to Environmental Engineering and Science*, Prentice-Hall, Englewood Cliffs, New Jersey, 1991.

6-14. G. K. Batchelaor, *An Introduction to Fluid Mechanics*, Cambridge University Press, 1967.

6-15. L. H. Keith, *Energy and Environmental Chemistry, Acid Rain*, Vol. 2, Ann Arbor Science, Ann Arbor, Michigan, 1982.

6-16. J. A. Roberson and C. T. Crowe, *Engineering Fluid Mechanics*, 6th ed. Wiley, New York, 1997.

6-17. R. K. Linsely and J. B. Franzini, *Water Resources Engineering*, 3rd ed., McGraw-Hill, New York, 1978.

6-18. I. J. Higgins and R. G. Burns, *The Chemistry and Microbiology of Pollution*, Academic Press, London, 1975.

PROBLEM SET

1. Calculate the hydraulic gradient if the vertical drop is 50 m over a distance of 200 km. Also find the Darcy velocity in m/yr and the intrinsic permeability in m^2 of the medium if the hydraulic conductivity is 10^{-2} cm/sec

2. Five hundred kilograms of n-propanol ($CH_3CH_2CH_2OH$) is accidentally discharged onto a body of water containing 10^8 L. By how much is the COD (in milligrams per liter) of this water increased? Assume the following reaction:

$$C_3H_8O + 9/2\ O_2 \rightarrow 3\ CO_2 + 4\ H_2O$$

3. A water sample whose BOD versus time data for the first five days has the following data:

Time(days)	BOD(Y, mg/L)
2	10
4	16
6	21

Calculate the kinetic constants k, k' and L_0

NATURAL WATER AND POLLUTION

*I*n this chapter, we will discuss various natural water systems including oceans, estuaries and fjords, rivers and streams, and lakes. For the ocean, it is important to understand that it acts as the sink for weathered salts, the buffer for CO_2 and the Sillen's brake for atmospheric oxygen. The forms of estuaries and fjords are perfect examples of sediment traps in which minerals and petroleum are formed. Next, the self-purification of rivers and streams with the understanding of stream management or zoning is emphasized. The principle is guided by Gibb's diagram. Finally, the formation and life cycles of lakes are explored especially in reference to the eutrophication of nutrients, fertilizers, sewage, and industrial tailings. In general, the principles of marine and fresh water chemistry are the theme of this chapter. For the pollution of natural water systems, we will use the oil pollution in open waters and the tributyltin pollution in marine environments as examples.

7.1 CHEMICAL OCEANOGRAPHY

The Oceans represent a chemical system covering 71% of the earth's surface and accounting for 77% of the water in the hydrogeological cycle, as discussed in Chapter 6. The massive body has an average depth of 3.37 km, whereas in deep sea the depth will exceed 10 km. Seawater is a solution of gas and solids containing both organic and inorganic compounds. It is a well-oxygenated solution and is buffered at pH = 8 containing almost all elements. It is considered as a sink for all the salts that are present in sediments from the weathering process of the lithosphere.

Thermodynamically, oceans form an open system where both energy and mass can be exchanged across the boundaries. They may be considered as a homogenous equilibrated system, yet this is an over-simplification. At best, oceans may be treated as a steady-state system. The major soluble elements in seawater are summarized in Table 7-1. Traditionally, those elements exhibiting concentrations greater than 1 mg/kg, with the exclusion of silicon are major soluble elements. When the ratios of the concentrations are found invariant with time and locations, the term **nonconservative** is employed, because concentration is governed by chemical and biological process only.

Table 7-1. The Major Elements in Seawater

Element	Chemical Species	Concentration for S = 35‰ $(mol\ dm^{-3})$	$(g\ kg^{-1})$	Ratio to Chlorinity (Cl = 19.374‰)
Na	Na^+	4.79×10^{-1}	10.77	5.56×10^{-1}
Mg	Mg^{2+}	5.44×10^{-2}	1.29	6.66×10^{-2}
Ca	Ca^{2+}	1.05×10^{-2}	0.4123	2.13×10^{-2}
K	K^+	1.05×10^{-2}	0.3991	2.06×10^{-2}
Sr	Sr^{2+}	9.51×10^{-5}	0.00814	4.20×10^{-4}
Cl	Cl^-	5.59×10^{-1}	19.353	9.99×10^{-1}
S	$SO_4^{2-}, NaSO_4^-$	2.89×10^{-2}	0.905	4.67×10^{-2}
C (inorganic)	HCO_3^-, CO_3^{2-}	2.35×10^{-3}	0.276	1.42×10^{-2}
Br	Br^-	8.62×10^{-4}	0.673	3.47×10^{-3}
B	$B(OH)_3, B(OH)_4^-$	4.21×10^{-4}	0.0445	2.30×10^{-3}
F	F^-, MgF^+	7.51×10^{-5}	0.00139	7.17×10^{-5}

Based on Dyrssen & Wedborg (1974).

Table 7-2. The Major Components in the Model Ocean Relative to 1 Liter of Water

Substance	Amount (moles)	Comments
H_2O	54.9	(1 liter)
Si	6.06	mostly solid silicates and SiO_2
Al	1.85	mostly solids
Na	0.76	0.48 moles in solution
Ca	0.56	0.01 moles in solution
Cl	0.55	mostly in solution
C	0.55	see C cycle
Fe	0.55	mostly solids
Mg	0.53	0.05 moles in solution
K	0.41	0.01 moles in solution

After Sillen, 1961.

Salinity, S, is defined as the total salt content; that is, the mass of the total dissolved solid expressed in g/kg or 0/00. The factor termed **chlorinity** (Cl 0/00) is determined by silver nitrate titration and cannot differentiate the halides present. The relationships can be expressed by:

$$S\ 0/00 = 1.80655\ Cl\ 0/00 \qquad [7\text{-}1]$$

Because the dissolved species in the ocean are controlled by the mineral solids and atmosphere present in the boundaries of the ocean, different physicochemical models have been set up to account for the behaviors of some of the major elements. **Sillen's model** will be explored here. The model is based on using 1 liter of water as indicated by Table 7-2. In this model where silicon is most abundant, the solution process is

$$SiO_2(\text{crystal}) + 2H_2O = Si(OH)_4\ (aq),\ K = 2\ 7 \times 10^{-4}$$

$Si(OH)_4$ is a weak acid; for example,

$$Si(OH)_4\ (aq) + H_2O = Si(OH)_4^-\ (aq) + H_3O^+,\ K = 3.5 \times 10^{-10}$$

$$Si(OH)_4\ (aq) = Si_4O_6(OH)_6^{2-}\ (aq) + 2H_3O^+ + 2H_2O$$

$$Al(OH)_3\ (s) + OH^- = Al(OH)_4^-\ (aq),\ K = 10^{-1}$$

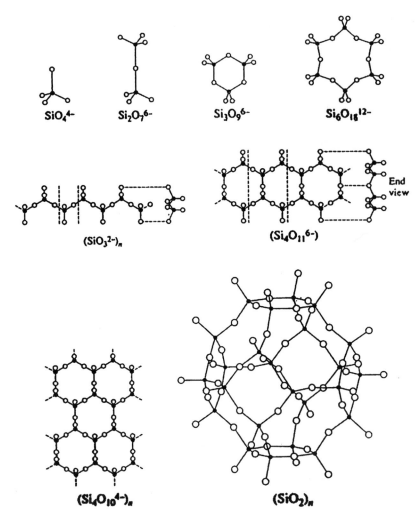

Figure 7-1. The common units in silicate minerals. (Source: Redrawn from R. W. Raisewell, P. Brimblecombe, D. L. Dent, and P. S. Liss. *Environmental Chemistry*. London: Edward Arnold Publishers. 1980)

This weak acid aluminum dissolved species can interact with $Si(OH)_4$ to form a basic solution.

$$Al(OH)_4^- \text{ (aq)} + Si(OH)_4 \text{ (aq)} = \frac{1}{2} Al_2Si_2O_5(OH)_4 \text{ (s)} + OH^- + 2\frac{1}{2} H_2O,$$

$$K = 10^{-6}$$

Because there are more cations than anions written in Table 7-2, some of these must form oxides or hydroxides.

$$3Al_2Si_2O_5(OH)_4 \text{ (s)} + 4SiO_2 \text{ (s)} + 2K^+ + 2Ca^{2+} + 15H_2O$$

$$= 2KCaAl_3SiO_6(H_2O)_6 \text{ (s)} + 6H_3O^+$$

The H_3O^+ is the controlling reaction for pH because the concentration of Al and Si are large. In sediments, K^+ is more abundant than Na^+, Ca^{2+} is more abundant than Mg^{2+}, and $K^+/Na^+ = 1.4$; whereas in seawater, $K^+/Na^+ = 0.0026$. Thus, for the previous equation,

$$K_{eq} = (H_3O^+)/(K^+)^2(Ca^{2+})^2$$

and pH of seawater is influenced by cationic concentration.

Other species (such as carbonate) will also enter the Sillen's model; for example, 0.46 moles of $CaCO_3$ and 0.09 moles of $MgCO_3$. The carbonate will enter into the pH-controlling reactions, but the amount is small compared to the proton capacity available from aluminosilicates. When all minerals are considered, the buffering capacity approaches 1 mole/L; therefore, the CO_2 in the atmosphere does not affect the pH of seawater much. Nevertheless, the seawater still maintains the balance of atmospheric CO_2 and the carbonate in sediments.

Oxygen and iron are important for the model. Most oxygen is in the atmosphere, but some is in the ocean according to Henry's law. Goethite may be the "brake" for an atmospheric increase of oxygen for global photosynthesis activity.

In natural water, both iron and manganese control the phosphorous and sulfur in a redox system. A good illustration can be seen in Figure 7-2, where manganese nodules and pyrite are formed. Furthermore, it seems that iron can regulate the partial pressure of oxygen in this planet's atmosphere as a **Sillen's brake** for global photosynthesis activity.

$$12FeOOH(s) = 4Fe_2O_3(s) + 6H_2O + O_2 \text{ (g)}$$

Assuming a steady state condition, the input from weathering products brought in by rivers, volcanic activity, atmospheric phenomena, and so on, is balanced by removal through sedimentation, ion exchange, biological production of inert materials, and so on.

$$(\frac{dC}{dt})_{in} = (\frac{dC}{dt})_{out} \qquad [7\text{-}2]$$

$(\frac{dC}{dt})$ is rate of change of concentration. The residence time, τ, is defined as

$$\tau = \frac{C}{(dC/dt)} \qquad [7\text{-}3]$$

where C is total dissolved concentration. The residence time calculated out is displayed in Table 7-3.

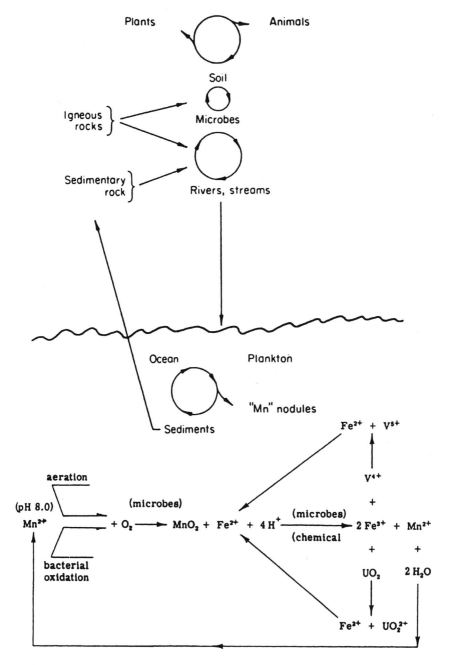

Figure 7-2. Biogeochemical manganese cycle and possible role of manganese in the oxidation of uranium and vanadium.

Table 7-3. The Residence Times of Some Elements in Seawater

Element	Principal Species	Concentration (mol dm^{-3})	Residence Time (y)
Li	Li$^+$	2.6×10^{-5}	2.3×10^6
B	B(OH)$_3$, B(OH)$^-_4$	4.1×10^{-4}	1.3×10^7
F	F$^-$, MgF$^+$	6.8×10^{-5}	5.2×10^5
Na	Na$^+$	4.68×10^{-1}	6.8×10^7
Mg	Mg^{2+}	5.32×10^{-2}	1.2×10^7
Al	Al(OH)$^-_4$	7.4×10^{-8}	1.0×10^2
Si	Si(OH)$_4$	7.1×10^{-5}	1.8×10^4
P	HPO$^{2-}_4$, PO$^{3-}_4$	2×10^{-6}	1.8×10^5
Cl	Cl$^-$	5.46×10^{-1}	1×10^8
K	K$^+$	1.02×10^{-2}	7×10^6
Ca	Ca^{2+}	1.02×10^{-2}	1×10^6
Sc	Sc(OH)$_4$	1.3×10^{-11}	4×10^4
Ti	Ti(OH)$_4$	2×10^{-8}	1.3×10^4
V	H$_2$VO$^-_4$, HVO$^{2-}_4$	5×10^{-8}	8×10^4
Cr	Cr(OH)$_3$, CrO$^{2-}_4$	5.7×10^{-9}	6×10^3
Mn	Mn^{2+}, MnCl$^+$	3.6×10^{-9}	1×10^4
Fe	Fe(OH)$^+_2$, Fe(OH)$^-_4$	3.5×10^{-8}	2×10^2
Co	Co^{2+}	8×10^{-10}	3×10^4
Ni	Ni^{2+}	2.8×10^{-8}	9×10^4
Cu	CuCO$_3$, CuOH$^+$	8×10^{-9}	2×10^4
Zn	ZnOH$^+$, Zn^{2+}, ZnCO$_3$	7.6×10^{-8}	2×10^4
Br	Br$^-$	8.4×10^{-4}	1×10^8
Sr	Sr^{2+}	9.1×10^{-5}	4×10^6
Ba	Ba^{2+}	1.5×10^{-7}	4×10^4
La	La(OH)$_3$	2×10^{-11}	6×10^2
Hg	HgCl$^{2-}_4$, HgCl$_2$	1.5×10^{-10}	8×10^4
Pb	PbCO$_3$, PbOH$^-$	2×10^{-10}	4×10^2
Th	Th(OH)$_4$	4×10^{-11}	2×10^2
U	UO$_2$(CO$_3$)$^{4-}_2$	1.4×10^{-8}	3×10^6

Based on Brewer (1975).

It is a common hypothesis that seawater is derived from the earth's crust while the inorganic ions may be derived by the interaction of major spheres, as shown in Figure 7-3. The seawater can also be lost by **evaporite** formation.

Deposits of evaporites generally consist of gypsum, rock salt, or a mixture of NaCl and KCl. If a narrow channel (such as a strait) is connected by oceans or semiclosed oceans, evaporite formation will take place. For example, the salinity of the Red Sea is 4.1%, whereas the average salinity is 3.5%.

A few words on oceans should be clarified here. The word **neritic** versus **oceanic** pertains to the closeness to shore. Also the **enphotic** zone (epipalagic) is shallow in depth; following that is the **disphotic** zone (mesopelagic); still deeper is the **aphotic** zone (bathypelagic); usually, the bottom of the ocean is called the **benthic** zone.

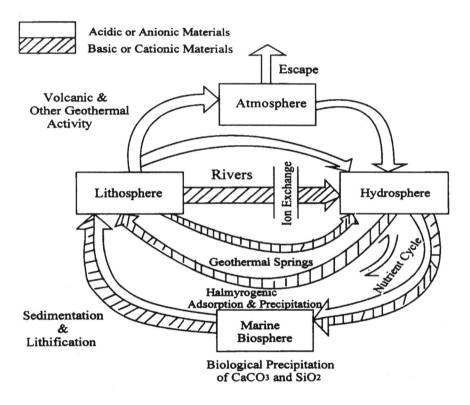

Figure 7-3. The cycle of cationic and anionic species in seawater.

7.1.1   Oil Pollution

One aspect of ocean pollution is contamination by petroleum. In 1975, the U.S. National Academy of Science workshop estimated that circa 6 million tons of petroleum hydrocarbons enter the ocean yearly. This flux is composed of natural seeps (10%), inputs from the atmosphere (10%), urban and river runoff and coastal industrial wastes (40%), and tanker operations (especially wasting of cargo tankers) (40%). It is clear that the last two sources cause 80% of the total flux to the ocean. Tanker operation is the major single source of oil pollution in the ocean. Figure 7-4 gives the world shipping lanes and major ocean currents.

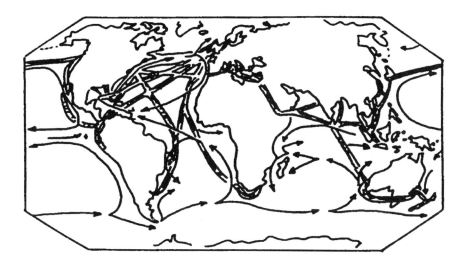

Figure 7-4. World shipping lanes before the closure of the Suez Canal (Stippled) and major ocean currents (arrows). (After A. Nelson-Smith, 1973.)

As early as 1922, the resulting oil pollution of British shores was harmful enough to prompt a law to prohibit the discharge of oil or oily waste in territorial waters. The present position in international law recognizes the absolute prohibition of visible oil discharge at sea for new ships over 20,000 tons.

The entry of petroleum hydrocarbons into the aquatic food web has been clearly demonstrated. Experiences with oil spills have been particularly revealing in regard to the vulnerability of the marine environment to petroleum. Certain petroleum hydrocarbons have been shown to interfere with the processes of chemoreception through the blocking of receptive organs. Reproductive processes also may become impaired as a result.

When oil spills occur, the oil spreads on the water. At the same time, components of low boiling points ("light ends") evaporate rapidly, leading to successively higher-boiling fractions. Significant amounts of compounds up to C_8 are carried off in this way. Figure 7-5 gives normal paraffin profiles of crude oil residues above 343°C. In the case of the Kuwait crude oil spill at sea, the crude soon lost most of its fractions of boiling point up to 300°C, diminishing by more than one-third of its mass or about 43% of the total volume.

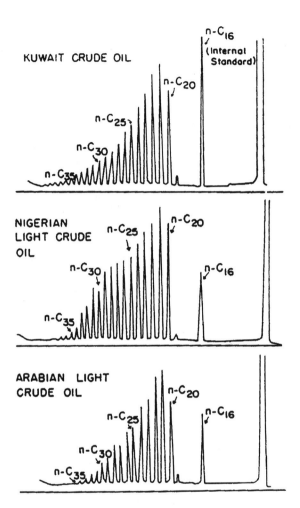

Figure 7-5. Normal paraffin profiles of crude oil residues above 343°C.

A model has been proposed to estimate the rate of oil slicking on water:

$$\frac{\pi\left(r_t^3 - r_o^3\right)\rho_\omega}{3V\left(\rho_\omega - \rho_o\right)\rho_o} = k_r t \qquad [7\text{-}4]$$

where

r_0 = slick radius at t = 0 (cm)

r_t = slick radius at t = t (cm)

V = volume of oil spilled (cm^3)

k_r = Blokker's constant (rate constant)

ρ_o = density of oil (g/cm^3)

ρ_w = density of water (g/cm^3)

t = time of spreading (sec)

Slick thickness

$$h_t = \frac{k}{t^{\frac{2}{3}}} \quad \text{(cm)} \tag{7-5}$$

$$k = \left(\frac{V}{\pi}\right)^{\frac{1}{3}} \left[\frac{\rho_w}{3\rho_o(\rho_w - \rho_o)k_r}\right]^{\frac{2}{3}} \tag{7-6}$$

The following are some of the Blokker's constants for petroleum k_r (at 9°C).

Libyan	1085
Iranian heavy	750
Kuwait	1480
Iraq	975
Venezuela	1340

Figure 7-6 gives the slick radius as a function of time for various crudes. Within 30 seconds, the slick radius can reach 7 feet for Kuwait crude. Figure 7-7 shows the variation of $(r_t^3 - r_0^3)$ with time, as an oil slick spreads.

Although spectacular oil spills have been highly publicized over the years, the smaller day-to-day inputs in the coastal waters and harbors of the world produce chronic pollution that is much larger in total volume. Municipal and industrial effluents, as well as runoff from the land and rivers, all contribute significantly. Offshore oil drilling will be increasingly important as the search for oil in the continental shelf area intensifies, adding more pollution to the ocean. Preventive measures of oil spills are shown below along with examples:

- Dispersant — $-(CH_2OCH_2)_n -$ amines

- Sinking agent — Siliconized sand

- Sorbent medium — For Bunker C oil, grams of oil absorbed per gram of sorbent medium is 72.7 for polyurethane and 5.8 for straw

- Combustion promoter — Silane treated silica

- Biodegradation organic   using prepackaged bacterial inoculate with soluble N and P compounds as nutrients

- Gelling agents   organic Ca salts

- Magnetic liquid   ferromagnetic properties of Fe_3O_4

- Beach cleaners   powdered enzymes and oxidizers

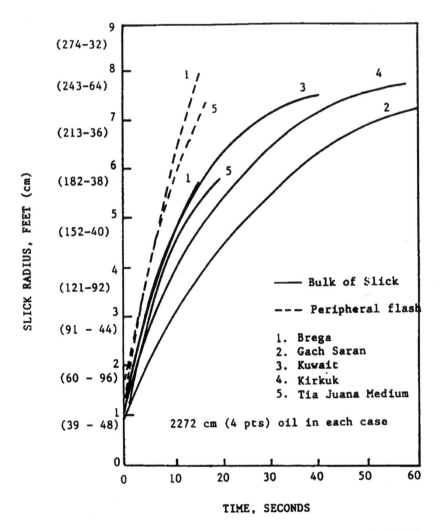

Figure 7-6. Oil slick spreading on water. (After Berridge et. al., 1965.)

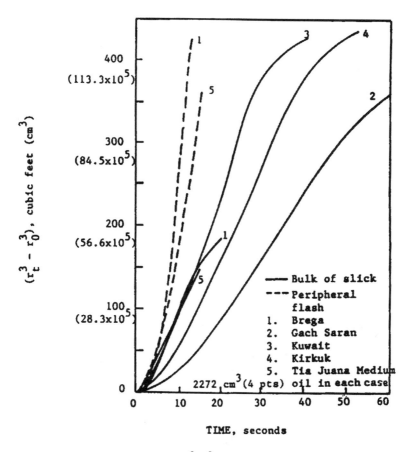

Figure 7-7. Variation of $(r_t^3 - r_0^3)$ with time as oil slick spreads. (Ref. Berridge et. al., 1965.)

7.1.2   Tributyltin in a Marine Environment

The anthropogenic addition of a number of chemical species in water (or the contamination) can be best illustrated with a metal such as lead. Suppose a soluble lead Pb^{2+} is present in water (this could originate from lead-based solder joints or scales). There are many soluble species whose formations depend on pH. (See Figure 7-8.)

   Tributyltin (TBT) is used as an antifouling marine paint for large ships as well as for recreation crafts. Although it is effective in preventing the attachment of barnacles to the surface of a vessel in such a manner as not to increase the dead weight of that vessel while traveling, TBT paint will slowly hydrolyze in marine water and is toxic to various nontarget organisms including native mollusks. The following is a list of derivations commonly used in the literature:

$$R'COOSnR_3 + H_2O \rightarrow R_3SnOH + R'COOH$$

Paint (copolymer)   TBE salt

$$R_3SnOH \xrightarrow[\text{TBT}]{} R_3Sn^+ + OH^-$$

(degradation scheme)

$$R_3SnX \rightarrow R_2SnX_2 \rightarrow RSnX_3 \rightarrow SnX_4$$

TBT   DBT   MBT   T

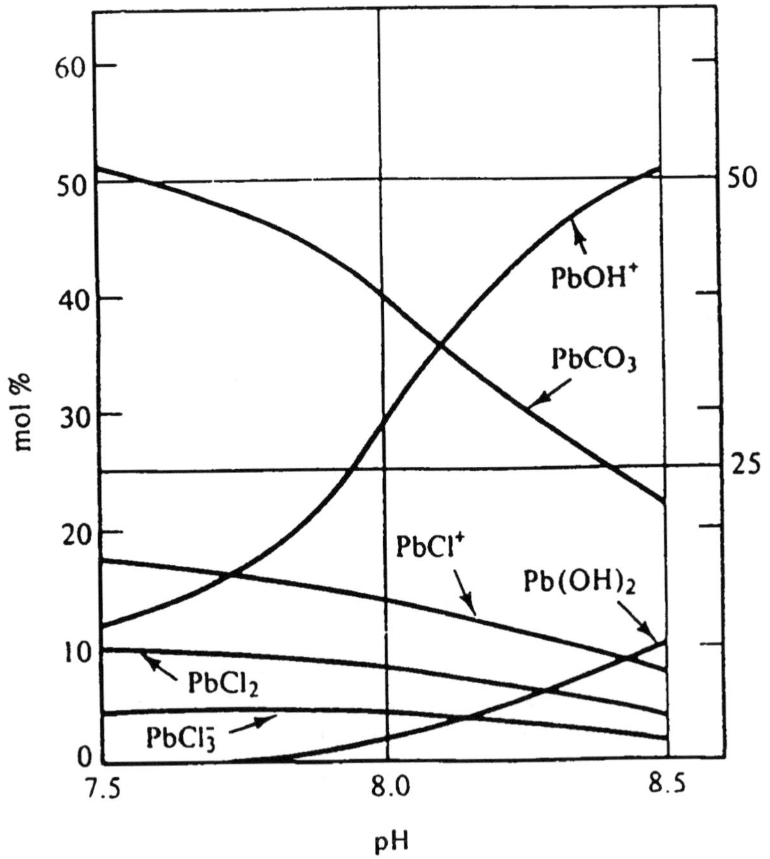

Figure 7-8. The calculated distribution of Pb (II) in seawater at 25°C and 1 atm. (From Bilinsky & Stumm, 1973.)

When these tributyltin ions transport across mitochondrial membranes for eukaryotic organisms, they will preferentially bind the imidazole system of the histidine residue from heme proteins. Because imidazole is essential in the coordination for binding oxygen hemoglobin, its interference will reflect on the sequence of electron carriers for mitochondria and will consequently inhibit phosphorylation, interrupting the basic energy process. This will explain the toxicity of the biocide activity of TBT. The scheme of oxygen transport is illustrated in Figure 7-9.

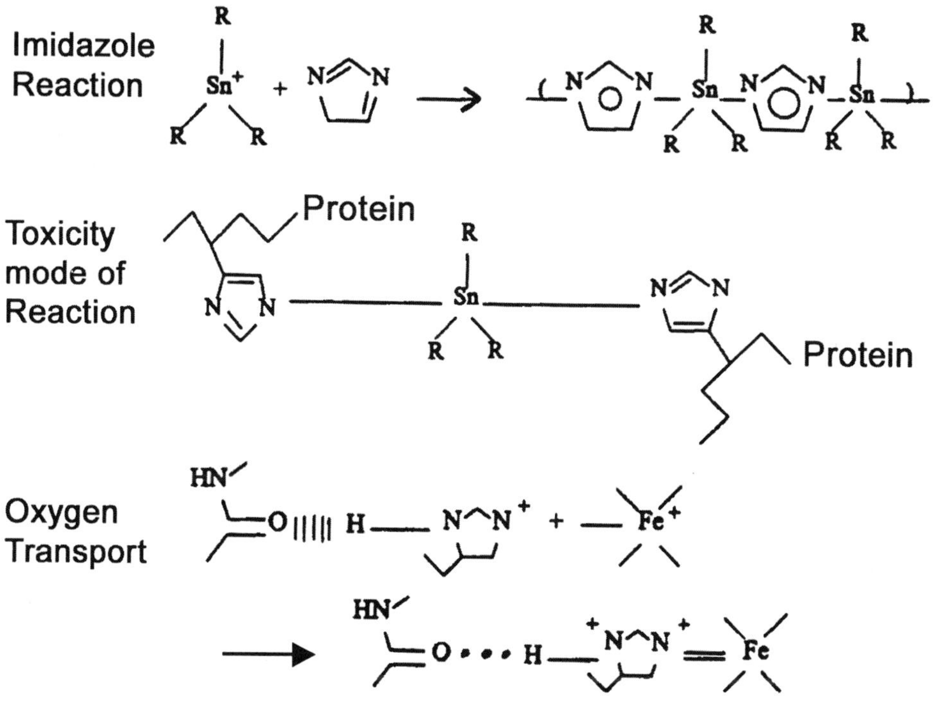

Imidazole Reaction

Toxicity mode of Reaction

Oxygen Transport

Figure 7-9. Imidazole for oxygen transport and blockage.

In 1976, the annual world consumption of organotin was 55 million lbs. Of this, the United States used almost half. The trend of consumption is upwards, with an annual growth rate of 12% for the last decade, as shown in Figure 7-10. The annual production of alkyltin compounds in the United States was 84 million lbs/yr in the year 2000. Of the total amount, about 30% ended up in the marine environment. No one knows what will be the consequence of the increase in potential loading and its impact on assimilative capacity. At this time, there is no simulated modeling for TBT in existence. Even submodels, such as hydrographic input of diffusive leaching, cannot be integrated into a sequestration because degradation, detoxification, and so on, are taking place simultaneously, as shown in Figure 7-11.

Increased use of organotin compounds is not limited to biocides. Another major use is for polyps stabilization. The tin content in sediment of Narragansett Bay is increasing logarithmically, as shown in Figure 7-12. Environmental stresses both in magnitude and duration are certainly real. The description or re-suspension in water has to be considered, as shown in Figure 7-13.

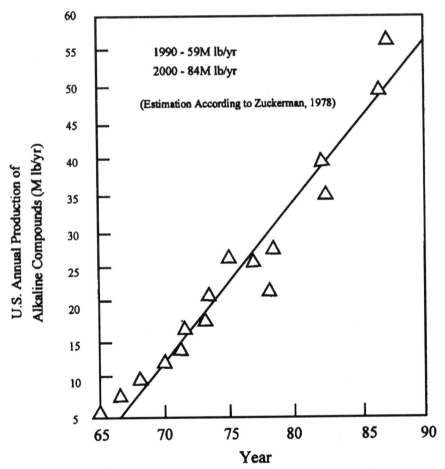

Figure 7-10. Projected annual vs. production of alkaline compounds.

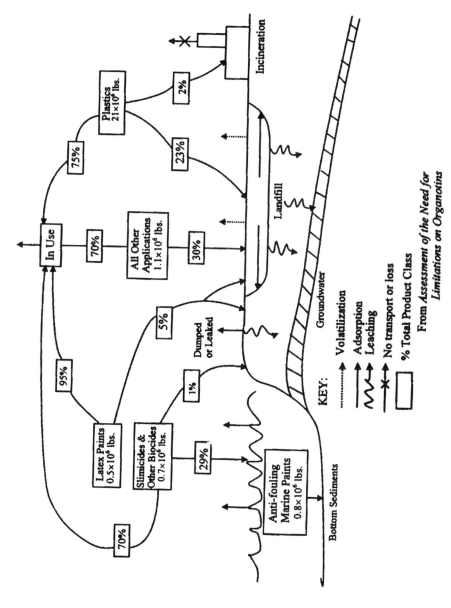

Figure 7-11. Geochemical cycle of TBT in the environment (After EPA- OTS Draft Report).

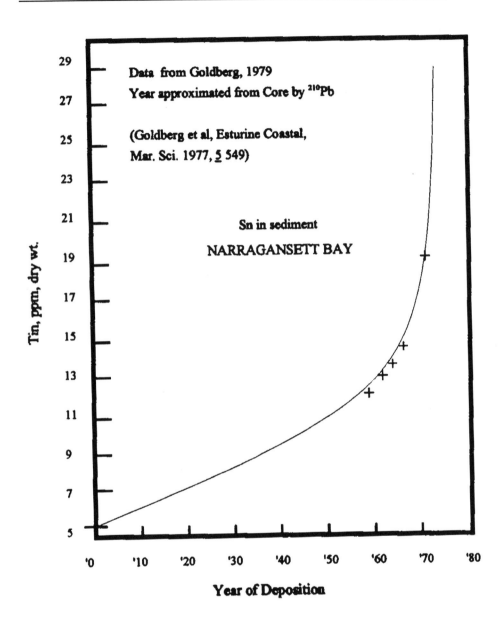

Figure 7-12. Environmental accumulation of tin in sediments is exponential.

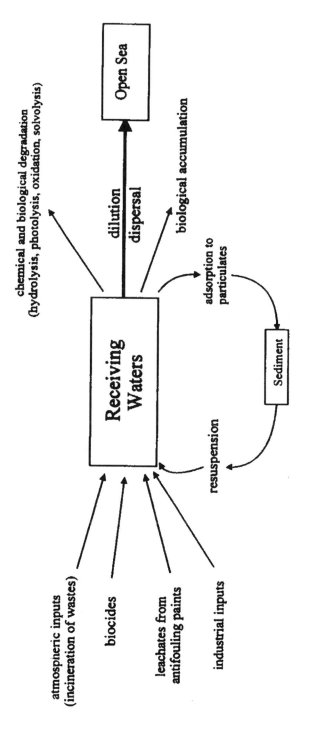

Figure 7-13. Input and uptake by TBT (After Stebbing, 1985).

7.2 RIVERS AND STREAMS

Riverine transport is the predominant way by which materials are inputted from the continents to the oceans. The average concentrations of elements in rivers and superficial rocks are summarized in Table 7-4. Usually the inputs carried by riverine waters are:

- salts in rain water

- continental material in drainage basins from weathering and erosion

- anthropogenic in origin

Table 7-4. A Comparison of the Concentration of Major Elements in "Average" Riverine Particulate Material and Surficial Rocks

Element	Concentrations (g kg^{-1})	
	Riverine Particulate Material	Surficial Rocks
Al	94.0	69.3
Ca	21.5	45.0
Fe	48.0	35.9
K	20.0	24.4
Mg	11.8	16.4
Mn	1.1	0.7
Na	7.1	14.2
P	1.2	0.6
Si	285.0	275.0
Ti	5.6	3.8

From Martin & Meybeck (1979).

Some rivers are linked with oceans; the important indicator for world surface water is Na^+ (for seawater) and Ca^{2+} (freshwater). A diagram expressing this is developed by Gibbs and can be located in Figure 7-14. A number of rivers are located. The **Gibbs' river diagram** can be further separated into precipitation dominance, rock dominance, and evaporation/crystallization dominance regions.

The global river influx (aq) of dissolved solids to oceans is 4.2 Pg/yr. Yet the particulate matter contribution, which is referred to as the **riverine particulate material** (RPM), is more than four times higher. The RPM is often reported by a 0.4-0.5 μm filter from the dissolved solid. Chemical fractionation has been used to sort out the anthropogenic inputs in RPM. A chemical extraction method has been developed to separate the anthropogenic elements from the natural weathering elements. By successive extraction with four different reagents, the following four different fractions are obtained:

- NH_4OAc — exchangeable
- NH_2OH/HCl — associated with Mn-Fe oxide surface coatings
- H_2O_2/HCl — organically associated
- $HF/HClO_4$ — resistant

The first two types are associated with anthropogenic sources; the latter two types are associated with natural weathering sources. The distribution of the four fractions will be an indication of the degree of pollution of a river. In Figure 7-15, the copper speciation of some rivers illustrates this point.

Chemically, if there is an increase in the concentration or mass transport of certain chemical species, it is known as pollution. Conversely, if there is a decrease, then it is termed **self-purification**. The self-purification of streams involve the following:

- transport and incorporation into deposits
- reaction within the water mass or suspended matter
- exchange reaction of volatiles with the atmosphere
- chemical and biochemical oxidation within the sediments

Amount of self-purification of rivers, S_m, in mole/sec can be expressed as

$$S_m = Q(C_o - C_u) \qquad [7\text{-}7]$$

where Q is flow in m^3/s and C_o and C_u is respectively the concentrations in the upstream and downstream in mole/m^3. The rate of self-purification can be expressed as

$$S_r = \frac{dC}{dt} = \frac{(C_0 - C_u)}{t} \qquad [7\text{-}8]$$

For ecological models, the rate can be expressed in terms of biomass in the river (mole/g/s)

$$S_e = \frac{S_m}{G} = \frac{Q(C_0 - C_u)}{t(g'Pv + g''Q)} \qquad [7\text{-}9]$$

where G is the total biomass in g, g' and g'' are respectively attached and suspended biomass in g/m^2 and g/m^3, P is the length of wetted cross profile in m, and v is flow velocity in m/s or

$$S_r = S_e \left(\frac{g'}{R} + g''\right) \quad \text{mol/m}^3/s \qquad [7\text{-}10]$$

where R is hydraulic radius in m.

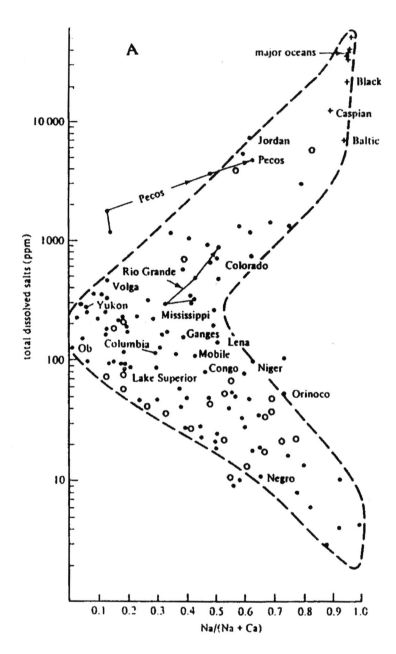

Figure 7-14. A variation of the weight ratio Na/ (Na + Ca) as a function of the total dissolved salts for several surface waters. (From Gibbs, 1970.)

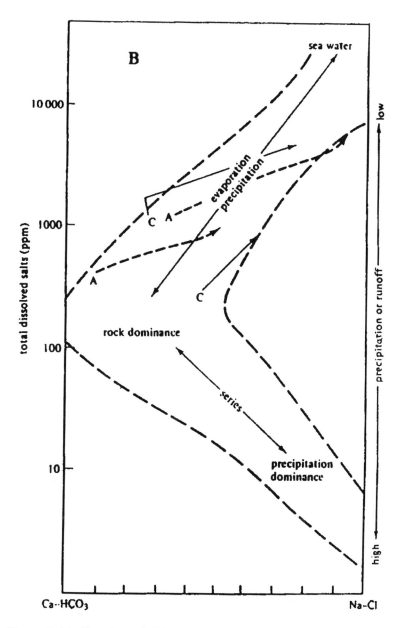

Figure 7-14. (Continued). Diagrammatic representation of processes controlling the chemistry of the world surface waters. (From Gibbs, 1970.)

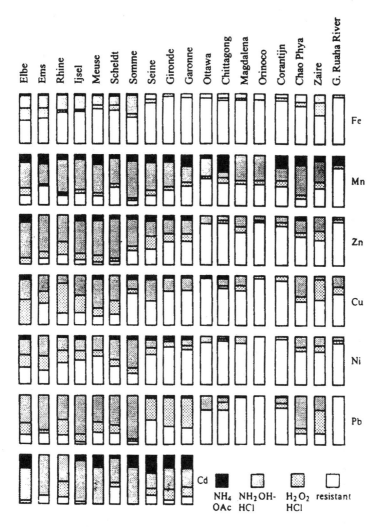

Figure 7-15. The speciation of trace metals in 18 different river sediments, arranged according to their approximate geographic position from north to south. Most tropical contained low cadmium levels, and no reliable data were obtained. For the Rio Magdalena and Orinoco River, insufficient material was available for determination of the 'Exchangeable' (NH_4OAc) fraction, and this is contained in the hydroxylamine extract. (From Salomons and Forstner, 1980.)

Different substrates added to model rivers confirm that the imported energy from substrate organics, E_s, is important because it can be correlated with heterotroph or prototroph to compete for dominance of growth space. The ratio of E_s/E_L is plotted versus P/H in Figure 7-16, where E_L is light energy, and P is phototrophic, and H is heterotrophic biomass.

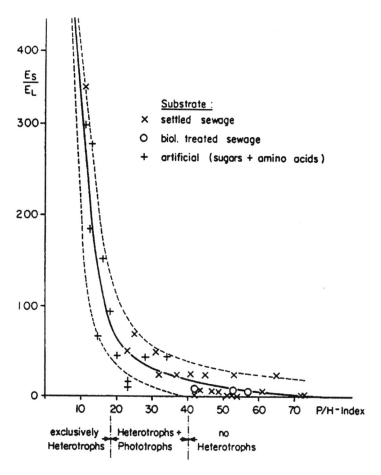

Figure 7-16. P/H Index of phytocenoses in model rivers at comparable hydraulic conditions as a function of the ratio E_8/E_L. River water: groundwater with the addition of various pollutants. E_8 and E_L in kcal/(dm^2) (day). (Data compiled from numerous independent model river studies from 1965-1970 by K. Wuhrmann.)

Volatilization of the organics released into the air is a major event, especially when both blowing wind and howling water are acting. This is similar to the release of volatile organics or petroleum to open water bodies as previously described in oil spills. The mass exchange of a given chemical across the air- water interface can be computed by the "two film" theory. If K is the liquid film coefficient (L/T) and K_g is the gas film coefficient (L/T), then the overall volatilization transfer rate, k_l (L/T), can be obtained from

$$(k_l)^{-1} = (K_l)^{-1} + (K_g H_e)^{-1} \qquad [7\text{-}11]$$

where H_e is the Henry's constant (dimensionless), representing the partitioning of the chemicals between the water and atmosphere phases.

$$H_e = \frac{H'_e (\text{atm} \cdot \text{m}^3 / \text{mole})}{RT (\text{atm} \cdot \text{m}^3 / \text{mole})} = \frac{\left[\dfrac{P(\text{atm})}{C_w (\text{mole/m}^3)} \right]}{RT} \qquad [7\text{-}12]$$

where P is partial pressure and C_w is water solubility concentration.

Furthermore, the liquid film coefficient can be estimated from the oxygen transfer coefficient, K_L, by

$$K_L = \left(\frac{32}{M} \right)^{\frac{1}{4}} K_L \qquad [7\text{-}13]$$

where M is the molecular weights of the spilled chemicals and K_L can be approximated from re-aeration coefficient

$$K_a = \frac{K_L}{H} = \frac{(D_L U)^{\frac{1}{2}}}{H^{\frac{3}{2}}} \qquad [7\text{-}14]$$

or

$$K_L = \left(\frac{D_L U}{H} \right)^{\frac{1}{2}}$$

where D_L is the oxygen diffusivity at 20°C (8.1×10^{-5} ft²/ hr), U is average stream velocity, and H is average depth. On the other hand, the gas film coefficient can be estimated empirically from

$$K_g = 168 \left(\frac{18}{M} \right)^{\frac{1}{4}} U_w \qquad [7\text{-}15]$$

where U_w is the wind speed in m/s.

[Example 7-1] Chlorobenzene is discharged into a stream where the wind speed is 5 m/s and water temperature is 20°C. The depth of the stream is 0.4 m and the water velocity is 0.60 m/s. Compute the volatilization rate of chlorobenzene.

Oxygen transfer coefficient can be estimated by

$$K_L = \left(\frac{D_L U}{H}\right)^{\frac{1}{2}}$$

$$= (1.81 \times 10^{-4} \text{ m}^2/\text{ day})^{1/2} (0.60 \text{ m/s})^{1/2} (8.64 \times 10^4 \text{ s/day})^{1/2}/ (0.4 \text{ m/s})^{1/2}$$

$$= 4.84 \text{ m/day}$$

Liquid film coefficient,

$$K_l = (32/113)^{1/4} (4.84 \text{ m/day}) = 3.53 \text{ m/day}$$

Gas film coefficient,

$$K_g = 168 (18/113)^{1/4} (5) = 531 \text{ m/day}$$

Dimensionless Henry's constant,

$$H_e = \frac{H'_e}{RT} = \frac{0.0037 \text{ atm - m}^3/\text{mole}}{8.206 \times 10^{-3} \text{ atm - m}^3/\text{mole °K} \times 293 \text{°K}} = 0.154$$

$$\frac{l}{k_l} = \frac{1}{K_l} + \frac{1}{K_g H_e} = \frac{1}{3.53} + \frac{1}{531(0.154)} = 0.283 + 0.0122$$

or

$$k_l = 3.38 \text{ m/day}$$

7.3 CHEMICAL LIMNOLOGY

The study of the physical, chemical, and biological characteristics of rivers and lakes are in the domain of **limnology**. For the counterpart, the study of water in the ocean is termed as oceanography. For rivers, because the turbulent flow causes mixing, there will be no temperature gradients developing. However, for lakes, due to the nonmixing, the top layer is heated up more than the bottom layer. In this manner there is vertical stratification. As shown in Figure 7-17, the upper layer is called **epilimnion** and the bottom layer is called **hypolimnion**. The middle layer is called **thermocline** (or metalimnion), or the transition zone and the temperature changes rapidly over a short distance of depth. Chemical factors affect the pattern within the limnolocal domain, vastly including all the biological communities. For example, phosphorous will affect both algae and fish as shown in Figure 7-18. The properties of lakes will be modified accordingly, including nutrients, as shown in Figure 7-19.

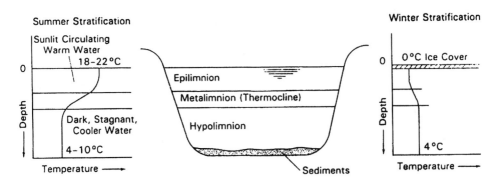

Figure 7-17. Thermal stratification of a deep lake.

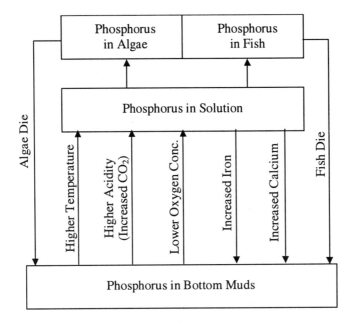

Figure 7-18. Chemical factors that effect phosphorus equilibria in lakes.
[Source: H.R. Jones, *Detergents and Pollution, Problems and Technical Solutions* (Park Ridge, N.J.; Noyes Data Corporation, 1972), p. 8.]

As an ecological unit, often the lake and river will also include the drainage basin, which is also known as catchment area or watershed. Because vegetation and soil will be surrounding the body of water, organic peat will be formed with bog vegetation as a sponge. It should be remembered that the peat would eventually become coal via the coalification process, for example, via lignite.

Heavy metals can settle from receiving waters (such as tributaries) together with biogenic and other particles as transport carriers. Actually, many sediments depth profiles serve as indicators of heavy metal pollution. For lakes, the residence time of heavy ions is drastically reduced due to scavenging actions by sedimentation. For example, in Lake Greiffensee, the sedimentation rate is 0.37 Tg/yr. The computation of the soluble metals is best performed by the mass balance model, assuming that a steady state is attainable.

For example, the lake can be comparable to a reactor as indicated by Figure 7-20. Let the input and output fluxes be J_i ($i=4$), then

$$V \frac{dC}{dt} = J_1 + J_2 - J_3 - J_4 \qquad [7\text{-}16]$$

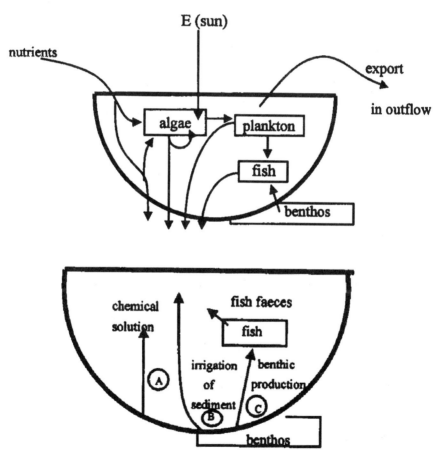

Figure 7-19. (a) Common view of nutrient pathways. (b) Potential sources of return from sediment.

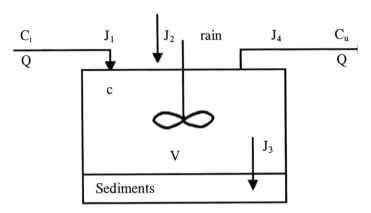

Figure 7-20. Mass balance model of a lake at steady state.

This equation is also used for the total phosphorous concentration calculation as seen in Equation [7-25].

Here J_1 is the annual input from rivers and streams, J_2 is the annual rainfall to the lake, J_3 is the annual sedimentation amount and J_4 is the annual outflow to other water body systems. Because precipitation is approximately equal to evaporation, the lake is a closed system.

Now the influx concentration of any metal sphere, C_i, is carried in by

$$J_1 = QC_I \qquad\qquad [7\text{-}17]$$

assuming the rain does not affect C_i. The final ultimate concentration is the exit concentration, C_u, and can be effected as

$$J_3 = SP = K_d C_u P \qquad\qquad [7\text{-}18]$$

Here, P is the sedimentation rate and S is the concentration of metal species in suspended matter. Because the distribution coefficient, k_d, is defined as

$$k_d = \frac{S}{C_u} \qquad\qquad [7\text{-}19]$$

Finally, the exit flux is

$$J_4 = QC_u \qquad\qquad [7\text{-}20]$$

For steady state,

$$C_u = C = \underline{C} \qquad\qquad [7\text{-}21]$$

Substituting Equations (7-17), (7-18), and (7-20) into Equation (7-16) for differentiation,

$$\frac{dC}{dt} = \frac{QC_i + J_2}{V} - C\left(\frac{Q + k_d P}{V}\right) \qquad [7-22]$$

or after integration

$$C(t) = C_0 \exp\left(\frac{Q + k_d P}{V}t\right) + \frac{QC_i + J_2}{Q + k_d P}\left[1 - \exp\left(\frac{-Q + k_d P}{V}t\right)\right] \qquad [7-23]$$

At $t = \infty$ and $C(t) = \underline{C}$, Equation [7-23] becomes

$$\underline{C} = \frac{QC_i + J_2}{Q + k_d P} \qquad [7-24]$$

Stumm and Morgan used Equation [7-24] to calculate the heavy metal concentration in the water leaving the lake. For example, in Greiffensee, assuming the volume is 0.125 Gm^3, Q is 89 M m^3/yr, P is 0.37 Tg/yr, and J_2 (rain) = 1300 kg/ yr. For metal ions $C_i(\text{Zn})$ = 1908 mg/m^3, $C_i(\text{Pb})$ = 3.2 mg/m^3, $k_d(\text{Zn})$ = 25 m^3/kg, and $k_d(\text{Pb})$ = 120m^3/kg. After using Equation [7-24], the results are $\underline{C}(\text{Zn})$ = 3.1 mg/m^3 and $\underline{C}(\text{Pb})$ = 0.4 mg/m^3. The field data collected are Zn = 4.1 mg/m^3 and Pb = 0.6 mg/m^3. The critical field data are close to the calculated values.

7.4 EUTROPHICATION

Natural waters acquire their chemical characteristics by dissolution and by chemical reactions with solids, liquids, and gases with which they have come into contact during the various parts of the hydrological cycle. In some instances, biological activities also play a role.

The water carried in streams is considered to consist of two fractions: one which is made up of subsurface water and groundwater that reenters the surface water, and the other which is a surface runoff fraction that enters the drainage system during and soon after the precipitation period. The relative proportions of those components and the concentration of dissolved species in each, as influenced by the interactions of rainwater with minerals and vegetation and by the evaporation and transpiration from plants, largely determine the composition of river waters. Figure 7-21 gives the dissolved solids of rivers as a function of runoff. In the modeling work, streams are usually assumed to be fixed-length plug flow chemical reactors with the superimposition of the influences of pollution and waste disposal. Rivers are also considered to possess the ability of self-purification.

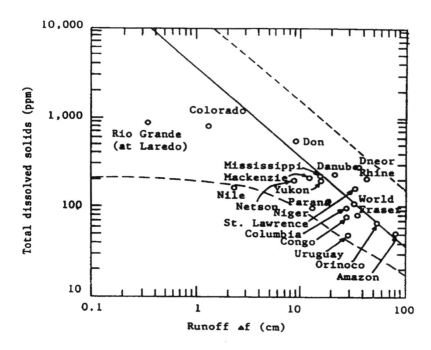

Figure 7-21. Dissolved solids of rivers as a function of runoff (After H. D. Holland).

As discussed before, the sea is an open, dynamic system with variable inputs and outputs of mass and energy for which the state of equilibrium is a constraint. Because the sea has remained constant during the recent geological past, it may be well justified to interpret the ocean as a **steady-state model**. Input is balanced by output in a steady-state system. The system considered is a single box model of the sea, that is, an ocean of constant volume, temperature, pressure, and uniform composition.

The situation in lakes, as mentioned before, is more complicated than that of the sea in many regards. Most substances entering lakes are **nonconservative** (i.e., have a residence time different from that of water). In most lakes, the input rates of many substances have increased, and consequently, concentrations of many constituents are not time-variant. If the input rates of nutrients are too high, lakes may also have some of the eutrophication problems that are discussed in Section 7.4.2.

The aquatic environments are further complicated by the interaction with the biosphere. There is a constant production, decomposition, and sedimentation of biomass, as illustrated by Figure 7-22. Another factor that should be pointed out is that agricultural irrigation will not only add dissolved solids to the rivers or lakes, but also put a burden on underground aquifers. For example, one of the largest aquifers in North America, the Ogallala aquifer, which crosses eight states in the United States, suffers from this; even the rate of use is three orders of magnitude greater than the rate of recharge.

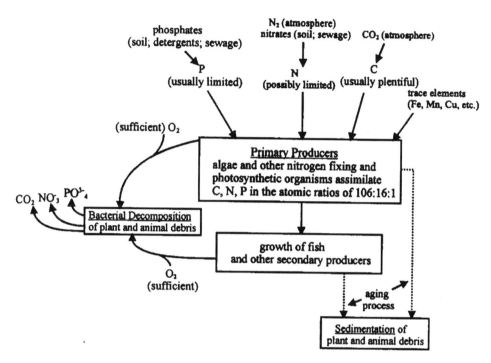

Figure 7-22. Factors that affect aquatic production, decomposition, and sedimentation.

7.4.1 Photosynthesis and Respiration

Energy-rich bonds are produced as a result of photosynthesis, thus, distorting the thermodynamic equilibrium. Bacteria and other respiring organisms catalyze the redox processes that tend to restore chemical equilibrium. In a simplified way, we may consider a stationary state between **photosynthetic production** P (rate of production of organic material) and **heterotrophic respiration** R (rate of destruction of organic material) and chemically characterize this steady state by a simple stoichiometry equation.

$$106 \, CO_2 + 16 \, NO_3^- + HPO_4^= + 122 H_2O + 18 H$$

$$P\downarrow \; R\uparrow$$

$$(C_{106}H_{263}O_{110}N_{16}P_1) + 138 \, O_2$$

Algal protoplasm may also be conveniently expressed as $(CH_2O)_{106}(NH_3)_{16}(H_3PO_4)$. The stoichiometric formulation of the equation reflects, in a simple way, **Liebig's law of minimum**, which states that plant growth is controlled by the availability of a single nutrient, the **limiting nutrient**, in this case, phosphorous.

Figure 7-23 shows the nitrogen cycle. Nitrifying bacteria have evolved to use reduced nitrogen as a fuel source in a process called **nitrification**. Some of them (organisms of the genus *Nitrosomonas*) oxidize ammonia to nitrite, as shown here:

$$NH_4^+ + OH^- + 1.5\ O_2 \rightarrow NO_2^- + H^+ + 2H_2O$$

While others (*Nitrobacter*) oxidize nitrite further to nitrate

$$NO_2^- + 0.5\ O_2 \rightarrow NO_3^-$$

The nitrogen cycle is completed by denitrifying bacteria in the nitrification process, which converts nitrate back to N_2.

$$NO_3^- \rightarrow NO_2^- \rightarrow NO \rightarrow N_2O \rightarrow N_2$$

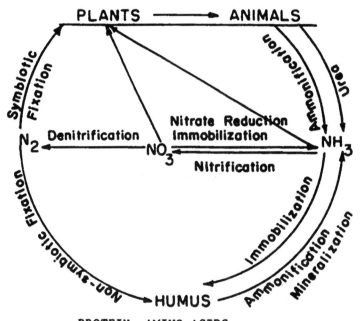

Figure 7-23. The nitrogen cycle.

Nitrate, which has converted to nitrite, will cause methemoglobieumia (blue blood) problems. Table 7-5 tabulates nitrate concentration in some vegetables, and it shows that the nitrate concentration in spinach is the highest.

Table 7-5. Average ppm Nitrate in a Variety of Foods

	Average ppm of Nitrate	Number Samples Analyzed
Mixed vegetables	88	2
Carrots	101	8
Green beans	163	3
Garden vegetables	180	5
Graham crackers	211	1
Squash	282	5
Wax beans	444	2
Beets	977	6
Spinach	1373	5

Elemental sulfur is chemically stable in the presence of oxygen in most environments but is readily oxidized by sulfur-oxidizing bacteria; for example, *Thiobacilles thiooxidans* results in the formation of sulfate and hydrogen ions. The energy generated can be used for building up the biomass.

$$S + 3/2\ O_2 + H_2O \rightarrow H_2SO_4$$

$$CO_2 + H_2O \rightarrow (CH_2O) + O_2$$

A wide variety of organisms can use sulfate as a sulfur source and carry out assimilatory sulfate reduction, for example, *Desulovibrio desulfuricans*.

$$CaSO_4 + 2\ (CH_2O) \rightarrow CaS + 2H_2O + 2CO_2$$

or

$$CaSO_4 + 2(CH_2O) \rightarrow CaCO_3 + H_2S + CO_2 + H_2O$$

7.4.2 Eutrophication and Its Control

Figure 7-24 illustrates the thermal stratification of a lake through the four seasons. In a stratified lake, excessive production of algae and oxygen in the upper layers (P>>R) may be paralleled by anaerobic conditions at the bottom (R>>P). This is because most of the photosynthetic oxygen escapes into the atmosphere and does not become available to the deeper water layers, and eventually the algae sink to the bottom of the lake.

Over-nutrition of bodies of water caused by inputs of phosphates, nitrogen compounds, or other nutrients is commonly called **eutrophication**. Technically, eutrophication is simply the natural process of providing a body of water with the nutrients for the aquatic life it supports. A lake starts its life cycle as a clear body of water, which is described as **oligotrophic**. As nutrients enter the lake through land runoff, and as aquatic life grows and dies, the water acquires a high content of organic debris. At this stage, the lake is considered **mesotrophic**. Eventually, it fills in completely, forming a marsh and then dry land. The Green River basin in Utah and Colorado was a lake basin (Lake Uinta) that lasted 4 million years and then dried out to be replaced by the Colorado River, which has already lasted for 40 million years. Lake Gosiate in Wyoming also dried out eventually. This basin is rich in oil shale deposits.

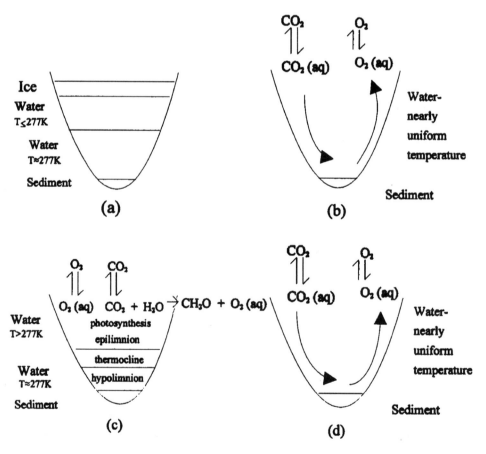

Figure 7-24. Thermal stratification of a lake through four seasons:
(a) Winter, (b) Spring (turnover), (c) Summer, and (d) Fall (turnover).

The rate of eutrophication establishes the balance between the production of aquatic life and its destruction by bacterial decomposition. Under natural conditions, the rate of decomposition is nearly equal to the rate of production, and little sedimentation occurs. Where there are large inputs of nutrients from human sources, bacterial decomposition cannot keep pace with productivity, and sedimentation increases. Table 7-6, which summarizes amounts of nitrogen and phosphorus reaching surface water, may serve as an example of the ratio of natural and man-made sources of loading. Figure 7-25 illustrates that there is a close relationship between the fertilizer used and the excess riverine nitrate concentration.

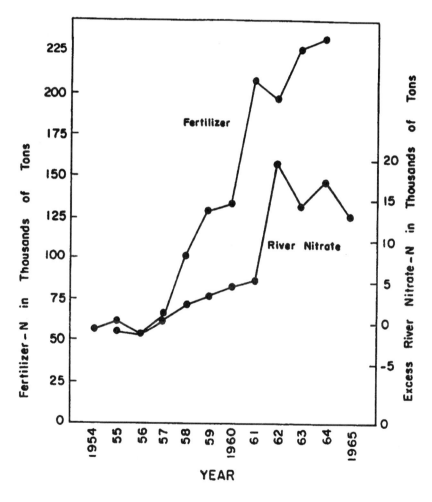

Figure 7-25. Trend of river nitrate to fertilizer use.

Table 7-6. Estimated Nitrogen and Phosphorus Reaching Wisconsin Surface Waters

Source	N	P	N	P
	Lbs. per year (% of total)			
Municipal treatment facilities	20,000,000	7,000,000	24.5	55.7
Private sewage system	4,800,000	280,000	5.9	2.2
Industrial wastes[a]	1,500,000	100,000	1.8	0.8
Rural sources				
manured lands	8,110,000	2,700,000	9.9	21.5
other cropland	576,000	384,000	0.7	3.1
forest land	435,000	43,500	0.5	0.3
pasture, woodlot and other lands	540,000	360,000	0.7	2.9
ground water	34,300,000	285,000	42.0	2.3
Urban runoff	4,450,000	1,250,000	5.5	10.0
Precipitation on water areas	6,950,000	155,000	8.5	1.2
Total	81,661,000	12,557,500	100.0	100.0

[a] excludes industrial wastes that discharge to municipal systems. Table does not include contributions from aquatic nitrogen fixation, waterfowl, chemical deicers, and wetland drainage.

To remedy the problem, people have applied methods to control the input of nutrients to lakes, for example, by restricting the usage of phosphate detergents and removing phosphate at sewage treatment plants. Phosphate removal can be done fairly simply by adding lime, calcium oxide, aluminum sulfate, or ferric chloride to the sewage. Figure 7-26 predicts promising results for the application of phosphate control. Some of the control methods are as follows:

- diverting nutrients from lakes — especially diverting of sewage.
- removing nutrients from sewage.
- controlling availability of nutrients within lakes — for example, flocculation of nutrients from euphoric zone, prevention of thermocline formation in summer.
- removing nutrients from lakes — for example, removing macrophytes and large quantity of fish, and so on.
- relieving symptoms of eutrophication — for example, using algicide such as copper sulfate and mechanical harvesting.
- improving agricultural practices — for example, using barrier for groundwater flow, grafting fertilizer molecules of biomass on humin on structure in soil.

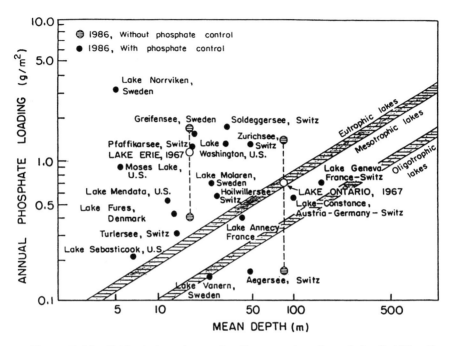

Figure 7-26. Critical phosphorus loading as a function of depth (After R. Vollen Weider).

The level of eutrophication is directly related to water quality variables such as T_p (μg/L), which is the total **phosphorous concentration** in a lake. Chlorophyll a concentration (μg/L), the **Secchi depth** (m), and hypolimnetic oxygen (DO) in % saturation are shown in Table 7-7. The Secchi depth usually employs a disk by lowering it in the water and recording the depth at which the disappearance of the disk by eye occurs. This empirical measurement, however, can be correlated to the extinction coefficient by solar radiation at the lake. The depth at which 1% of the surface remains is of use in eutrophication studies; for example, $z_1 = 4.61/ K_e$ where $I / I_o = \exp(-K_e z)$.

Total phosphorous concentration still remains an important factor related to the eutrophication problem. Usually, the phosphorous mass balance for a completely mixed lake (assumption) with steady state conditions, representing a seasonal/annual average is used. As indicated by Equation 7-16 and Figure 7-27,

$$V\frac{dP_T}{dt} = J_e - u_s A_s P_T - QP_T \qquad [7\text{-}25]$$

$$V\frac{dP_T}{dt} = J_e - k_s P_T V - QP_T \qquad [7\text{-}26]$$

For $k_s = u_s/H$, where

V = volume of lake (L^3)

P_T = total phosphorous concentration in lake (M/L^3); for example, $\mu g/L$

Q = outflow

A_s = lake surface area (L^2)

J_e = external flux of phosphorous, (M/T); for example, g/s

k_s = overall loss rate of total phosphorous ($1/T$)

H = depth of lake (L)

at a steady state,

$$P_T = \frac{J_e}{Q + u_s A_s} \qquad [7\text{-}27]$$

If an area loading rate is used

$$J_e' = J_e/A_s \qquad [M/L^2 \cdot T; \text{ for example, g/m}^2 \cdot yr] \qquad [7\text{-}28]$$

then

$$P_T = J_e' / (q + u_s) \qquad [7\text{-}29]$$

$$q = Q/A_s = \text{hydraulic overflow rate (L/T)}$$

or

$$P_T = J_e' / H(t_d^{-1} + k_s) \qquad [7\text{-}30]$$

where $t_d = V\!/Q$ = detention/time of the lake, k_s is difficult to determine, so an estimation has been used,

$$k_s = 10/H \qquad [7\text{-}31]$$

Table 7-7. Trophic Status of Lakes

Water Quality Variable	Oligotrophic	Mesotrophic	Eutrophic	Reference[a]
Tp ($\mu g/1$)	<10	10-20	>20	1
Chlorophyll ($\mu g/1$)	<4	4-10	>10	2
Secchi depth (m)	>4	2-4	<2	1
Hypolimnetic oxygen (% saturation)	>80	10-80	<10	1

a References:
1. USEPA (1974). National Eutrophication Survey Working Paper, No. 23.
2. NAS, NAE (1972). Water Quality Criteria, A Report of the Committee on Water Quality.
(After R.V. Thomann and J.A. Mueller, 1987).

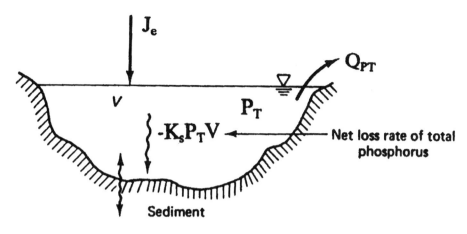

Figure 7-27. Schematic of phosphorus mass balance- completely mixed lake. (Modified after Thomann and Mueller, 1987.)

[**Example 7-2**]   A lake in Pennsylvania has a surface area of 7.77×10^7 m^2 and an average depth of 8m. This lake receives 30 in/yr of rainfall. (a) There is a sewage treatment plant close to the city, which serves a population of 50,000. The water usage is 150 gcd, the influent phosphate for the plant is 6 mg/L, and the plant can only remove 20% of the phosphorous. (b) The combined sewers, which serve 6 mi^2, have a runoff coefficient of 0.45. The overflow of phosphorous concentration is 4 mg/L and is captured by the sewage treatment plant of 5%. (c) The storm drains have a runoff coefficient of 0.27 in a service area of 4 mi^2. The concentration of phosphorous from the storm drain is 0.7 mg/L. (d) At the hillside of the lake, there is an upstream gage that measures the annual average flow of 500 cfs coming from virgin land and contains a phosphorous concentration of 20 ppb. (e) To the north of the lake, there is agricultural land with a drainage area of 60 mi^2, which has a phosphorous loading of 0.5- lb/Mi^2- day and the runoff is 30% of rainfall. (f) To the south of the lake, there is a forest with 80 Mi^2 of drainage area, which carries a phosphorous loading of 0.15 lb/mi^2- day, and the runoff is 30% rainfall (after Thomann and Mueller).

Calculate the total phosphorous concentration of the lake. What is the trophic state of this lake?

In analysis, the lake geometry is as follows:

$$A = 7.77 \times 10^7 \text{ m}^2$$

$$H = 8 \text{ m}$$

$$V = AH = 6.22 \times 10^8 \text{ m}^3$$

The upstream flow plus the sum of the instrumental flow from the drainage areas equals the outflow.

Q_a (sewage treatment plant)

$$= 50{,}000 \text{ cap} \times 150 \text{ g cal} \times \text{MGD}/10^6 \text{ gal} \times 1.548 \text{ cfs}/ \text{MGD} = 11.6 \text{ cfs}$$

Q_b (combine sewers)

= C (runoff coefficient) I (rainfall rate in/hr) A (acres) (1 – capture)

= 0.45 (30 in/yr × 1 yr/365 days × 1 day/24 hr) (6 mi² × 640 acre/mi²) (1 – 0.05)

= 5.61 cfs.

$$Q_c = 0.27 \left(\frac{30}{3.64 \text{ x } 24}\right)(4 \times 640) = 2.36 \text{ cfs}$$

$$Q_d = 500 \text{ cfs}$$

$$Q_e = (30 \times 0.3) \text{ in/yr} \left(\frac{0.07367 \text{ cfs}/ \text{ mi}^2}{\text{in/yr}}\right)(60 \text{ mi}^2) = 39.8 \text{ cfs}$$

$$Q_f = (30 \times 0.3)(0.7367)80 = 53.0 \text{ cfs}$$

$$\sum Q = Q_a + Q_b + Q_c + Q_d + Q_e + Q_f = 612 \text{ cfs}$$

$$= 612 \text{ cfs} \left(\frac{1 \text{ m}^3/\text{s}}{35.4 \text{ cfs}}\right) = 17.3 \text{ m}^3/\text{s}$$

For the lake total phosphorous loading ($J = Q_c$),

$$J_a = 11.6 \text{ cfs} [\, 6\,(1 - 0.20)\ \text{mg/L}\,]\,(5.39\,\frac{\text{lb/day}}{\text{mg/L - cfs}}) = 301 \ \text{lb/day}$$

$$J_b = 5.61 \times 4 \times 5.39 = 121 \ \text{lb/day}$$

$$J_c = 2.36 \times 0.7 \times 5.39 = 9 \ \text{lb/day}$$

$$J_d = 500 \times 0.02 \times 5.39 = 54 \ \text{lb/day}$$

$$J_e = 0.5 \ \text{lb/mi}^2 \ \text{day} \times 60 \ \text{mi}^2 = 30 \ \text{lb/day}$$

$$J_f = 0.15 \times 80 = 12 \ \text{lb/day}$$

$$\Sigma J = J_a + J_b + J_c + J_d + J_e + J_f = 527 \ \text{lb/day}$$

This is similar to Equation [7-16].
Thus, the area loading is

$$J = 527 \ \text{lb/day} \times 365 \ \text{day/yr} \times 454 \ \text{g/lb} = 8.73 \times 10^7 \ \text{g/yr}$$

or

$$J' = J/A = 8.73 \times 10^7 / 7.77 \times 10^7 = 1.12 \ \text{g/ m}^2 \text{yr}$$

The hydraulic detention time is

$$t_d = V/Q = 6.22 \times 10^8 \ \text{m}^3/ \ 17.3 \ \text{m}^3/\text{s} \times 1 \ \text{yr}/ \ 3.154 \times 10^7 \text{s} = 1.14 \ \text{yr}$$

$$q = \text{overflow rate} = Q/A = (V/H \cdot 1/Q)^{-1} = H/t_d = 8\text{m}/ \ 1.14 \ \text{yr} = 7.02 \ \text{m/yr}$$

Assuming $K_s = 1.55$ or $U_s = 12.4 \ \text{m/yr}$ $(K_s = U_s/H)$

or

$$P_t = J'/ (q + U_s) = \left(\frac{1.12q/ \ \text{m}^2 \ \text{yr}}{7.02 \ \text{m/yr} + 12.4 \ \text{m/yr}} \right) = 0.058 \ \text{g/ m}^3 = 58 \ \mu\text{g/L}$$

The lake is in entrophic status.

An important note should be made here. For hydraulic and water-related calculation, the concentration expressed in English system is MGD, cfs, or lb/day, but for the metric systems, the concentration is always expressed in mg/L, ppm, and so on. There are two formulae. Equations [7-32] and [7-33] are commonly used to accommodate both systems in practice.

$$J = 8.34\,QC \hspace{6cm} [7\text{-}32]$$

$$\text{lb/day} = (\frac{\text{lb}}{\text{MG - mg/L}})\,(\text{MGD})\,(\text{mg/L})$$

or

$$J = 5.39\,QC \hspace{6cm} [7\text{-}33]$$

$$\text{lb/day} = (\frac{\text{lb/day}}{\text{cfs - mg/L}})\,(\text{cfs})\,(\text{mg/L})$$

In actual case for metric system, the flow is expressed in m^3/s. Also the approximation can be made

$$1\ \text{mg/L} = 1\ \text{g/m}^3 = 10^{-3}\ \text{kg/m}^3$$

Thus,

$$W = QC$$

where W is in g/s, Q is m^3/s and C is mg/L.
The preceding equations [7-32] and [7-33] can be simplified as

$$8.34\ \text{lb/day} = \text{MGD} - \text{mg/L} \hspace{4cm} [7\text{-}32]$$

$$5.39\ \text{lb/day} = \text{cfs} - \text{mg/L} \hspace{4cm} [7\text{-}33]$$

The formulae are essential in the calculation problems later in environmental processes.

REFERENCES

7-1. J. P. Riley and G. Skirrow, *Chemical Oceanography*, Academic Press, London, 1975.

7-2. G. Sillen, "The Physical Chemistry of Seawater" in *Oceanography* (M. Sears, ed.), American Association of Science, Publication No. 67, Washington, DC, 1961.

7-3. H. Brichert, J. P. Riley, and G. Skirrow, *Chemical Oceanography*, Academic Press, London, 1965.

7-4. A. Lerman, *Geochemical Processes: Water and Sediment Environments*, Wiley-Interscience, New York, 1979.

7-5. A. Nelson-Smith, *Oil Pollution and Marine Ecology*, Plenum Press, New York, 1973.

7-6. H. D. Holland, *The Chemistry of the Atmosphere and Oceans*, Wiley-Interscience, New York, 1978.

7-7. S. A. Berridge, R. A. Dean, R. G. Fellows, and A. Fish, "The Properties of Persistent Oil at Sea" in *Proceedings of the Symposium Scientific Aspects of Pollution of the Sea by Oil* (P. Hepple, ed), Institute of Petroleum, London, 1965.

7-8. R.M. Harrison, S.J. deMora, S. Radsomanikis, and W.R. Johnston, *Introductory Chemistry for the Environmental Sciences*, Cambridge University Press, Cambridge, 1991.

7-9. R.V. Thomann and J.A. Mueller, *Principles of Surface Water Quality Modeling and Control*, Harper and Row, New York, 1987.

7-10. R. M. Harrison, *Understanding Our Environment: An Introduction to Environmental Chemistry and Pollution*, 2nd ed., Royal Society of Chemistry, Cambridge, 1995.

7-11. W. Stumm and J. J. Morgan, *Aquatic Chemistry*, 2nd ed., Wiley-Interscience, 1981.

7-12. R. V. Thomann and J. A. Maeller, *Principles of Surface Water Quality Modeling and Control*, Harper and Row, New York, 1987.

7-13. E. J. Middlebrooks, D. H. Falkenborg, and T. E. Maloney, *Modeling the Eutrophication in Process*, Ann Arbor Science, Ann Arbor, Michigan, 1974.

7-14. A. V. Kaffka, *Sea-Dumped Chemical Weapons: Aspects, Problems and Solutions*, Kluwer Academic, Dororecht, 1996.

7-15. S. J. deMora, *Tributylin, Core Study of an Environmental Contaminant*, Cambridge University Press, Cambridge, 1996.

7-16. L. C. Wrobel, *Water Pollution, 2. Modeling, Measuring and Prediction, Computational Mechanics*, Southampton, 1993.

7-17. G. Tchohanogloeus and T. G. Schroeder, *Water Quality*, Addison Wesley, 1985.

PROBLEM SET

1.   If Venezuelan oil was spilled off shore 2 miles to Long Beach Harbor of 100,000 gal quantity, what time would the first oil slick reach the shore? What is the thickness of the slick?

2.   The Cu^{2+} concentrations of a river upstream and downstream are respectively 4 and 2 moles/m^3/day. If the attached and suspended biomass is about 0.5 g/m^3 respectively and the hydraulic radius is 50 m, find the rate of self-purification and the amount of ecological self-purification of the river.

WATER TREATMENT

*T*he chemicals present in water affect the water quality for its end use. We must learn that not only the key parameters for describing the water quality are essential, but also the chemistry of various spheres including the interactions thereof are equally important. Prior to even developing a treatment technology, the constituents (as well as the amount of chemical species in a given wastewater) must be identified. In general, the development of various combinations of schematics from chemical unit processes and observations is essential for development of a useful, multi-stage treatment.

This chapter consists of four sections. The first section will exemplify pollution and how it affects water quality. We will review the chemistry of DO, BOD, and COD and their role in water quality criterion. In the second section, we will review why staged water treatment is necessary. In the third section, odor and taste in water will be addressed. Then, in the fourth section, industrial wastewater will be discussed, with regard to its characteristics and treatments.

8.1 WATER QUALITY CRITERIA

The ratio of pollutant fluxes to natural fluxes increases with the increasing activity of civilization. Thus the quality of water bodies generally reflects the range of human activity within the catchment area. In a broad sense, the potential perturbation of lakes, rivers, estuaries, and coastal areas may be related to population density and energy dissipation in the drainage area of these water bodies. Figure 8-1 shows the relationships among per capita energy consumption, population density, and energy consumption per unit area for various countries. As Figure 8-1 shows, in most countries of the Northern hemisphere, the energy flux by civilization markedly exceeds the biotic energy flux.

Pollutional loading may be related to the population density and to the per capita waste production in a drainage area. The potential loading, J, of various rivers and estuaries may be estimated by

$$J = \frac{\text{inhibitants}}{\text{drainage area}} \times \frac{\text{drainage areas}}{\text{runoff}} \times \frac{\text{waste production}}{\text{capita}} \times (1-\eta) \qquad [8\text{-}1]$$

where η is the **effectiveness of environmental protection measures** such as recycling, waste retention, and waste treatment. The higher the effectiveness, the lower is the loading factor, J. Similarly, the loading of a lake can be formulated as

$$J_L = \frac{\text{inhabitants}}{\text{drainage area}} \times \frac{\text{drainage areas}}{\text{lake area}} \times \frac{1}{\text{lake depth}} \times \frac{\text{waste production}}{\text{capita}} \times (1-\eta) \quad [8\text{-}2]$$

The gross national product per time within the drainage area may be used to estimate the potential waste production because it measures economic production, i.e., the value of material goods and services for private and public consumption. Table 8-1 illustrates the comparison of some loading parameters of a few lakes. The six lakes at the top of the list in the table are or have been eutrophied prior to treatment or waste diversion.

Water pollution consists of a variety of material flows that depend on population density, lifestyle, and cultural activities. The resulting water composition is determined by the interacting chemical, physical, and biological factors, which are **intensity factors** (activity, concentration, redox potential, temperature, and velocity gradient). These intensive variables, above all the activities of the chemical constituents, primarily determine the type of community of organisms present in the water.

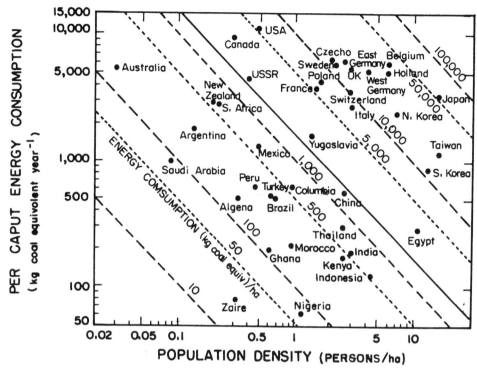

Figure 8-1. The relationships among per-person energy consumption, population density, and energy consumption per unit area (or population potential) for different countries in 1972. [Note: the area of each country includes only FAO-defined agricultural areas. For comparison, solar input at Earth surface is ca. 10^6 kg coal eq year^{-1} ha^{-1}. (1 kg coal eq year^{-1} = 1.1 x 10^{-4}W m^{-2}).] (After Y. H. Li ref. 8-9.)

Water quality criteria are scientifically established requirements concerning intensity factors. These criteria form the basis for judgments with respect to the compatibility of a water composition with ecological objectives or designated water uses. Standards are tolerance levels established by governmental authorities in programs for water pollution abatement. Table 8-2 lists criteria ranges for raw water sources of domestic water supplies in California.

For drinking water quality, there is the maximum contaminant level (MCL). To reach the MCL goal, the U.S. EPA established the primary standard concerning the synthetic organic chemicals (SOCs), the trihalomethanes (THMs), the volatile organic chemicals (VOCs), and the microbiological contaminant of coliform ranges from 10^6/100 mL to 1/100 mL, as shown in Table 8-3. The 1994 MCL goal from the EPA also includes the elimination of *Giardia*, *Legionella*, and viruses.

Table 8-1. Comparison of Some Loading parameters of Some Lakes

Lake	Country	Surrounding Factor	Mean depth (m)	Inhabitants Per km^{-2}	Inhabitants per m^3 lake volume	Energy Consumption per lake volume (W) m^{-3}
Greifensee	Switzerland	15	19	441	348	1.81
Plattensee	Hungary	10	3	~60	200	0.97
Lake Washington	United States	~15	18	~50	42	0.48
Lake Constance	Switzerland-Germany-Austria	19	90	114	24	0.12
Lake Lugano	Switzerland- Italy	11	130	264	22.3	0.11
Lake Biwa	Japan	4.5	41	~150	16	0.07
Lake Winnipeg	Canada	35	13	~3	8.1	0.07
Lake Titicaca	South America	14	~100	~40	5.6	0.001
Lake Victoria	Africa	3	40	~70	5.1	0.002
Lake Baikal	USSR	17	730	~5	0.6	0.0005
Lake Tanganiika	Africa	4	572	~50	0.3	0.0001
Lake Inari	Lapland	12	~50	0.5	0.1	0.0005
Lake Superior	Canada-United States	1.5	145	~5	0.05	0.0005

Table 8-2. Criteria Ranges for Raw Water Sources of Domestic Water Supply in California

Constituent	Excellent Source of Water Supply, Requiring Disinfection Only As Treatment	Good Source of Water Supply, Requiring Usual Treatment Such As Filtration and Disinfection	Poor Source of Water Supply, Requiring Special or Auxiliary Treatment and Disinfection
BOD (5-day), mg/ L			
Monthly average:	0.75-1.5	1.5-2.5	over 2.5
Maximum day, or sample:	1.0-3.0	3.0-4.0	over 4.0
Codiform MPN per 100 ml			
Monthly average:	50-100	50-5,000	over 5,000
Maximum day, or sample:	Less than 5% over 100	Less than 20% over 5.000	Less than 5% over 20,000
Dissolved oxygen			
mg/ L average:	4.0-7.5	4.0-6.5	4.0
% saturation:	75% or better	60% or better	---
pH (average)	6.0-6.5	5.0-9.0	3.8-10.5
Chlorides, max. mg/ L	50 or less	50-250	over 250
Fluorides, mg/ L	Less than 1.5	1.5-3.0	over 3.0
Phenolic compounds,			
max, mg/ L	None	0.005	over 0.005
Color, units	0-20	20-150	over 150
Turbidity, units	0-10	10-250	over 250

The preservation of fresh water as a supply of potable water and the maintenance of most natural waters as life preservation systems such as production of aquatic food and reservoirs for genetic diversity) are among the most important goals of water pollution control. It is difficult to evaluate objectively and codify water quality because (1) the effects of water composition on the various ecological consequences are not well understood and are difficult to quantify, and (2) it is difficult to define a reference state (a hypothetical pristine state) of the water.

Two water quality criteria have been used most commonly in previous decades: (1) the concentration of **dissolved oxygen** as a pollution and the biochemical oxygen demand as a loading parameter; and (2) **indicator organisms** that are indicative of the existence of certain pollution conditions. This concept, which is referred to as biotic index, is illustrated in Figure 8-2.

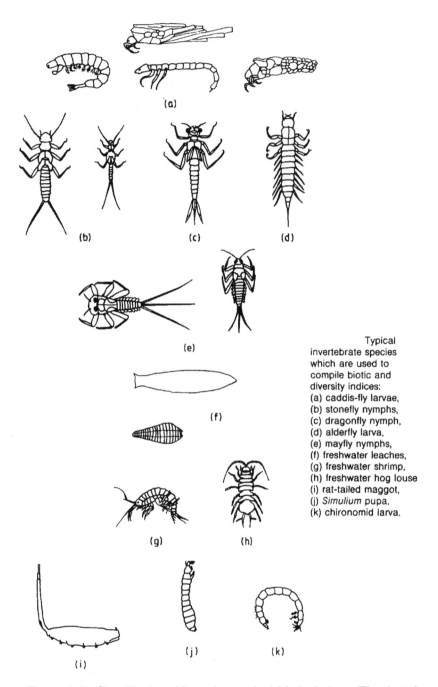

Typical invertebrate species which are used to compile biotic and diversity indices: (a) caddis-fly larvae, (b) stonefly nymphs, (c) dragonfly nymph, (d) alderfly larva, (e) mayfly nymphs, (f) freshwater leaches, (g) freshwater shrimp, (h) freshwater hog louse (i) rat-tailed maggot, (j) *Simulium* pupa, (k) chironomid larva.

Figure 8-2. Simplified ranking of a typical biotic index. (The last few, i.e.(i), (j) and (k), are always found in oxygen-deficient zones.)

There are some collective parameters, such as **chemical oxygen demand** (COD), **biological oxygen demand** (BOD), and **total organic carbon** (TOC). One or more of these parameters are often used to estimate the quantity of organic matter present in water bodies. COD is obtained by measuring the equivalent quantity of an oxidizing agent (usually permanganate or dichromate in acid solution), necessary for oxidation of the organic constituents. The amount of oxidant consumed is customarily expressed in equivalents of oxygen. The BOD test measures the oxygen uptake in the microbiologically mediated oxidation of organic matter directly. In both tests, not all the organic matter reacts with the oxidants. In determinations of TOC, the carbon oxide produced in the oxidation or combustion of a water sample is measured.

Table 8-3. Drinking Water Standards

Contaminant	United States EPA	Canada NHW	International WHO
PRIMARY STANDARDS (Health) MCL			
Total Coliforms (Membrane Filter)	Avg 1/100 mL	2/100 mL	0
	Max 4/100 mL	3/100 mL	—
Turbidity	1-5 TU	1-5 TU	<1
Inorganic Chemicals (mg/ L)			
Arsenic (As)	0.05	0.05	0.05
Barium (Ba)	1.0	1.0	—
Cadmium (Cd)	0.010	0.005	0.005
Chromium (Cr)	0.05	0.05	0.05
Fluoride (F)	1-2 (15°C)	1.5	1.5
Lead (Pb)	0.05	0.05	0.05
Mercury (Hg)	0.002	0.001	0.001
Nitrate (N)	10.0	10.0	10.0
Selenium (Se)	0.01	0.01	0.01
Silver (Ag)	0.05	0.05	—
Organic Chemicals (mg/ L)			
Endrin	0.0002	0.0002	—
Lindane	0.004	0.004	0.003
Methoxychlor	0.1	0.01	0.030
Toxaphene	0.005	0.005	—
2-4-D	0.1	0.1	0.1
2,4,5 TP	0.01	0.01	—
Trihalomehanes	0.10	0.35	—

Table 8-3. (Continued)

SECONDARY STANDARDS (Aesthetics)	RCL		
Chloride (Cl)	250 mg/ L	250 mg/ L	250 mg/ L
Color	15 color units	15 color units	15 color units
Copper (Cu)	1.0 mg/ L	1.0 mg/ L	1.0 mg/ L
Iron (Fe)	0.3 mg/ L	0.3 mg/ L	0.3 mg/ L
Manganese (Mn)	0.05 mg/ L	0.05 mg/ L	0.1 mg/ L
Odor	3 (Threshold)	Nil	Nil
pH	7.5 ± 1	7.5 ± 1	7.5 ± 1
Sulfate (SO4)	250 mg/ L	500 mg/ L	400 mg/ L
Total Diss. Solids	500 mg/ L	500 mg/ L	1000 mg/ L
Zinc (Zn)	5.0 mg/ L	5.0 mg/ L	5.0 mg/ L

Where two values are noted (turbidity, fluorides), the lower one indicates the recommended contaminant level (RCL); the higher one indicates the maximum contaminant level (MCL) acceptable.

Sources: U.S. Environmental Protection Agency (EPA), *National Interim Primary Drinking Water Regulations,* EPA 570/9-76-003, 1976; Dept. of National Health and Welfare Canada (NWH), *Guidelines for Canadian Drinking Water Quality,* 1978; World Health Organization (WHO), *International Standards for Drinking Water,* 1983.

8.1.1   Chemistry of Dissolved Oxygen (DO)

Any gas equilibrated with water is governed by Henry's law. This law states that the amount of gas that dissolves is proportional to the partial pressure of the gas. If X is any gas, then

$$X_{(aq)} = \text{const.} \, P_{X(g)}$$

or

$$K_H = \frac{X_{(aq)}}{P_{X(g)}} \qquad [8\text{-}3]$$

The proportional constant is Henry's law constant. Some common values are listed, all expressed in units of mole/L/atm.

N_2	6.5×10^{-4}
CO_2	3.4×10^{-2}
CO	9×10^{-4}
O_2	1.3×10^{-3}
O_3	1.3×10^{-2}

In the case of oxygen, the atmosphere contains 0.21 atm O_2, so the solubility of oxygen in water can be evaluated.

$$O_{2(aq)} = K_H \, P_{O_2 \, (g)}$$

$$= (1.3 \times 10^{-3} \text{ mole L}^{-1} \text{ atm}^{-1})(0.21 \text{ atm})$$

$$= 2.7 \times 10^{-4} \text{ mole L}^{-1}$$

or

$$= (2.7 \times 10^{-4} \text{ mole L}^{-1})(32 \text{ g mole}^{-1})(1000 \text{ mg/g})$$

$$= 8.7 \text{ mg L}^{-1} = 8.7 \text{ ppm}$$

(Notice that the difference of ppm in air is only by volume basis. In dissolved species in water, the unit can be expressed on weight basis.)

Dissolved oxygen (DO) is essential for aquatic life, and this value can be altered by thermal pollution and decomposition of biomass; examples include algal blooms and any oxidizable substances in water such as sewage. The measurements of DO can be carried out as follows:

Winkler's Method of Titration

The sample is treated with manganese sulfate in an alkaline solution. The precipitated manganese dioxide is used to oxidize I^- to I_2, which is back-titrated with standard sodium thiosulfate solution until no I_2 end point can be seen.

$$Mn^{2+} + 2 \, OH^- + \tfrac{1}{2}O_2 \rightarrow MnO_2(s) + H_2O$$

$$MnO_2(s) + 4H^+ + 2I^- \rightarrow I_2 + Mn^{2+} + 2H_2O$$

$$I_2 + 2Na_2S_2O_3 \rightarrow Na_2S_4O_6 + 2NaI$$

Spectrophotometry

Methyltene blue and indigo carmine, which are dyes, can be oxidized by O_2 to the forms leaving different colors.

Makareth Oxygen Electrode

Oxygen can diffuse in the cell through a thin, disposable polyethylene membrane.

Cathode (made of Ag)

$$O_2(g) + 2H_2O(e) + 4e^- \rightarrow 4OH^-(aq)$$

Anode (made of Pb)

$$4OH^-(aq) + 2Pb(s) \rightarrow 2Pb(OH)_2(s) + 4e^-$$

Overall,

$$O_2(g) + 2H_2O(l) + 2Pb(s) \rightarrow 2Pb(OH)_2(s)$$

The potential across this cell, which depends on P_{O_2}, can be measured as follows:

$$E_{cell} = E^0 - \frac{RT}{nF} \ln\left(\frac{1}{P_{O_2}}\right) \qquad [8\text{-}4]$$

This voltametric device is calibrated with constant potential, and the current flow is directed proportional to DO.

8.1.2   Biochemical Oxygen Demand (BOD)

In a water solution, the organics can be utilized by microorganisms. The amount of oxygen consumed during microbial utilization of the organics is termed **BOD**. Laboratory determination is based on the initial and final DO concentrations. Most natural and municipal wastewaters contain a population of microorganisms that will consume the organics. In sterile waters, microorganisms must be added and the material containing the organisms determined and subtracted from total BOD of the mixture. Usually, the BOD_5 represents the oxygen consumed in five days. The total BOD or DOD at any given time period can be determined.

The rate at which organics are utilized by microorganisms is assumed to be first-order. Mathematically, this can be expressed as

$$\frac{dL_t}{dt} = -kL_t \qquad [8\text{-}5]$$

where L_t is the oxygen equivalent of organics at time t and k is the reaction rate constant. Usually the units of L_t are mg per liter, and k is expressed as day^{-1} or

$$\frac{d L_t}{L_t} = -kdt$$

$$\int_{L_0}^{L} \frac{dL_t}{L_t} = -k \int_0^t dt \qquad [8\text{-}6]$$

$$\ln\left(\frac{L_t}{L_0}\right) = -kt \tag{8-7}$$

$$L_t = L_0\, e^{-kt}$$

Here, L_0 represents total oxygen equivalent of organics at time zero. Graphically, the oxygen equivalents, L_t and BOD, expressed as mg/L of O_2 can be seen in Figure 8-3. We define

$$y_t = L_0 - L_t = L_0 - L_0 e^{-kt} = L_0\left(1 - e^{-kt}\right) \tag{8-8}$$

where y_t represents BOD_t of the water. As y_t approach L_o, the BOD_t becomes BOD_u, the ultimate BOD. Therefore,

$$BOD_u = L_0 \tag{8-9}$$

Usually, the common logarithm of base 10 is used.

$$\frac{L_t}{L_o} = e^{-kt} = 10^{-k't} \tag{8-10}$$

where $k' = k/2.3$. To evaluate k and L_o, Thomas has developed a graphical method, from

$$y_t = L_o\left(1 - 10^{-k't}\right) \tag{8-11}$$

Rearrange to read:

$$\left(\frac{t}{y_t}\right)^{\frac{1}{3}} = \left(2.30k'L_0\right)^{-\frac{1}{3}} + \left(\frac{k'^{\frac{2}{3}}}{3.43L_0^{\frac{1}{3}}}\right)t \tag{8-12}$$

If one plots $(t/y_t)^{1/3}$ versus t, a straight line will result with slope b and intercept a, and

$$k' = 2.61\frac{b}{a} \tag{8-13}$$

$$L_0 = \frac{1}{\left(2.3\,k'a^3\right)} \tag{8-14}$$

and

$$k = 2.3\,k' \tag{8-14a}$$

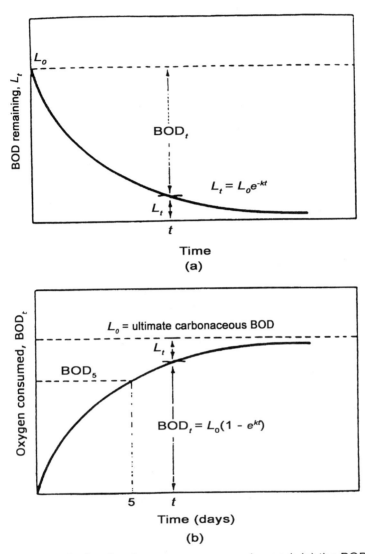

Figure 8-3. Idealized carbonaceous oxygen demand: (a) the BOD remaining as a function of time, and (b) the oxygen consumed.

8.1.3 Chemical Oxygen Demand (COD)

This test employs potassium dichromate in boiling sulfuric acid (150°C) in the presence of a silver catalyst to oxidize the organics into CO_2 and H_2O. The hexavalent chromate is reduced to trivalent chromium ion. Taking monopotassium salt of phthalate as an example, the balanced equation is

$$2\,KC_8H_5O_4 + 10\,K_2Cr_2O_7 + 41\,H_2SO_4 = 16\,CO_2 + 46\,H_2O + 10\,Cr_2(SO_4)_3 + 11\,K_2SO_4$$

In the preceding equation each molecule of potassium dichromate has the same oxidizing power as 1.5 molecules of oxygen. The reason is that Cr^{6+} to Cr^{3+} requires three electrons, but oxygen only requires two electrons. Therefore, two moles of potassium phthalate consumes 15 moles of oxygen, which is in this case, equivalent to 10 moles of dichornate. The COD of the sample is determined by titrating the remaining dichromate with ferrous sulfate

$$6Fe^{2+} + Cr_2O_7^{2-} + 14H^+ = 6Fe^{3+} + 2Cr^{3+} + 7H_2O$$

In general, the balance of equation of organic molecules to have the chromate can be simplified by the following:

$$C_n H_a O_b + c\,Cr_2O_7^{2-} + 8\,cH^+ = nCO_2 + \frac{(a+8c)}{2} H_2O + 2c\,Cr^{3+}$$

and

$$c = \frac{2n}{3} + \frac{a}{6} - \frac{b}{3}$$

Using this equation to figure out the ratio of dichromate to phthalate, the ratio is still 5 to 1 because c = 5.

An oxidation-reduction indicator such as ferroin (ferrous 1,10-phenonithroline sulfate) can indicate when all the dichromate can be reduced by ferrous ions. It gives sharp color changes in spite of the green color produced by Cr^{3+} ion.

8.1.4 Theoretical Oxygen Demand (ThOD) and Other Tests

Similar to potassium dichromate oxidation as mentioned in Section 8.1.3, potassium permanganate has been selected to obtain similar result. This is termed **permanganate value (PV)**. A rapid automated test has been developed to oxidize the sample in the presence of a catalyst at 900°C with a stream of air. This is termed **total oxygen demand (TOD)**. A total carbon analyzer can also determine the total organic carbon (TOC) by removing the inorganic carbon through acidification or through dual combustion tube in the analyzer. The TOC can correlate well with COD result. Stumm and Morgan have provided a chart that the oxidation state of organic compound can be estimated by

$$\frac{4(TOC - COD)}{TOC} = \text{oxidation state} \qquad [8\text{-}15]$$

Figure 8-4 is such an example. Linear relationships are also generally found to exist between each of four assays, with relative strength being in the following order:

$$PV < BOD < COD < TOD$$

ThOD can be approximated from simple calculation. For urea, CH_4N_2O, oxygen is still needed.

$$CH_4N_2O + \frac{9}{2}O_2 = CO_2 + 2H_2O + 2NO_3$$

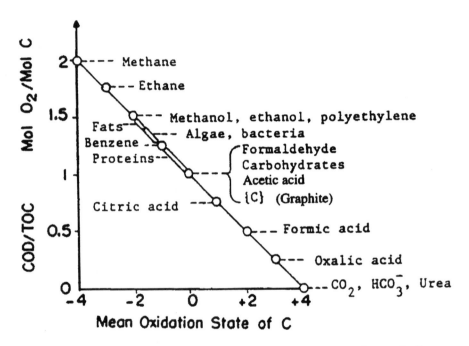

Figure 8-4. Oxygen demand and mean oxidation state of organic C. (Stumm and Morgan, 1981. Ref. 8-1.)

Thus, 1 g urea requires 1.2 mg of O_2, or in solution the ThOD value is 1200 mg O_2/L. Implicit in water quality criteria is a chemical model. With the help of the model tolerance levels compatible with designated water uses can be rationalized and quantified. In measuring and quantifying the diverse chemical variables (including pollutants), one first encounters the analytical problems of sensitivity and specificity. Analytical chemistry has made remarkable progress in improving the sensitivity of detection.

The environmental effects of a substance, such as the toxicological effects and chemical and geochemical reactivity, is structure specific, as shown in

Figure 8-5. The other factor making the environmental interpretation more complicated is its dependence on the knowledge of pathways and of biogeochemical parity.

Remember that even the water we drink is not pollutant-free. Chromatographic results for the analysis of volatile substances found in ground, river and drinking water within the same polluted watershed indicated numerous contaminating organic compounds. Some of the more refractory substances (some hydrocarbons and chlorinated hydrocarbons) detected in the river water may also typically occur in groundwater and drinking water. Recently a number of emerging chemicals have been found (Figure 8-6). These chemicals are of concern from the standpoint of human health and ecological risk.

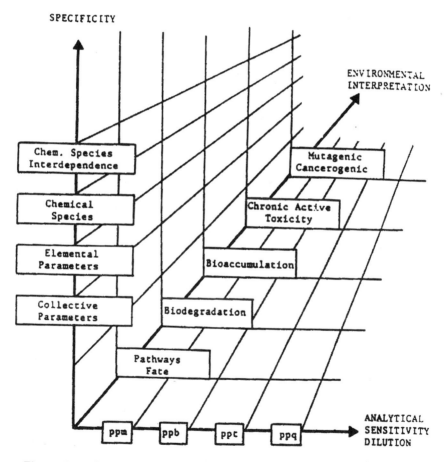

Figure 8-5. Sensitive analytical techniques and a knowledge of individual chemical species and their interdependence are prerequisites for water quality interpretations. (Stumm and Morgan, 1981, Ref. 8-1.)

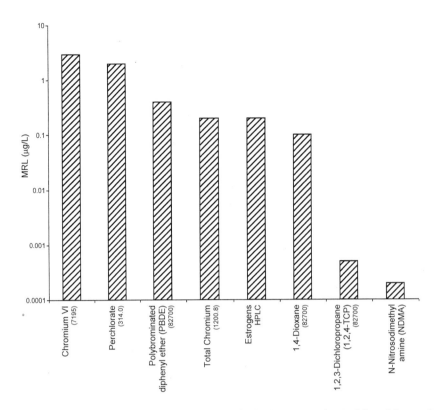

Figure 8-6. Examples of emerging chemicals in water, the achievable method reporting limit (MRL) in Mg/L is indicated by the y-axis. Numerals in parentheses are EPA's method. Data are from Columbia Analytical Service, Kelso, WA, Technology Bulletin, Spring 2004.

8.2 STAGED WASTEWATER TREATMENT

To the environmental engineer's concern, water treatment can be divided into sewage treatment of wastewater and drinking water supplies. Potable water treatment plants are generally simpler than sewage treatment plants. The unit operations in the potable water treatment plants can also be found in the practice of wastewater treatment. Here, we will focus on sewage and industrial wastewater treatment.

Figure 8-7 illustrates the fate of industrial wastewater constituents in treatment processes. One can also see from the figure that various operations can be applied. The processes selected in a treatment plant depend heavily on the quality of the source water and the quality of the effluent required. Economic consideration and regulation standards of the effluents are usually the controlling factors of water treatment. The main objective of conventional wastewater treatment processes is the reduction of the biochemical oxygen demand, suspended solids, and pathogenic organisms. In addition, it may be necessary to remove nutrients, toxic components, nonbiodegradable compounds, and dissolved solids. Conventional wastewater treatment processes are often classified as primary treatment, intermediate treatment, secondary treatment, tertiary treatment, and quaternary treatment. Most of the common operations are shown in Figure 8-7.

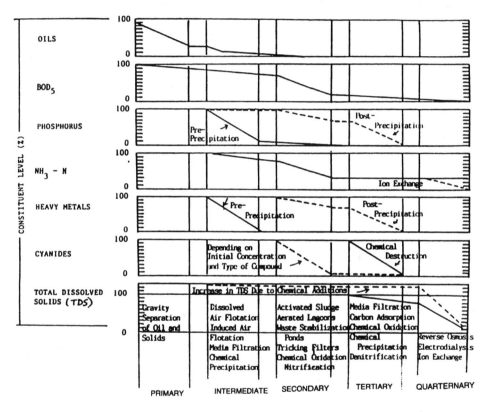

Figure 8-7. Fate of industrial wastewater constituents in treatment processes.

Because many different combinations of these operations are possible, each situation must be evaluated in order to select the best combination. Remember that in each of the treatments, there are at least a half dozen unit operations that have evolved. Each unit operation again depends on a particular contaminant (chemicals) in the water. The order of the unit process may be discharged according to (1) the original chemical composition (2) the by-product formed during the treatment, and (3) the precursors for potential and anticipated products.

8.2.1   Primary and Intermediate Treatment

Primary treatment processes are used to screen out coarse solids, to reduce the size of solids, to eliminate floating oil and grease, and to equalize fluctuations in flow or concentration through short-term storage. **Intermediate treatment processes** include dissolved air flotation (DAF), induced air flotation, media filtration, and chemical precipitation.

Impurities in water vary in size by about six orders of magnitude, from a few angstroms for soluble substances to a few hundred microns for suspended materials. The removal of a large proportion of these impurities in wastewater treatment is accomplished by sedimentation. In a sedimentation unit, solid particles are allowed to settle to the bottom of the tank under quiescent conditions. However, because many of the impurities are too small for gravitational settling alone, to be an effective removal process, the aggregation of these particles into large, more readily settleable aggregates is essential for successful separation by sedimentation. This process of aggregation is termed **"coagulation."** Chemicals are often used in the coagulation process to destabilize the colloidal particles and thus to increase the rate of aggregation. Common chemicals used are alum, ferric chloride, or lime. Table 8-4 summarizes the chemical compounds used in the coagulation process.

The reduction of solids in these stages reduces oxygen requirements in a subsequent biological step and also reduces the solids loading into the secondary sedimentation tank.

8.2.2   Secondary Treatment

Secondary treatment generally involves a **biological process** to remove organic matter through biochemical oxidation. The operations include activated sludge, aerated lagoons, waste stabilization ponds, trickling filters, chemical oxidation, and nitrification. The particular process selected depends upon such factors as quantity of wastewater, biodegradability of waste, and availability of land. Activated sludge reactors and trickling filters are the most commonly used biological processes.

In the activated sludge process, wastewater is fed to an aerated tank where microorganisms consume organic wastes for maintenance and for generation of new cells. The resulting microbial floc (activated sludge) is settled in a sedimentation vessel called a clarifier, or thickener. A portion of the thickened biomass is usually recycled to the reactor to improve the performance through higher cell concentrations. Trickling filters are beds packed with rocks, plastic structures, or other media. Microbial films grow on the surface of the packing and remove soluble organics from the wastewater flowing over the packing.

Table 8-4. Chemical Compounds Used in Coagulation Process

Compounds	Formula	Commercial Strength	Grades Available	Weight (lbs/ft)	Remarks
Coagulants:					
Aluminum sulfate	$Al_2(SO_4)_3 \cdot 18H_2O$	17 % Al_2O_3	lump / powder granules	Powder: 38-45 / Other: 57-67	Coagulation and sedimentation; prior to pressure, filters for removal of suspended matter and oil
Sodium aluminate	$Na_2Al_2O_3$	55 % Al_2O_3	crystals	50-60	Usually added with soda ash to softeners
Ammonium alum	$Al_2(SO_4)_3(NH_3)_2SO_4 \cdot 24N_2O$	11 % Al_2O_3	lump powder	60-68	Coagulation system – not widely used
Potash alum	$Al_2(SO_4)_3 \cdot K_2SO_4 \cdot 24N_2O$	11 % Al_2O_3	lump powder	64-68	Coagulation system – not widely used
Copperas	$FeSO_4 \cdot 7H_2O$	55 % $FeSO_4$	crystals granules	63-66	Suitable coagulant only pH range of 8.5-11.0
Chlorinated copperas	$FeSO_4 \cdot 7H_2O + 1/2CO_2$	48 % $FeSO_4$	—	—	Ferrous sulfate and chlorine are fed separately
Ferric sulfate	$Fe_2(SO_4)_3$	90 % $Fe_2(SO_4)_3$	powder granules	60-70	Coagulation – effective over wide range of pH, 4.0-11.0
Ferric chloride hydrate	$FeCl_3 \cdot 6H_2O$	60 % $FeCl_3$	Crystals	—	Coagulation – effective over wide range of pH, 4.0-11.0
Magnesium oxide	MgO	90 % MgO	powder	25-35	Essentially insoluble – fed in slurry form
Coagulant Aids:					
Bentonite	—	—	powder	60	Essentially insoluble – fed in slurry form
Sodium silicate	$Na_2O(SiO_2)_{3 \cdot 25}$	40 Be solution	solution	86	—
pH Adjusters:					
Lime, hydrated	$Ca(OH)_2$	93 % $Ca(OH)_2$	powder	25-50	pH adjustment and softener
Soda ash	Na_2CO_3	99% Na_2CO_3	powder	34-52	pH adjustment and softener
Caustic soda	$NaOH$	98% $NaOH$	flake / solid / ground / solution	—	pH adjustment, softening, oil removal systems
Sulfuric acid	H_2SO_4	200 % H_2SO_4	liquid	—	pH adjustment

8.2.3   Tertiary Treatment

Many effluent standards require wastewater treatment to remove particular contaminants or to prepare water for reuse. **Tertiary treatment** is also called the **polishing step.** Some common tertiary operations are removal of phosphorous compounds by coagulations with chemicals, removal of nitrogen compounds by ammonia stripping with air or by nitrification — denitrification in biological reactors, and removal of residual organic and color compounds by adsorption on activated carbon. The effluent water is often treated with chlorine or ozone to destroy pathogenic organisms before discharge into the receiving waters.

8.2.4 Quaternary Treatment

In cases where softer water is desired, **quaternary treatment** is necessary, and an ion exchange column is added to the process. Also, reverse osmosis or electrodialysis are commonly used if the removal of dissolved solids is required.

8.3 OLFACTION AND TASTE IN WATER

Both taste and odor are important perceptions of an aesthetic quality. The primary test qualities are sour (e.g., HCl), salty (e.g., NaCl), sweet (e.g., sucrose), and bitter (e.g., caffeine). The primary olfactory qualities are camphoraceous (moth repellent), peppermint (mint candy), floral (roses), ethereal (dry-cleaning fluid), pungent (vinegar), and putrid (H_2S). There is a **threshold concentration** for various substances that cause taste and odor. It should be pointed out that the substances in water causing olfaction are different from those in air, as shown in Table 8-5 and 8-6.

A technique for detecting taste and odor-producing organic compounds in water or sediments has been developed. This is called **close-loop stripping** analysis (CLSA), which strips semi-volatile organics from water for identification by GC-MS. The odor thresholds and detection limits are summarized in Tables 8-6 and 8-7.

It is widely known that odor, taste and tongue sensations in water are caused by various **algae** and **actinomycetes** present in groundwater and surface water (Tables 8-6 and 8-8). The associated biochemical transformations including decaying and putrefaction can be important. Chlorination of water, which produces all types of chlorophenols, will give bitter tastes. Various organic compounds isolated from odor-causing aquatic organisms are identified in Table 8-9. The taste thresholds are listed in Table 8-10.

Table 8-5. Odor Thresholds of Various Substances in Air

Substance	Description	Concentration Causing Faint Odor (mg/L)
Allyl sulfide	Garlic odor	0.00005
Amyl acetate (iso)	Banana odor	0.0006
Benzaldehyde	Odor of bitter almonds	0.003
Chlorine	Pungent and irritating odor	0.010
Coumarine	Vanilla odor, pleasant	0.00034
Crotyl mercaptan	Skunk odor	0.000029
Diphenylchlorarsine	Shoe polish odor	0.0003
Hydrogen sulfide	Odor of rotten eggs, nauseating	0.0011
Ozone	Slightly pungent, irritating odor	0.001
Phenyl isothiocyanate	Cinnamon odor, pleasant	0.0024
Phosgene	Odor of ensilage or fresh-cut hay	0.0044

Source: Adapted from J. Olishifski, *Fundamentals of Industrial Aggiene*, National Safety Council, Chicago, (1971)

Table 8-6. Odor Thresholds of Various Substances in Water

Compound	Threshold Odor Concentration (mg/ L)
2 – Octanol	0.13
Styrene	0.05
Ethylbenzene	0.1
Nathalene	0.007
p-Dichlorobenzene	0.15
Chloroform	20.0
Nonanal	0.001
Methyl sulfide	0.003
Geosmin	0.000005
Methylisoborneol (MIB)	0.000005
Trichloroethene	2.6
Tetrachloroethene	2.8
Dichloromethane	24.0
Toluene	0.14

Table 8-7. Odor Thresholds and CLSA Detection

Compounds	CLSA Detection Limit (ng/ L)	Lowest Reported Threshold Odor Concentration (ng/ L)
Geosmin	2	10
2 – Methylisoborneol	2	29
2 – Isopropyl – 3 – methoxy pyrazine	2	2
2 – Isobutyl – 3 – methoxy pyrazine	2	2
2,3,6 - Trichloranisole	5	7

Source: M.J. JAWWA *73*, 530-537, McGuire et al. (1981). Reported with permission.

Table 8-8. Odors, Tastes, and Tongue Sensations Association with Various Algae

Algal Genus	Algal Group	Odor When Algae Are: Moderate	Odor When Algae Are: Abundant	Taste	Tongue Sensation
Anabaena	Blue-green	Grassy, nasturtium, musty	Septic	—	—
Anacystis	Blue-green	Grassy	Septic	Sweet	—
Aphanizomenon	Blue-green	Grassy, nasturtium, musty	Septic	Sweet	Dry
Asterionella	Diatom	Geranium, spicy	Fishy	—	—
Ceratium	Flagellate	Fishy	Septic	Bitter	—
Dinobryon	Flagellate	Violet	Fishy	—	Slick
Oscillatoria	Blue-green	Grassy	Musty, spicy	—	—
Scenedesmus	Green	—	Grassy	—	—
Spirogyra	Green	—	Grassy	—	—
Synura	Flagellate	Cucumber, muskmelon, spicy	Fishy	Bitter	Dry, metallic slick
Tabellaria	Diatom	Geranium	Fishy	—	—
Ulothrix	Green	—	Grassy	—	—
Volvox	Flagellate	Fishy	Fishy	—	—

Source: Adapted from C.M. Palmer, Algae in Water Supplies, U.S. Dept HEW, Washington D.C., (1962).

Table 8-9. Structure of Various Compounds Isolated from Odor-Causing Aquatic Organisms

Compound	Structure	Associated Organisms
Methylisoborneol (MIB)	CH_3 CH_3 CH_3 OH CH_3	*Actinomycetes* *Oscillatoria curviceps* *Oscillatoria tenuis*
Geosmin	OH CH_3 CH_3	*Actinomycetes* *Sympioca muscoum* *Oscillatoria tenuis* *Oscillatoria simplicissima* *Oscillatoria scheremetievi*
Mucidone	CH_3—O O CH_3 CH_3	*Actinomycetes*
Isobutyl mercaptan	CH_3 $CHCH_2$ — SH CH_3	*Microcystis flos-aquae*
N–Butyl mercaptan	$CH_3(CH_2)_3$–SH	*Microcystis flos-aquae* *Oscillatoria chalybea*
Isopropyl mercaptan	CH_3 CHSH CH_3	*Microcystis flos-aquae*
Dimethyl disulfide	CH_3–S–S–CH_3	*Microcystis flos-aquae* *Oscillatoria chalybea*
Dimethyl sulfide	CH_3–S–CH_3	*Oscillatoria chalybea* *Anabaena*
Methyl mercaptan	CH_3SH	*Microcystis flos-aquae* *Oscillatoria chalybea*

Table 8-10. Taste Thresholds for Selected Materials

Material	Taste Threshold (mg/L)
Zn^{2+}	4-9[b]
Cu^{2+}	2-5[b]
Fe^{2+}	0.04-0.1[b]
Mn^{2+}	4-30[b]
2 – Chlorophenol[a]	0.004
2,4 – Chlorophenol[a]	0.008
2,6 – Chlorophenol[a]	0.002
Phenol	>1.0
Fluoride	10[b]

[a]Created by the action of chlorine on phenol.
[b]Threshold detected by 5 percent of panel.
Source: J. Cohen, et al., JAWWA 52,660 (1960).

Table 8-11. Results of Screening for Specific Anosmics

Compounds	Odor Character	Percent of Anosmics	Anosmic Defect Factor Between Mean Anosmic Threshold and Mean Threshold for Normal Observers
Isovaleric acid	Sweaty	3	42
1 – Pyrroline	Spermous	16	39
Trimethylamine	Fishy	6	830
Isobutyraldehyde	Malty	36	340
5α-Androst–16–en–3–one	Urinuous	47	770
ω - pentadecalactone	Musky	12	13
4 - chloroaniline	Mixed	41	—

Source: H. Van Langenhore and N. Schamp in Encyclopedia of Environmental Control Technology, Vol.2, by P.N. Cheremisinoff, editor.

Both odor and taste thresholds do not have individual Gaussian distributions. Individuals may be different and there are anosmic defects (less sensitive). The results of screening for specific anosmics are in Table 8-11. The acceptability of olfactometric results obtained have to be agreed on by different panels, as shown in Figure 8-8.

Prevention and control is always done at the source. The use of algicides (such as copper sulfate) together with citric acid as a complexing agent is often the practice. The use of chlorine and the destratification of reservoirs by mechanical mixing is also possible. The use of biological control is more efficient, as with the use of *Bacillus subtilis* and *B. cereus* for the oxidation of the many metabolites of actinomycetes. For water supplies, activated carbon and ozonation is usually employed.

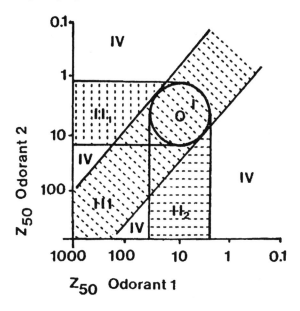

Figure 8-8. Evaluation of olfactometric results obtained by different panels with different olfactometers. I. Both measurements are acceptable. II. One or both measurements is unacceptable. III. Systematic errors have been made. IV. Both results are unacceptable. (V. Thiele and H. Bahnmuller, Staub, Reinhalt Luft, <u>45</u>, 200, 1985.)

8.4 INDUSTRIAL WASTEWATER

Domestic wastewater primarily contains human excreta which can be easily decomposed. Even so-called **Graywater** (excluding toilet waste) can be easily treated. On the contrary, industry wastewater contain largely xenobiotics (not from nature) which are often resist to treatment.

8.4.1   Chemical Perspective of Industrial Wastewater

It is difficult to meaningfully describe the wide variety of wastewater derived from every industrial sector. However, a few remarks can be made. First, industrial wastewater is different from municipal wastewater and stormwater. When compared with municipal wastewater, the industrial wastewater possesses a greater variety of chemical compositions due to various industrial processes and the larger amount of production. An attempt at a comparison is shown in Table 8-12. According to the output of the gross chemical species, the general breakdown of the major categories are as follows:

- Metal plating wastes, painting and derusting wastes, electronic and printing wastes, mining and metal processing wastes, etc. — These wastes have high metal contents as well as chelating agents such as cyanides, and citric acids.

- Dairy and canning wastes, fruit and sugar wastes, brewery and distillery wastes, stockyard, slaughterhouse, packinghouse and poultry-plant wastes, agriculture wastes, and so on. — These wastes are characterized by the high protein or carbohydrate content. One remarkable feature is the recoverable fraction of the relatively nontoxic wastes that can be reused in the food industry. The principle of recovery of whey or casin from dairy wastes is no different from the recovery of albumin from modern abattoirs. So is the recovery of valuable amino acids or specific saccharides from cane-sugar manufacturers.

- Paper and pulp wastes, textile waste, leather wastes, and dyeing and painting wastes — Various chemicals are used, including the vat dyes. The recoverable material includes lignin sulfonate, tannic acid, etc.

- Energy and power producing wastes, including nuclear and geothermal plants, refinery wastes, oilfield wastes, heavy chemical industry wastes, steel mill, metal recovery, and finishing wastes — any high temperature operation will generate a number of toxic materials. The bulk of water is used in cooling operations, ballast water blowdowns and miscellaneous discharges.

- Various fine chemical industry wastes, munitions and weaponry wastes, etc. — These wastes are generally hazardous in nature.

Regarding the wastewaters from the industry sectors, there are also some inherent problems with them. The following list is a survey of these wastewaters. In general, they are mere difficulties to cope with. In many instances, they are described as "alive", meaning that the water quality parameters (such as BOD and COD) change from day to day. This is due to either the photochemical change, air oxidation, or microorganism content.

Table 8-12. Representative Values of Contaminants in Wastewater

Waste Parameter	Municipal (R + C + I)			Industrial (Process)				Stormwater (Annual Runoff)
	Small	Medium	Large	Food	Meat	Plating	Textile	Small/Medium/Large
Volume (L)								
/capita/ day	400	500	600					—
/tonne prod.				10,000	12,000		100,000	—
% runoff								30/35/45
MPN (10⁶/100/mL)	100	80	70	0		0	0	0.008
BOD₅	190	240	300	1,200	640	0	400	14
COD	320	400	500					100
TOC	135	170	215					—
Susp. Solids	225	300	350	700	300	0	100	170
Diss. Solids	450	600	700		200		1,900	170
Total N	40	30	25	0	3	0	0	3.5
Total P	10	8	7	0		0	0	0.35
PH	7.0	7.0	7.0		7.0	4 or 10	10	—
Copper	0.14	0.17	0.21	0.29	0.09	6	0.31	0.46
Cadmium	0.003	0.010	0.016	0.006	0.011	1	0.03	0.025
Chromium	0.04	0.08	0.16	0.15	0.15	11	0.82	0.16
Nickel	0.01	0.06	0.11	0.11	0.07	12	0.25	0.15
Lead	0.05	0.1	0.2					—
Zinc	0.19	0.29	0.38	1.08	0.43	9	0.47	1.6

Note: Concentrations of constituents are in mg/L. Values for all parameters may vary widely from those noted.

Small   Small residential (R) community

Medium   Medium-sized diversified municipality, residential (R), commercial (C), and industrial (I) areas with separate sewers

Large   Large industrialized city (R + C + I) with combined sewers

Food   Food waste (canning factory: pickles, beets, tomatoes, pears)

Meat   Meat processing (poultry plant with no manure or blood recovery)

Plating Plating shop (wastes are acidic with chromate baths, alkaline with cyanide baths, and are less than 2,000 m³/d for most plants

Textile Textile mill (spun cotton yarn processed into cotton goods, sized with starch)

Adapted in part from P. G. Collins, and J. W. Ridgeway, Journal Environmental Engineering Division, American Society of Civil Engineers, Vol. 106, Reading, Mass: EE-1, 1980; L. A. Klein et al., Journal, Water Pollution Control Federation, Vol. 46:12, 1974; N. L. Nemerow, *Industrial Water Pollution*, Addison-Wesley, 1978; I. Polls and R. Lanyon, Journal, Environmental Engineering Division, American Society of Civil Engineers, Vol. 106, EE-1, 1980; and D. H. Waller, and Novak, Z., Journal of Water Pollution Control Federation, Vol. 53:3, 1981.

Problems of Industrial Wastewater

High concentration loading > 1,000–100,000 TOC when compared with domestic wastes

Complexes and chelates

Unstable

May contain biodegradable components

Colloidal state

Highly colored

One of the characteristics of industry wastewater treatment is that all the available methods can be applied. Usually, a given process will include 10-20 stages of different unit processes and operations. If one intends to treat a given wastewater, one should be prepared to understand the entire chemical process. After that, the potential value and compatibility for a given stage or course can be realized, and the wastewater can be more effectively treated. Figure 8-9, a flow chart of cotton-textile finishing waste treatment, illustrates this point.

Furthermore, for a particular type of wastewater one must know exactly what the chemical composition is. The sample indication of COD, BOD, etc. is not sufficient. Unless the targeted compound, which one has to eliminate, is known, one cannot develop an antidote for it. Thus in the screening test, exact chemical analysis will be conducted to discover exactly what is an average chemical composition in a given wastewater. The following table lists some treatment plans for certain types of wastewater.

Type of Wastewater	Chemical Composition	Treatment Methods
Geothermal brine	H_3BO_3, HF, H_2S, Hydrocarbons	Stripping - GAC
Retort water	NH_4HCO_3, RCOOH, Nitrogen bases	Electro-oxidation
Coal conversion water	Phenol, NH_3, PAH	Pre-separation treatment, Biofilter
Postchlorination water	$CHCl_3$, THM	Ultrasound
Derusting water	Citric acid, Triethanolamine	AOP (UV-H_2O_2)
Pink water	TNT, RDX, NG, DOP, Polyacrylic acid	Anaerobic treatment

The first item in the preceding table illustrates the problem of boron in geothermal plant wastewater or geothermal brine. Boron is toxic to plant and vegetation because it interrupts the transportation metabolism of micronutrients. Therefore, efforts have been targeted towards removing boron in wastewater.

An efficient way of removal is to employ a highly selective boron resin, Amberlite IRA-743, to take up the borate from waste brine, followed by using strong acid for the removal of boron from exhausted resin and subsequently regenerating the resin by an alkali. The mechanism of the reaction is illustrated in Figure 8-10.

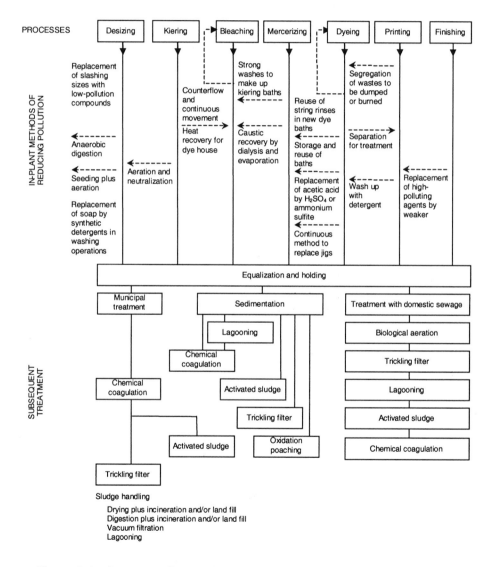

Figure 8-9. Cotton-textile finishing waste-treatment flow chart. (Taken from the chart prepared for the F.W.P.C.A.)

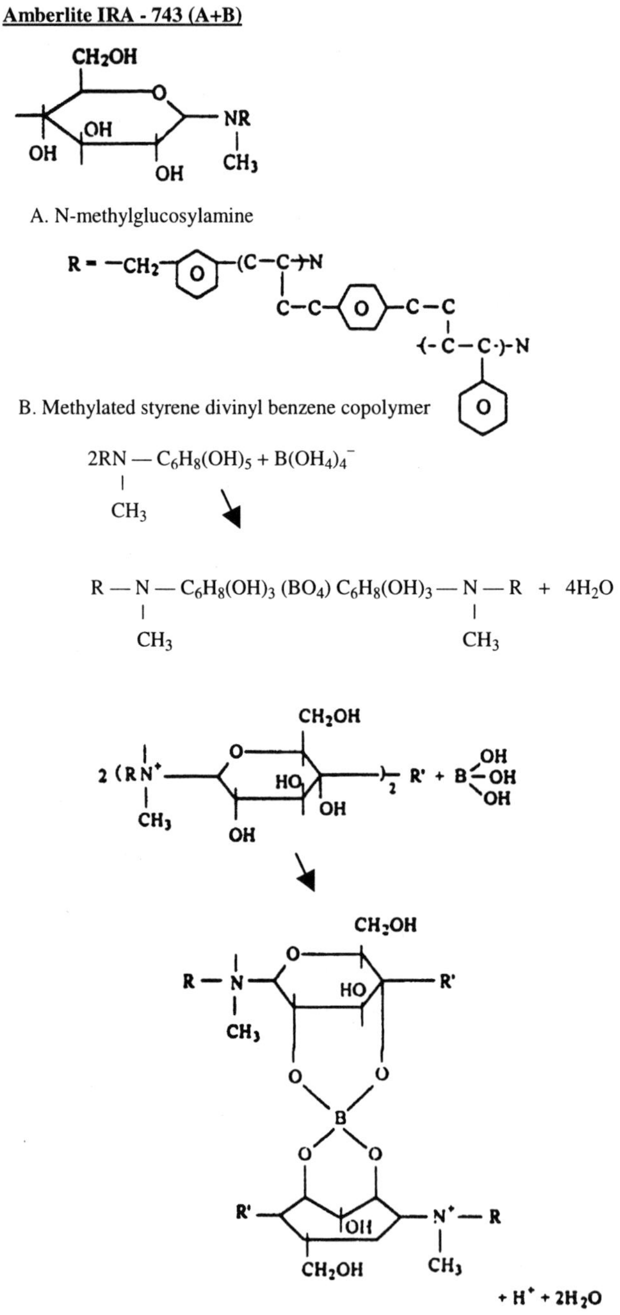

Amberlite IRA - 743 (A+B)

A. N-methylglucosylamine

B. Methylated styrene divinyl benzene copolymer

$$2RN - C_6H_8(OH)_5 + B(OH_4)_4^-$$
$$\quad\;\; |$$
$$\quad CH_3$$

$$R - N - C_6H_8(OH)_3 \, (BO_4) \, C_6H_8(OH)_3 - N - R \; + \; 4H_2O$$
$$\quad\;\; | \qquad\qquad\qquad\qquad\qquad\qquad\qquad | $$
$$\quad CH_3 \qquad\qquad\qquad\qquad\qquad\qquad CH_3$$

Figure 8-10. Reaction mechanism for remediation of Boron.

An efficient process in most cases is economically unsound and is not practical. For example, in the Kizilidere geothermal field in Turkey, the maximum permissible discharge to the nearby Büyük Menderes River is below 1 ppm. This river water is used for irrigation, and because the process to control the boron is costly (not even F and S), electricity production costs would increase by 7¢ (US)/ kWH^{-1}. This is not ideal because utilities for most places are only 10¢ / kWh^{-1}. Both the retort wastes and coal conversion water will be discussed in synfuel wastewater.

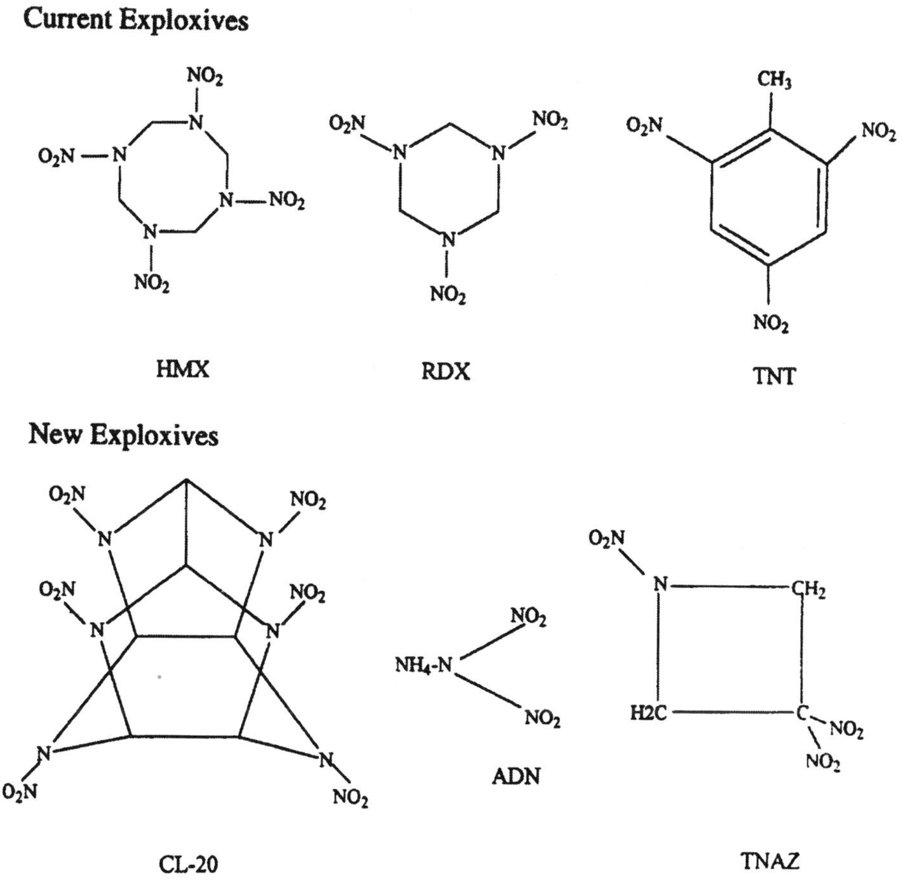

Current Exploxives

HMX   RDX   TNT

New Exploxives

CL-20   TNAZ

Figure 8-11. Common energetics available.

Pink water usually refers to military or munitions wastewater, which contains explosives or energetics. Most energetics or explosives are nitro compounds that can be produced by C-nitration (such as TNT and TNB); the N-nitration, which includes nitramines (such as RDX, HMX and ADN) and; the O-nitration, which includes nitroglycerin, nitrocellulose, etc. Figure 8-11 illustrates some of the current energetics available, which may be possible components of the pink water. At the moment, using mixed anerobic bacteria of digested sewage is the most effective treatment of pink water, as shown in Figure 8-12. The principle is based on the bacteria for the dentrification process, shown in Figure 8-13 and 8-14. Pesticides of nitrosamine functions can also be subjected to this treatment, as shown in Figure 8-15.

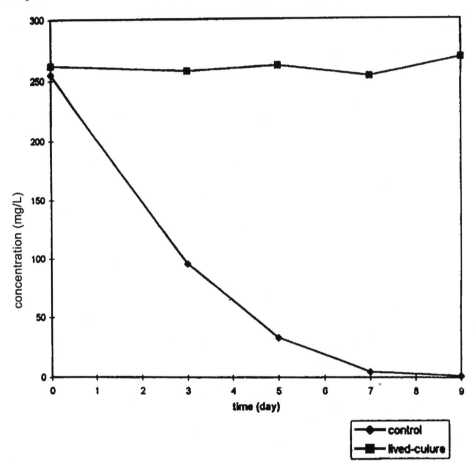

Figure 8-12. ADN degradation by digested sewage sludge. (Kwon et al. North Am. Water and Environmental Congress, Anaheim, CA, 1996.)

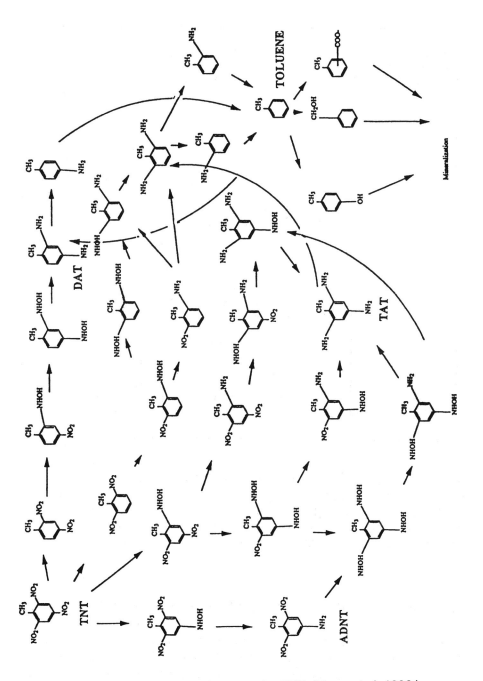

Figure 8-13. Plausible biopathway for TNT. (Kwon et al. 1996.)

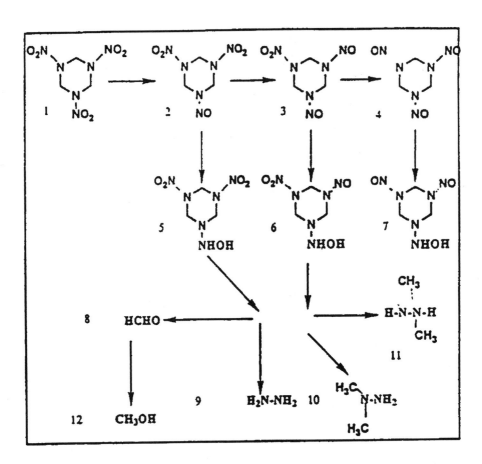

Figure 8-14. Proposed pathway for the anaerobic biodegradation of RDX.
Compounds: 1,RDX; 2, MNX; 3, DNX; 4, TNX; 5, 1-hydroxlamino-3,5-dinitro-
1,3,5-triazine;6, 1-hydroxylamino-3-nitroso-5-1,3,5-triazine; 7, 1-hydroxylamino-
3,5-dinitroso-1,3,5-triazine;8, formaldehyde;9,hydrazine; 10,1,1-
dimethyhyrazine; 11,1,2-dimethylhydrazine; 12, methanol. (McCormick, 1981.)

8.4.2 Chemistry Associated with Synfuel Wastewaters

In order for the synfuel industry to produce alternative liquid fuels for
transportation as well as pumpable (portable) fuels, chemical conversion of the
raw fuel and the subsequent refining and upgrading methods must involve high-
temperature treatment. As a result, the weak linkages in the fuel molecules will
be dissociated in many instances, producing products with heteroatoms such as
N, S, and O. These new species produced may have a significant influence on
our health.

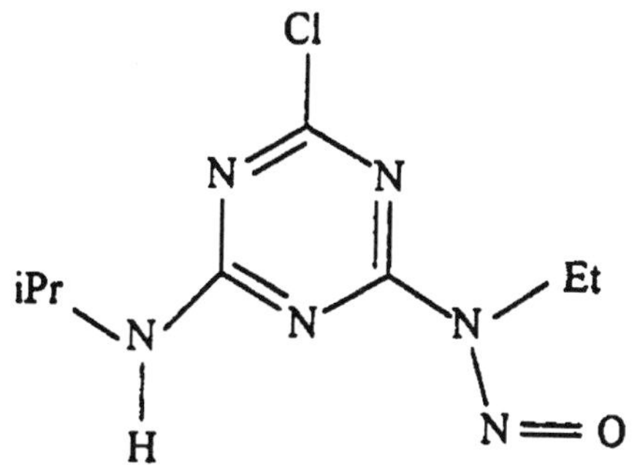

N - Nitrosoastrazine

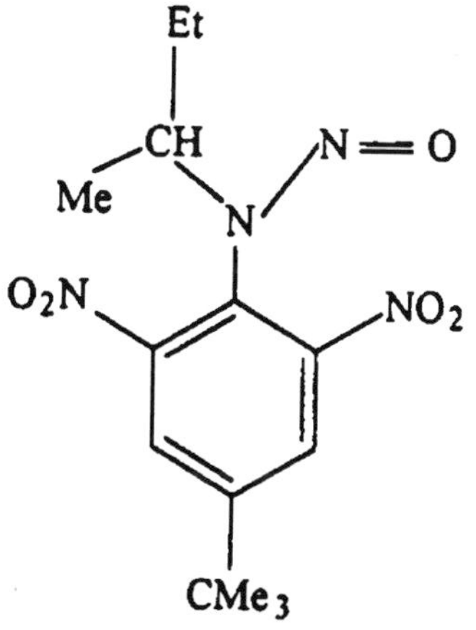

N - Nitrosobatralin 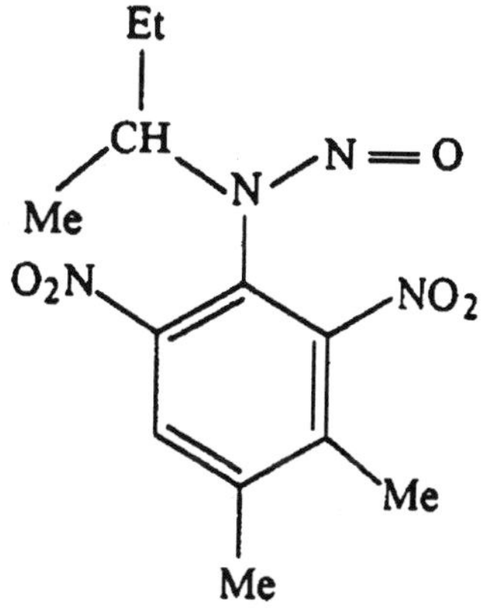   N - Nitrosopendimethalin

Figure 8-15. Example of Nitrosated Agrochemicals.

The sources of coal conversion may be various, from the cooling water to the actual scrubber water, recycles, and condensates. They may be derived either from gasification or liquefaction processes. For oil shale retort water, this is most interesting, because water is usually coproduced with the oil in the amount of equal quantities. The loading of chemical constituents of these waters is extraordinarily high; and, in many cases, they appear as black syrups. The typical composition of coal wastewater is in Table 8-13 and retort water is in Table 8-14. The trace elements in the in-situ retort water is in Tables 8-15 and 8-16. Similarly, the trace elements in coal process waters are listed in Table 8-17. It has been found that arsenic appeared in many organic species in retort water. Furthermore, uranium has also been identified in retort water.

Table 8-13. Comparison of Coal Wastewaters

	Coke Plant	Synthane	Fluidized Bed Scrubber Water	Pittsberg & Midway Liquefaction Recycle Water
pH	8.3-9.1	7.9-9.3	9.1	9.1
Alkalinity	1,200-2,700	—	3,200	3,360
NH₃	1,800-4,300	2,500-11,000	3,000	5,000
CN	10-37	0.1-0.6	—	—
SCN	100-1,500	21-200	—	—
Phenols	410-2,400	200-6,600	10,000	6,000
COD	2,500-10,000	1,700-43,000	18,000	12,000
Specific Conductance	11,000-32,000	—	5,300	4,700

Concentrations are expressed in ppm; conductivities are in μmhos/cm.

Table 8-14. Typical Retort Water Analyses

Parameter	Concentration (mg/L, Except pH)
pH	8.5-8.7
Alkalinity	15,000-32,900
NH₃ – N	4,800-16,800
Organic – N	733-17,000
Phenols	8.5-170
COD	11,000-18,000
BOD	350-12,000

Table 8-15. Trace Elements in Oil Shale Retort Water

Species	Concentration (mg/ L)	Land Application Limit for Irrigation Water
Boron	30 – 46	2 – 10
Fluorine	16 – 33	6 – 15
Lead	.04 – .19	4 – 10
Molybdenum	1 – 4	.02 – .05
Mercury	.001	—
Uranium	.064 – 1.08	—
Vanadium	.05 – .08	—

Table 8-16. Trace Elements in Water Extracted from an
Experimental In-Situ Retort

Species	Concentration (mg/L)	Est. Safe Concentration (mg/L)
Arsenic	.02 - .26	.05
Cadmium	.002 - .0035	.001
Copper	.042 - .087	.004
Manganese	.05 - .15	.10
Nickel	.10 - .30	.03
Selenium	.35 - .56	.01
Zinc	.09 - .77	.001 - .010

From Washburne and Yen, 1981.

Table 8-17. Trace Elements in Coal Process Waters

Species	Gasification Sample (mg/L)	Liquefication Sample (mg/L)
Ni	< 0.6	0.26
Pb	< 0.1	0.36
Fe	0.27	0.12
Mn	0.18	0.005
Cr	0.12	0.12
As	< 0.05	< 0.05
Be	0.2	0.2
Cd	0.07	0.06
Co	0.06	0.1
Cu	0.05	0.02
Hg	0.01	0.01
Zn	0.05	0.03
Se	1.3	1.2
Ag	0.75	0.08
Mo	0.6	< 0.5
V	< 100	< 100
Sn	3.5	5.0
Mg	3.4	0.03
Ca	3.3	0.4
B	0.8	8.0
Al	< 10	< 10
Ba	< 5	< 5

Perhaps one of the major properties of these waters is the appearance of a large variety of polycyclic aromatic hydrocarbons (PAH). Because the process waters are intimately in contact with coal liquid, the processing waters will have large numbers of PAH, as shown in Figure 8-16. No matter how oil shale is retorted, the coproduced waters also contain PAH, as shown in Figure 8-17.

The detection can be easily obtained by capillary gas chromatography, shown in Figure 8-17, or reversed-phase liquid chromatography, shown in Figure 8-18. Both benz(a)- and benz(e)- pyrens are found in these waters, as shown in Figure 8-19.

The nitrogen-rich fraction isolated from retort water by separation from the macroreticular resin method is especially important. A great variety of organic nitrogen-skeletons have been formed, some of which obviously can be related to carcinogenicity, as shown in Table 8-18 and Figures 8-20 and 8-21.

Table 8-18. Nitrogen Systems Identified in Retort Water

Skeleton	Functionality (Compared to Known Carcinogenic Compound)
1,2-Benzene dicarbonitrile	Conjugated dicyanide
4-Diethylamino benzaldehyde	Conjugated carbonyl amine
2,5-Pyrrolidinedione	Conjugated carbonyl amine
Aziridine	Small aza-arene
2H-Quinolizine	Bridge head amine
Pyramidinedione	Linked diamide
2-Aminoisoxazole	Aza-(O)-substituted amide
2-Pyridiamine	α,α-diamine
Octahdrotetrazocine	Cyclic poly α,α-diamine
2-Methyl pyrazine	Conjugated Diamine

Source: T. F. Yen, 1984, Ref. 8–10

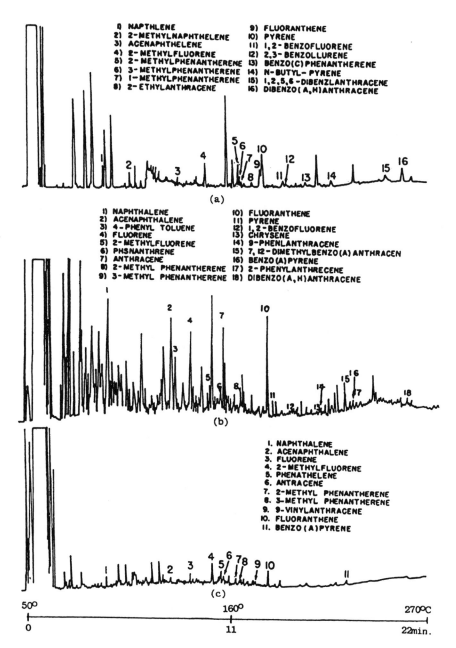

Figure 8-16. GC analysis of (a) coal wastewater (No. 3), (b) oil shale report water (Omega-9), and (c) oil shale report water (No. 16). [From Sadeghi et al., Fuel Sci. Tech. Int., 13, 1393-1412,(1994)].

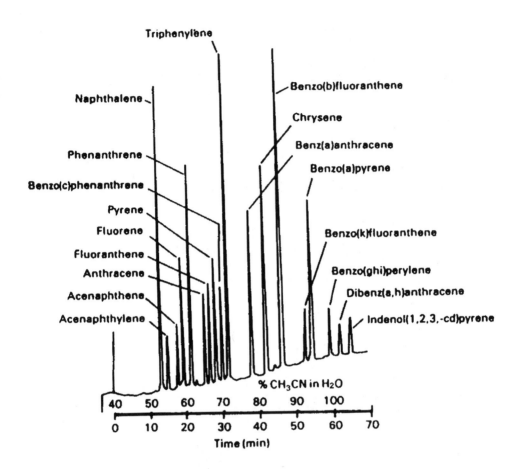

Figure 8-17. Reverse-phase liquid chromatographic separation of polycyclic aromatic hydrocarbons. Column: Vydac 201TP reverse phase; detection: ultraviolet absorbance at 254 NM; conditions: linear gradient from 40–100% acetonitrile in water at 1% min at 1 mL/min. (Adapted from Wise et al., 1980.)

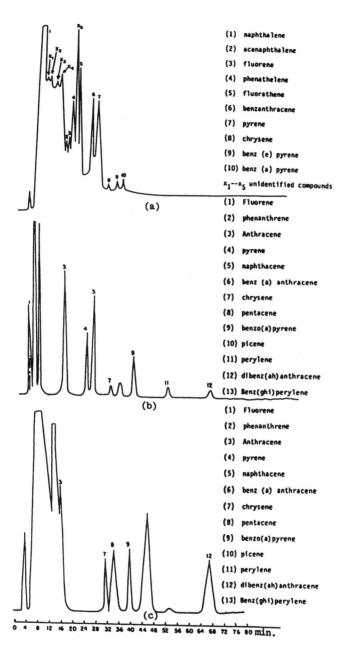

(a)

(1) naphthalene
(2) acenaphthalene
(3) fluorene
(4) phenathelene
(5) fluorathene
(6) benzanthracene
(7) pyrene
(8) chrysene
(9) benz (e) pyrene
(10) benz (a) pyrene
x_1--x_5 unidentified compounds

(b)

(1) Fluorene
(2) phenanthrene
(3) Anthracene
(4) pyrene
(5) naphthacene
(6) benz (a) anthracene
(7) chrysene
(8) pentacene
(9) benzo(a)pyrene
(10) picene
(11) perylene
(12) dibenz(ah)anthracene
(13) Benz(ghi)perylene

(c)

(1) Fluorene
(2) phenanthrene
(3) Anthracene
(4) pyrene
(5) naphthacene
(6) benz (a) anthracene
(7) chrysene
(8) pentacene
(9) benzo(a)pyrene
(10) picene
(11) perylene
(12) dibenz(ah)anthracene
(13) Benz(ghi)perylene

Figure 8-18. HPLC analysis of (a) coal wastewater (No. 3), (b) oil shale retort water (Omega-9), and (c) oil shale retort water (No. 16). [From Sadeghi et al. (1994). (Yen et al. ref 8-5.)]

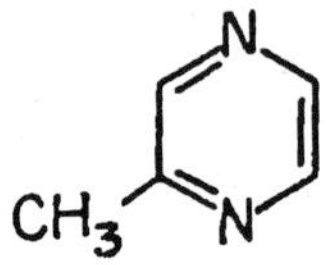

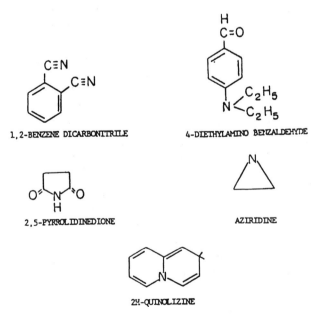

2-METHYL PYRAZINE

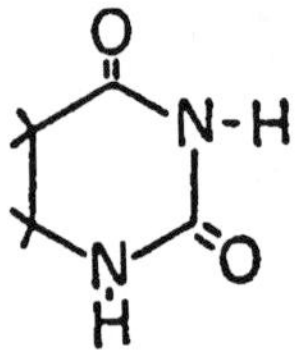

PYRAMIDINEDIONE

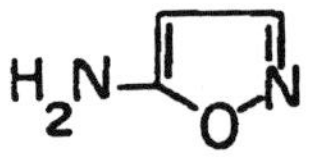

5-AMINOISOXAZOLE

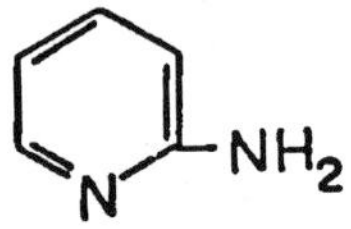

2-PYRIDINAMINE

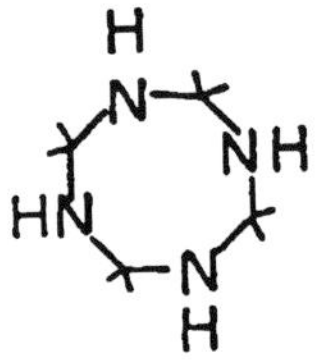

OCTAHYDROTETRAZOCINE

Figure 8-19. Some diazine and polyazine compounds identified from retort water. (Yen, ref. 8-10.)

1,2-BENZENE DICARBONITRILE

4-DIETHYLAMINO BENZALDEHYDE

2,5-PYRROLIDINEDIONE

AZIRIDINE

2H-QUINOLIZINE

Figure 8-20. Some nitrogen systems identified in retort water. (Yen, ref 8-10.)

REFERENCES

8-1. W. Stumm and J. J. Morgan, *Aquatic Chemistry*, Wiley, New York, 1981.

8-2. D. W. Sundstrom and H. E. Klei, *Wastewater Treatment*, Prentice-Hall, Englewood Cliff, New Jersey, 1980.

8-3. W. J. Weber Jr., *Physicochemical Processes for Water Quality Control*, Wiley, New York, 1972.

8-4. R. L. Sanks, *Water Treatment Plant Design*, Ann Arbor Science, Ann Arbor, Michigan, 1980.

8-5. T. F. Yen, J. I. S. Tang, M. Washburne, and S. Cohanim, *Analysis of Hazardous Organics Present in Liquid Wastes from Coal Conversion Processes*, EPA-R806 167-01-0, (NTIS), 1981.

8-6. T. W. Schultz and J. N. Dumont, "Cytotoxicity of Untreated Coal Liquefaction Process Water (And a Comparison with Gasification Process Water)," *J. Env. Sci. Health*, Al3(9), pp. 641–651, (1978).

8-7. T. F. Yen and G. C. Slawson Jr., *Compendium Reports on Coal Oil Shale Technology*, PB 293, 279, EPA-600/7-79-039 (NTIS), 1979.

8-8. R. H. Gray, H. Drucker and, M. J. Massey, *Toxicology of Coal Conversion Processing*, Wiley, New York, 1988.

8-9. Y. H. Li, *Environ. Conserv.*, 3, 171 (1976).

8-10. T. F. Yen, "Nature of Pollutants in Oil Shale and Coal Conversion Wastewaters," in *Environmental Engineering* (American Society of Civil Engineers, 1984 Specialty Conference) pp. 460–471, 1984.

8-11. H. Van Langenhore and N. Schamp, "Olfactometric Odor Measurement," in *Encyclopedia of Environmental Control Technology 2, Air Pollution Control* (P. N. Cheremisinoff, ed.), Gulf Pub., Houston, 1989, pp. 935–964.

8-12. P. C. G. Isaac, *Waste Treatment*, Pergamon, Oxford, 1960.

8-13. J. M. Montgomery, *Wastewater Treatment Principles and Design*, JMM, Pasadena, California, 1985.

8-14. S. D. Faust and O. M. Aly, *Chemistry of Water Treatment*, Butterworth, Boston, 1983.

8-15. I. J. Tinsley, *Chemical Concepts in Pollution Behavior*, Wiley, New York, 1979.

8-16. American Wastewater Association, *Water Quality and Treatment*, 4th ed., McGraw-Hill, New York, 1990.

PROBLEM SET

1. A water sample whose BOD versus time data for the first five days has the following data:

Time (days)	BOD (Y, mg/ L)
2	10
4	16
6	21

    Calculate the kinetic constants k, k', and L_0.

2. Determine the ThOD for glycine ($CH_2(NH_2)COOH$) using the following assumptions:
    a.) In the first step, the carbon is converted to CO_2, and the nitrogen is converted to ammonia.
    b.) In the second and third steps, the ammonia is oxidized to nitrite and nitrate.
    c.) The ThOD is the sum of the oxygen required for all three steps in grams of O_2 per mole.

3. Given the following results determined for a wastewater sample at 20°C, determine the ultimate carbonaceous oxygen demand, the ultimate nitrogenous oxygen demand (NOD), the carbonaceous BOD reaction-rate constant (K), and the nitrogenous NOD reaction-rate constant (K_n). Determine K ($\theta = 1.05$) and K_n ($\theta = 1.08$) at 25°C. (Courtesy of Edward Foree.)

Time, d	BOD, mg/ L	Time, d	BOD, mg/ L
0	0	11	63
1	10	12	69
2	18	13	74
3	23	14	77
4	26	16	82
5	29	18	85
6	31	30	87
7	32	25	89
8	33	20	90
9	46	40	90
10	56		

4. The following BOD results were obtained on a sample of untreated wastewater at 20°C:

t, d	0	1	2	3	4	5
y, mg/ L	0	65	109	138	158	172

Compute the reaction constant K and ultimate first-stage BOD using both the least-squares and the Thomas methods. (Courtesy of Metcalf and Eddy, Inc.)

THE PEDOSPHERE

The realm of soils and their inhabitants is the pedosphere. Soil is the medium in which crops grow, and it is the basis for nearly all forms of life on the solid earth. In addition, it acts as a buffer to control the water flow between sky, land, and sea. As the world becomes more industrialized, the role of soils becomes more important because they are a major sink for many of the xenobiotics. As with other natural, nonrenewable resources, it is important to ensure that soils are used conservatively so that the advantages we enjoy from soils now can be with us even in the future. To achieve this goal, it is necessary for us to understand soils and all the associated phenomena.

This chapter is composed of four sections. The first section is an introduction to soil, followed by a description of soil chemistry and its related erosion and weathering problems. These subjects are related to both agricultural and geotechnical applications. The third section deals with the contamination problems of soils. Current soil contamination regarding PCBs and related compounds is fully explored as well as DDT and other chloro-organics in use.

The last section of the chapter gives a discussion on recent advances in the insecticide industry, including an in-depth investigation of the nature of the third generation of insecticides. It also focuses on our duty to learn more about the chemical warfare stockpile in different locations worldwide.

INTRODUCTION TO SOIL

Soil is the end product of physical-chemical weathering and erosion in combination with biological actions on igneous rock. Soil is the interface, a thin layer mantle of the lithosphere that supports life. Soil is where the interactions between the lithosphere and the hydrosphere, atmosphere, and biosphere occur. Physically, soil is a porous mixture of inorganic particles, organic matter, air, and water. The three dominant inorganic particles in soil are **sand**, **silt**, and **clay**. They are usually classified according to their sizes, as shown in Table D-1. Clays are the finest particles, with an equivalent diameter smaller than 0.002 mm, and they have a large specific surface area.

Component	Sand	Silt	Clay
Particle Size (mm)	2	0.05	0.002
Surface Area (cm^2/g)	10	500	80×10^6

Table D-1. Size and Surface Area of Soil Particles

Particle Type	Diameter (mm)[a]	No. of Particles/ g	Surface area (sq. cm./ g)
Very coarse sand	2.00 - 1.00	90	11
Coarse sand	1.00 - 0.50	720	25
Medium sand	0.50 - 0.25	5,700	45
Fine sand	0.25 - 0.10	46,000	91
Very fine sand	0.10 - 0.05	722,000	227
Silt	0.05 - 0.002	5,780,000	454
Clay	0.002	90,500,000,000	80,000,000

[a]Assumed to have spherical shapes. Calculated on the basis of maximum diameters of the particle type.

Data from Foth and Turk (1972)

Particle size distribution is most often called **soil texture** and is probably the most important physical property of soil. The U.S. Department of Agriculture has developed a method for naming soils based on particle size distribution. The relationship between textural analysis and class names is shown in Figure D-1; it is often called a **textural triangle.**

Soil classification system differs from country to country. The U.S. system is very comprehensive; for example, there are 10 orders, 47 suborders, and then greatgroups, subgroups, and series. The frequently used 10 orders are listed in Table D-2. A common system for soil types is based on the **unified soil classification** system, shown in Table D-3, which is based on the **group symbols** shown in Table D-4.

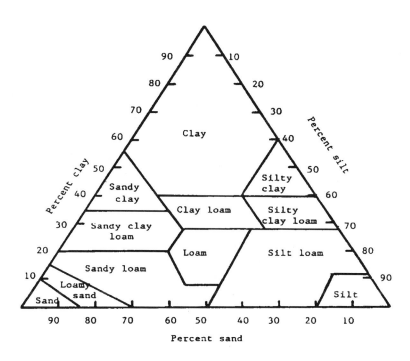

Figure D-1. The textural triangle from which the names of textural classes are obtained. There are a total of 12 classes of soil. (Source: U.S. Department of Agriculture.)

Table D-2. Soil Classification System - Key to Soil Orders

If soil has	Order
> 30% clay to 1 meter (40 in.) or to lithic/ paralithic contact gilgai or slickensides or wedge-shaped aggregates	Vertisols
No diagnostic horizon other than ochric, anthropic, albic, argic	Entisols
No spodic, argillic, natric, oxic, petrocalcic, plinthite; but has cambic or histic	Inceptisols
Ochric or argillic but no oxic or spodic and usually dry	Aridisols
Spodic	Spodosols
Mean annual soil temperature > 8°C (47°F), properties not placing it in one of above, percentage base saturation < 35 @ 1.25 m (50 in.) below top of argillic or 1.8 m (72 in.) below surface	Ultisols
Mollic but no oxic	Mollisols
All other mineral soils without oxic	Alfisols
Oxic horizon	Oxisols
> 30% organic matter to a depth of 40 cm (16 in.)	Histosols

This is a widely used approximate method for different types of coarsely-grained and finely-grained soils. In most soils, the particles tend to be grouped into aggregates or peds. Figure D-2 illustrates a hypothetical model of a soil aggregate. Building and maintaining a good structure depends on (1) the influence of organic matter, (2) the shrink-swell processes associated with wetting and drying or freezing and thawing, (3) the action of plant roots and soil microorganisms, and (4) the modifying effects of adsorbed cations. If the aggregates so formed are stable, various forms of pores such as transmission pores, storage pores, residual pores, and bonding space will exist in the interstices between them, as shown in Table D-5. Within these pore spaces, soil water can move freely, and the exchange of gases between soil and the atmosphere can proceed. Table D-6 gives a simple classification of soil aggregates by size. In summary, there exist the following:

- **Transmission pores** — air, seepage
- **Storage pores** — capillary effect
- **Residual pores** — ion exchange
- **Bonding spaces** — strength

     Clod → aggregate → microaggregate → domain

Table D-3. Unified Soil Classification Chart

Group symbols	Typical names	Information required for describing soils
GW	Well-graded gravels, gravel-sand mixtures; little or no fines	Given typical name; indicate approximate percentages of sand and gravel, maximum size, angularity, surface condition, and hardness
GP	Poorly graded gravels, gravel-sand mixtures; little or no fines	of the coarse grains; local or geologic name and other pertinent descriptive information; and symbol in parentheses.
GM	Silty gravels, poorly graded gravel-sand-silt mixtures	
GC	Clayey gravels, poorly graded gravel-sand-clay mixtures	For undisturbed soils add information on stratification, degree of compactness, and drainage characteristics.
SW	Well-graded sands, gravelly sands; little or no fines	
SP	Poorly graded sands, gravelly sands; little or no fines	Example:Silty sand, gravelly; about 20 % hard, angular gravel particles 12 mm maximum size;
SM	Silty sands, poorly graded sand-silt mixtures	rounded and subangular sand grains coarse to fine; about 15 % nonplastic fines with low
SC	Clayey sands, poorly graded sand-clay mixtures	dry strength; well compacted and moist in place; alluvial sand; (SM)
ML	Inorganic silts and very fine sands, rock flour, silty or clayey fine sands with slight plasticity	Give typical name; indicate degree and character of plasticity, amount and maximum size of coarse grains; color in wet condition, .
CL	Inorganic clays of low to medium plasticity, gravelly clays, sandy clays, silty clays, lean clays	odor if any, local or geologic name, and other pertinent descriptive information; and symbol in parentheses
OL	Organic silts and organic silt-clays of low plasticity	
MH	Inorganic silts, micaceous or diatomaceous fine sandy or silty soils, elastic silts.	For undisturbed soils add information on structure, stratification, consistency in undisturbed and remolded states, moisture and drainage conditions
CH	Inorganic clays of high plasticity, fat clays	Example: Clayey silt, brown; slightly plastic; small percentage of fine sand; numerous vertical root holes; firm and dry in place; loess; (ML)
Pt	Peat, muck, peat-bog, and so on.	

Table D-4. Group Symbols of Soils

Soil type	Prefix	Subgroup	Suffix
Gravel	G	Well graded	W
Sand	S	Poorly graded	P
		Silty	M
		Clay	C
Silt	M	WL<50%	L
Clay	C	WL>50%	H
Organic	O		
Peat	Pt		

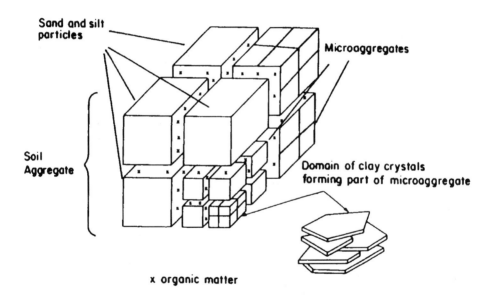

Figure D-2. A hypothetical model of a soil aggregate, illustrating the clustering of clay crystals to form domains, of domains to form microaggregates, and of microaggregates to form aggregates. Molecules of soil organic matter act as bonding agents between silt particles. [After Williams, Greenland and Quirk, Aust. J. Soil Research 5 77-83 (1967).]

The coarse-grained soil is:

GW, GP or SW, SP	≤ 5% passes No. 200 sieve
GW-GM, GP-GM, GW-GC, GP-GC or SW-SM, SP-SM, SW-SC, SP-SC	5% <passing No. 200 sieve ≤ 12%
GM,GC or SM, SC	> 12% passes No. 200 sieve

Table D-5. A functional classification of soil pores

Name	Function	Equivalent, Cylindrical Diameter, μm
Transmission pores	Air movement and drainage of excess water	50
Storage pores	Retention of water against gravity, and release to plant roots	0.5-50
Residual pores	Retention as well as diffusion ions in solution	0.5
Bonding spaces	Support major forces between soil particles	0.005

After D.J. Greenland, Phil Trans. Roy. Soc. (London) *B281* 193-208 (1977)

Table D-6. A Simple Classification of Soil Aggregates by Size

Name	Description	Size Range
Clod	Clusters of aggregates	> 5 mm
Aggregate	Clusters of microaggregates and sand particles	0.5 - 5 mm
Microaggregate	Domains, silt and fine sand particles bonded by organic polymers	5 - 500 μm
Domain	Oriented clusters of clay crystals	< 5 μm

Data from Greenland, Soil and Fertilizers, *34*, 237-251, 1971

SOIL CHEMISTRY

9.1 SOIL FORMATION AND SOIL PROPERTIES

The origin of most soils can be ultimately traced back to the alteration of igneous rocks from disintegration and decomposition processes. There are five major factors involved in soil formation: parental material (p), climate (cl), geological time (t), relief pressure (r), and natural organisms (o). This statement can be expressed as

$$\text{Soil formation} = f(p,cl,t,r,o)$$

which is called the **Dokuchaiev soil formation function**.

Soils are sometimes described in terms of their profile morphology. The differences in readily observable profile characteristics are reflections of chemical, mineralogical, physical, and biological differences in the soils. Figure 9-1 shows a typical **soil profile** in which the O,A,B,C horizon nomenclature is commonly used. The following scheme is often representative:

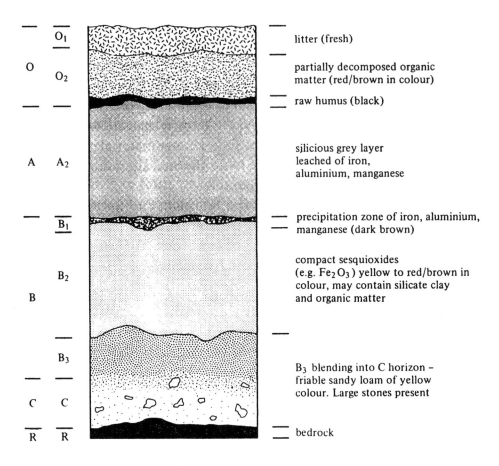

Figure 9-1. Representation of a podsol profile. The boundaries between O and A and A and B horizons are quite distinct, and good examples of podsols lack A_1 and A_3 zones which, in other soils, represent transitional areas between O and B horizons, respectively.

$$\text{OM}\downarrow$$
Igneous rock $\rightarrow$ Soil C Horizon C $\rightarrow$ Horizon B subsoil $\rightarrow$ Horizon A topsoil
$$\downarrow \text{loss of Ca, Mg, K, Na}$$

Silicon is one of the most abundant elements in the lithosphere, and because the pedosphere originates from the lithosphere, it is the main element in soil. Silicate minerals are the basis of igneous rocks, which result from the solidification of the earth's molten magma. Sedimentary rocks, such as sandstone, form igneous ones by the action of weathering (including water erosion). Another type of rock, the metamorphic, is created from the former two by heat, pressure, or solvent action (e.g., serpentine). The most abundant igneous rock is granite--a mixture of feldspar, mica, and quartz. Next to granite, the most important of the igneous rocks are ferromagnesium silicates, pyroxenes, and amphipoles. The continents appear to be slabs of granite about 20 miles thick, floating on the heavy basaltic material of the lithosphere.

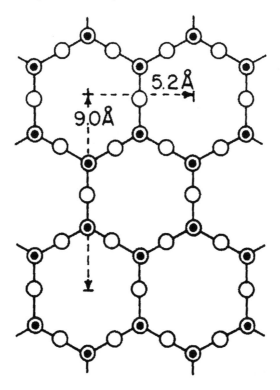

Figure 9-2. Diagram of phyllosilicate sheet (top view). The black circles represent silicon atoms and the open circles represent oxygen atoms. Each silicon atom is tetrahedrally bound to four oxygen atoms. The oxygen atoms shown superimposed on the silicons are directed upward and bound to a second parallel layer.

Silicon dioxide, or silica, is a polymeric solid with a network consisting of silicon atoms bound tetrahedrally to four oxygen atoms. The oxygen atoms are each in turn bound to two silicon atoms. In this manner, the tetrahedron structure can become leaflike if the fourth bond of silicon sticks up out of the sheet, as shown in Figure 9-2. If the bond center is an aluminum ion, then the formation of a phillosilicate sheet such as kaolinite, shown in Figure 9-3, takes place. Aluminum prefers to have an octahedral coordination resulting in the formation of clay sheets, which we will discuss further in the section on weathering.

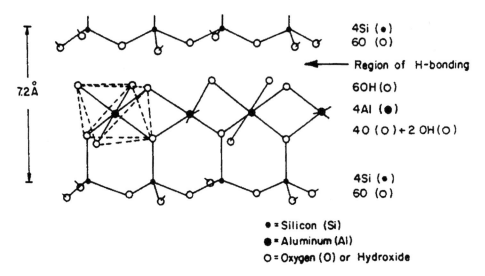

Figure 9-3. Structure of kaolinite, $Al_4Si_4O_{10}(OH)_8$. Figure shows phyllosilicate and octahedral layers. The distance between two successive plates is 7.2 Å. Dashed line shows sixfold coordinate positions in octahedral layer. Note that six oxygen atoms are associated with four silicon atoms in the phyllosilicate layer because silicon shares each of its three oxygens with another silicon. Hence, each Si is bonded to 3/2 oxygen atoms.

Due to the nature of the sheet type of stacking, a number of structures can be created from the many possible types of 3-dimensional tessellation. The regular spacing that remains can be a variety of regular cavities or canals acting as either residue or storage pores. This is the basis of heterogeneous catalysis in petro-chemical engineering that involves a number of faujasite structures such as zeolite or perovskite. Due to the interlayer polar characteristics within the clay layers, hydrogen bonding and interchelation molecules such as water, alcohol, and glycerine can expand or swell the clay minerals. This adsorption capacity can allow the organic molecules (such as humin) to be bound within the soil.

The structure of the clay soil, as observed by scanning electron microscopy, is quite complex, as shown in Figures 9-4 and 9-5. Usually, if the soil is under stress, such as overburden, the clay soil assumes a more-or-less compact structure. Because joints, fissures, root holes, vorves, silt or sand lenses, and other discontinuities will affect the total mechanical properties of soil, the orientation of various structures in an undisturbed soil is important. For example, the water content relating to different degrees of dispersion can effect the properties of cohesive soil, as shown in Figures 9-6 and 9-7.

There is oxygen and nitrogen in different locations of the pedosphere, and it is especially necessary for the **rhizosphere**. The rhizosphere is defined as the local soil environment greatly influenced by plant roots, as shown in Figure 9-8. The storage and transmission pores also greatly enhance the microbial growth, which is very common for any garden soil possessing different types of enzyme activists. The deterioration and biodegradation properties will be discussed later in detail.

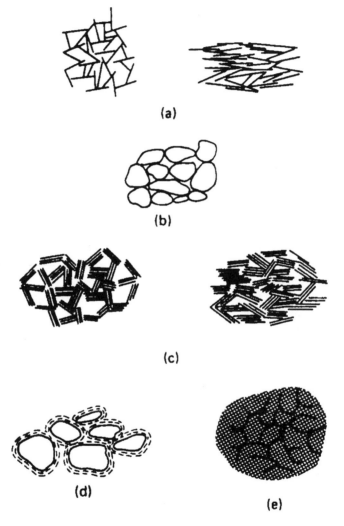

Figure 9-4. Schematic representations of elementary particle arrangements: (a) individual clat platelet interaction; (b) individual silt or sand particle interaction; (c) clay platelet group interaction; (d) clothed silt or sand particle interaction; (e) partly discernible particle interaction. [After K. Collins and A. McGown, "The Form and Function of Microfabric Features in a Variety of Natural Soils," *Geotechnique,* 24 (2), 223-254, 1974.]

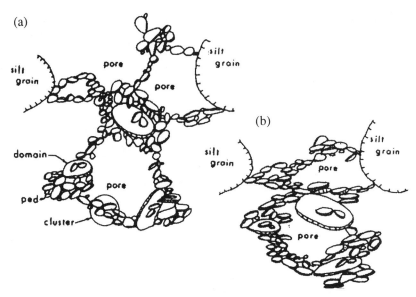

Figure 9-5. Structure of a clay soil: (a) a porous, flocculated sediment interspersed with silt grains; (b) the sediment after it has been subjected to overburden or other stresses that have resulted in a reorientation of the domains, clusters, and peds into a more parallel (dispersed) state.

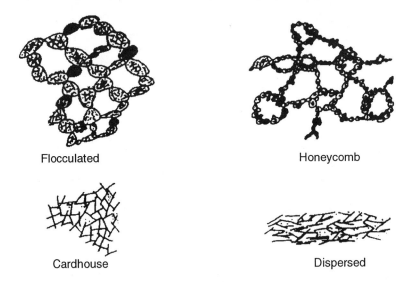

Figure 9-6. Structure of clay soil using earlier terms of structure orientation. The flocculent structure might be obtained from sedimentation in water with a low salt content. The honeycomb structure could be obtained from sedimentation in a marine (high-salt-content) environment. The cardhouse description was highly used prior to SEM studies. The dispersed state is a convenient description for reorientation from compaction.

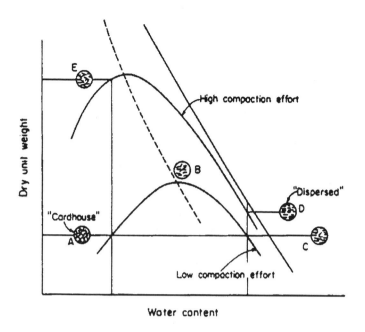

Figure 9-7. Qualitative effect of compaction on soil fabric and structure. [After T.W. Lambe, JSMFD, ASCE, *84* SM2, 1655 (1958).]

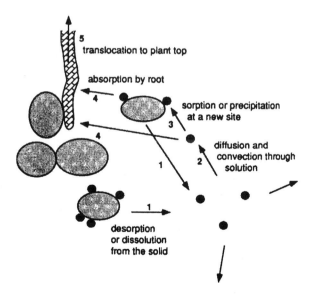

Figure 9-8. Rhizosphere - the space used by plant roots to transport solid soil to the top of the plant. There are 5 steps necessary for this. (Modified from McBride)

9.1.1   Soil Organic Matter

Although the organic components in soil only account for at most 5% by weight, they are essential in many aspects. Among all the organic components, humus is the most important. Humus (or the insoluble portion referred as **humin**) is the result, or residue, of the final degradation from biomass by microbiological actions, and as such, is resistant to biodegradation. Thus it is a recalcitrant. Some major classes of organic compounds in soils are listed in Table 9-1. Soil humus consists of a base soluble fraction composed of humic acid and fulvic acid, which are naturally occurring high molecular weighted macromolecules. A hypothetical formula can be seen with the presence of aryl, carboxylic, hydroxyl, aldehyde, ketone, etc., as shown in Figure 9-9. The insoluble fraction from soil humus is called **humin**, which in general contains nitrogen and sulfur. Young and Yen speculated that the formation of humin may have originated from melanoidin, which is a complex reaction product between amino acids and glucose. The electron spin resonance spectra of soil may be directly related to those of humus, as shown in Table 9-2.

Table 9-1. Major Classes of Organic Compounds in Soil

Compound Type	Composition	Significance
Humus	Degradation-resistant residue from plant decay, largely, C, H, and O	Most abundant organic component, improves soil physical properties and exchanges nutrients, is a reservoir of fixed N
Fats, resins, and waxes	Lipids extractable by organic solvents	Generally, they are only 1-5% soil organic matter may adversely affect soil physical properties by repelling water, and they may be phytotoxic
Saccharides	Cellulose, starches, hemicellulose, gums	Major food source for soil microorganisms, helps stabilize soil aggregates
N-containing organics	Nitrogen bound to humus, amino acids, amino sugars, and other compounds	Provides nitrogen for soil fertility
Phosphorous compounds	Phosphate esters, inositol phosphates (phytic acid), and phospholipids	Source of plant phosphate

Table 9-2. EPR Data of Some Naturally Occurring Materials

Sample	g-Value	Width (gauss)	Shape
Chlorophyll	7.81	420	+
	5.15	500	d*
	3.26	940	+
	2.43	520	d
	2.19	60	d
Humic acid	7.99	420	-
	5.09	570	d
	2.76	1040	+
	2.15	60	d
	2.08	620	+
Browning products	7.72	440	-
	5.00	540	d
	2.90	1300	-
	2.15	60	d
	1.94	1080	+
Hemoglobin	14.95	550	d
	4.18	1750	+
	2.98	30	d
Acetone extract of sediment	7.99	360	+
	5.08	560	d
	2.73	1800	+
	2.15	20	d

* d, derivati e, + and − represent upward and downward features, respectively.

Source: T.F. Yen, *Chemistry of Marine Sediments*, Ann Arbor Sci. Pub., 1977, p. 16.

There are various organics in soil (see a later discussion with Fig. 9-20). Some of them are based on the metabolites of microorganisms; for example, rocks can be weathered through fungi interacting with the oxalic acid, and citric acid, both of which can be good complexing agents. It should be emphasized here that the $\equiv$SiOH function can not only link with humus, extending their ability of bonding or complexing with a number of chemical agents as shown in Figure 9-10, which the ion exchange in the soil can also achieve. The concept of geo-polymer linking by biopolymer in this manner can be accomplished.

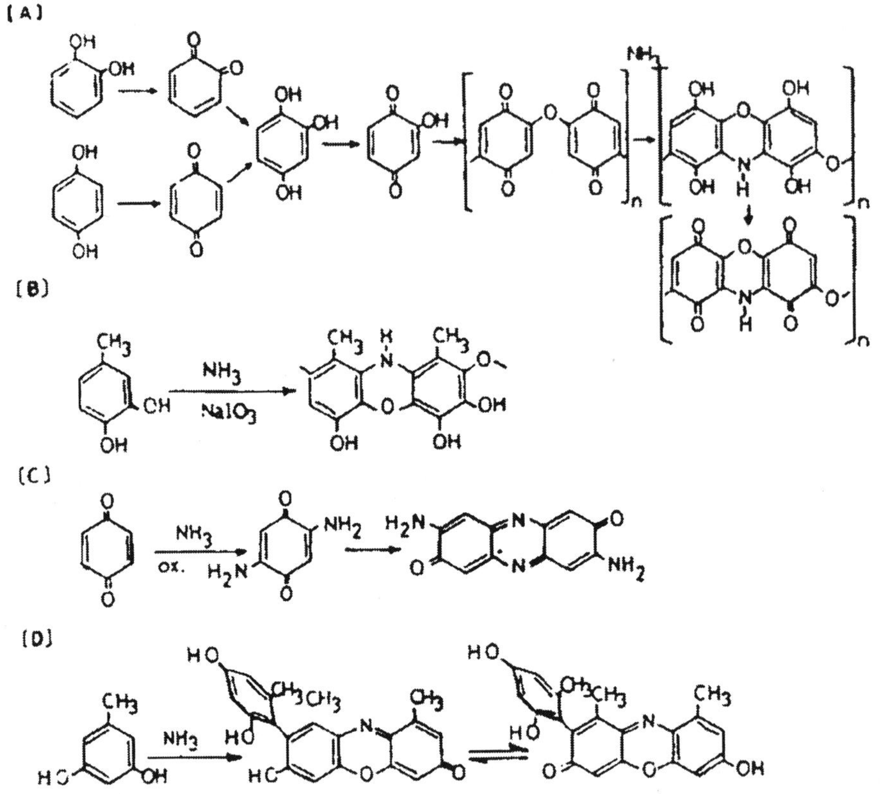

Figure 9-9. Precursors of soil humin. (After Bremner, Flaig, Manskaya, and Drosdova, Yen etc.)

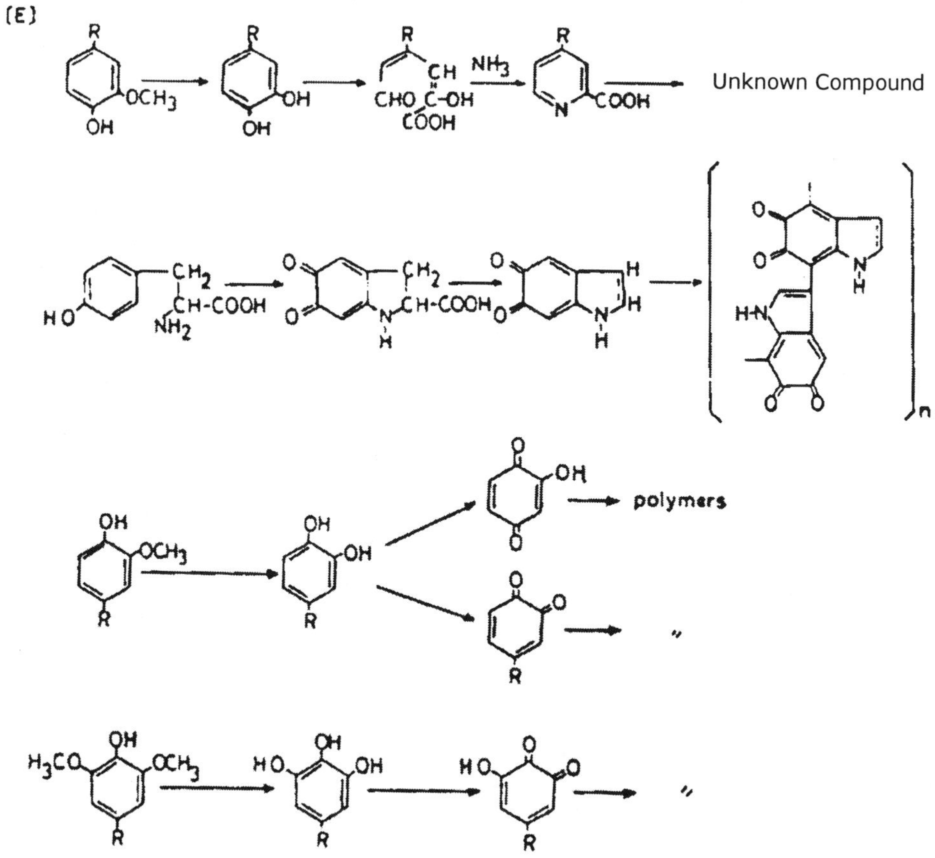

Figure 9-9. (Continued) Precursors of soil humin.

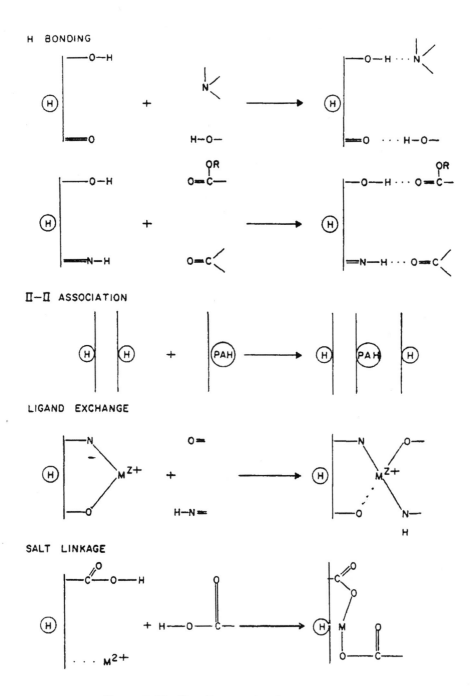

Figure 9-10. Bonding mechanisms

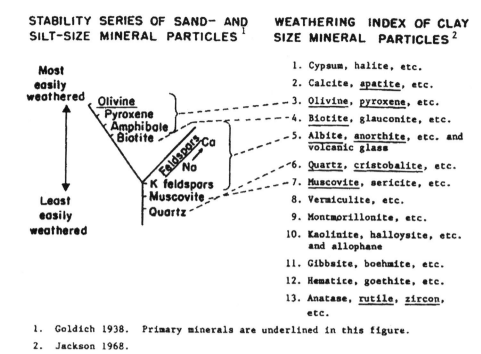

STABILITY SERIES OF SAND- AND SILT-SIZE MINERAL PARTICLES[1]

WEATHERING INDEX OF CLAY SIZE MINERAL PARTICLES[2]

1. Gypsum, halite, etc.
2. Calcite, apatite, etc.
3. Olivine, pyroxene, etc.
4. Biotite, glauconite, etc.
5. Albite, anorthite, etc. and volcanic glass
6. Quartz, cristobalite, etc.
7. Muscovite, sericite, etc.
8. Vermiculite, etc.
9. Montmorillonite, etc.
10. Kaolinite, halloysite, etc. and allophane
11. Gibbsite, boehmite, etc.
12. Hematite, goethite, etc.
13. Anatase, rutile, zircon, etc.

1. Goldich 1938. Primary minerals are underlined in this figure.
2. Jackson 1968.

Figure 9-11. The relative stability of coursely and finely grained minerals in soils. The first series consists of primary minerals arranged from top to bottom, in the order of their crystallization from molten material, and also in the order of decreasing ease of weathering. The second series consists of a condensed version of the first in which the positions of muscovite and quartz have been interchanged because of the greater stability of soils of clay-sized mica. At the top and in most of the lower part of this series are secondary minerals. (Reprinted by permission from *Soil Genesis and Classification* by S. W. Buol. F. D. Hole and R. J. McCraken. ©1972 by the Iowa State University Press, Ames. Iowa 50010.)

9.1.2 Chemical and Physical Weathering

Disintegration and decomposition are the two major weathering processes. Disintegration results from physical processes such as differential expansion of rock materials, expansion of water due to formation of ice in rock fissures, and abrasion and grinding action from hydrological flow. Decomposition is due to chemical processes such as hydrolysis, hydration, oxidation, reduction, carbonation, and solution effects. In addition to the physical and chemical processes, biological activity is also important in the weathering of rocks. For example, when moisture is present, lichens can grow on bedrock, which, in turn, can produce acids such as citric acid and acetic acid through their metabolism. These acids can erode the rock.

Certain minerals in the parent rock in the soil are more resistant than others. The more chemically acidic minerals (such as quartz, muscovite mica, and some feldspars) are, the more stable they are as they become predominant in soil. On the other hand, the more basic mineral components (such as pyroxenes) are more readily altered by chemical weathering, as shown in Figure 9-11. Although quartz (SiO_2) has a high degree of resistance to weathering, it does dissolve slightly in natural water to form silicic acid (H_4SiO_4). This slow dissolution process, as well as the chemical weathering of other minerals, follows the Arrhenius relationship,

$$k = A \exp(-E_a/RT).$$

The chemical weathering schemes are shown here:

$$\xrightarrow{\text{(hydrolysis)}} H^+, Mg^{2+}, Fe^{2+}, H_4SiO_4$$

$$\text{Hypersthene (Mg, Fe)SiO}_{3\,(s)} \xrightarrow{\text{H}_2\text{O(hydration)}} Mg^{2+}, OH^-, Fe_2SiO_4, H_4SiO_4$$

$$\xrightarrow{\text{O}_2\,\text{(oxidation)}} Fe_2O_3 \bullet 3H_2O, Mg_2SiO_4, H_4SiO_4$$

Therefore, the reaction is temperature dependent, with the weathering of rocks faster in tropical than in temperate areas. Incidentally, quartz is not stable under strong basic conditions. At high temperatures it will dissolve in quinoline at 360°C. Minerals formed at high temperatures and pressures also weather more rapidly than those formed at low temperatures.

Silicon chemistry is very complex; silica can be hydrolyzed into many species and some of them can be polymeric. If the aluminum ion can participate, layered materials termed **phyllosilicates** can be obtained.

$$SiO_2 \xrightarrow{2\,H_2O} H_4SiO_4 \rightarrow 4H^+ + (SiO_4)^{4-} \xrightarrow{H^+} Si(OH)_4 \xrightarrow{Al^{3+}} \text{Octahedral layers}$$

$$\xrightarrow{3\,H_2O} H_6Si_2O_7 \text{ (phyllosilicates)} \xrightarrow{4\,H_2O} H_8Si_3O_{10}$$

When hydrous oxide species of aluminum and silicon are present in appropriate concentrations in solution or are simultaneously released during weathering, the formation of silicoaluminum copolymers is thermodynamically favorable. Hence, clay minerals are formed. Clay minerals (phyllosilicates) are primarily crystalline alumina or magnesium silicates with stacked-layer structures. Each unit layer is in turn a sandwich of tetrahedral and octahedral sheets. In the tetrahedral sheets, each silicon atom is surrounded by four oxygen atoms in a tetrahedral arrangement. The octahedral sheet consists of the planes of oxygen in a hexagonal closest-packed arrangement, with alumina or magnesium atoms at the octahedral sites. Some clay minerals (e.g., kaolinite), which contain one tetrahedral sheet and one octahedral sheet in each of their layers, are termed **1:1 layer silicates**. Some three-sheet layer clays containing two outer tetrahedral sheets and one inner octahedral sheet are termed **2:1 layer silicates**. Both type of layered silicates are illustrated in Figure 9-12. Some common phyllosilicates with their interlay and composition are summarized in Table 9-3. The details of the crystal structures of clays were not elucidated until Pauling in 1930 proposed using X-ray diffraction analysis on these minerals.

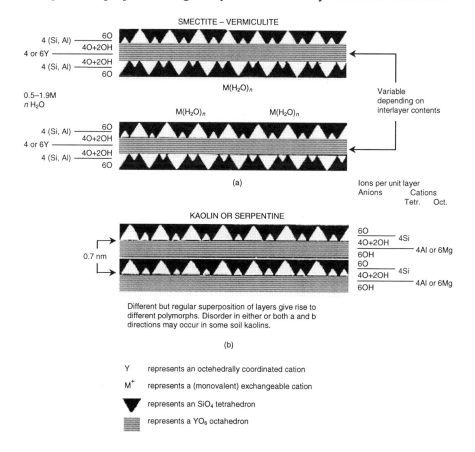

Figure 9-12. Schematic representation of structures of principle phyllosilicate minerals: (a) 2:1 minerals, and (b) 1:1 minerals. Pyrophyllite and talc are similar to the micas, except that the tetrahedral occupancy is 4Si, so that the layer change is zero and there are no interlayer cations.

Table 9-3. Classification and Generalized Structural Formulae of Phylosilicates

Layer Type	Group	Octahedral occupancy	Negative Charge per unit[a]	Unit formula			Layer thickness nm	Interlayer composition			Occurrence in soils
				Cations[b] Oct.	Tet.	Anions		Cations[c]	Hydroxide sheet[d]	Water	
1:1	Kaolinite	Di	0	Y_4	Z_4	$O_{10}(OH)_8$	0.7	None			Common
	(Halloysite)	Di	0	Y_4	Z_4	$O_{10}(OH)_8$	1.0-0.7	None		$0.6\text{-}4H_2O$	Common
	Serpentine	Tri	0	Y_6	Z_4	$O_{10}(OH)_8$	0.7	None			Rare
2:1	Pyrophyllite	Di	0	Y_4	Z_8	$O_{20}(OH)_4$	0.92	None			Rare
	Talc	Tri	0	Y_6	Z_8	$O_{20}(OH)_4$	0.9	None			Rare
	Micas	Di		Y_4	Z_8	$O_{20}(OH)_4$	1.00	X_2'			
		Tri	2	Y_6	Z_8	$O_{20}(OH)_4$	1.00	X_2'			Common
	Brittle Micas	Di		Y_4	Z_8	$O_{20}(OH)_4$	1.00	X_2''			
		Tri	4	Y_6	Z_8	$O_{20}(OH)_4$	1.00	X_2''			Rare
		Di		Y_4	Z_8	$O_{20}(OH)_4$	1.40		$A'''_4(OH)_{12}$		
	Chlorites	Di, tri	Variable	Y_4	Z_8	$O_{20}(OH)_4$	1.40		$A''_6(OH)_{12}$		Common
		Tri		Y_6	Z_8	$O_{20}(OH)_4$	1.40		$A''_6(OH)_{12}$		
	(Swelling chlorite)		Variable				≥ 1.40		$\sim A'''_4(OH)_{12}$	nH_2O	Common

Table 9-3 (Continued)

Layer Type	Group	Octahedral occupancy	Negative Charge per unit [a]	Unit formula Cations[b] Oct.	Tet.	Anions	Layer Thickness nm	Interlayer composition Cations[c]	Hydroxide sheet[d]	Water	Occurrence in soils
Smectites		Di	0.5-1.2	Y_4	Z_8	$O_{20}(OH)_4$	≥ 0.96	$X'_{0.5\text{-}1.2}$		nH_2O	Common
		Tri	0.5-1.2	Y_6	Z_8	$O_{20}(OH)_4$	≥ 0.96	$X'_{0.5\text{-}1.2}$			
Vermiculites		Di	1.2-1.9	Y_4	Z_8	$O_{20}(OH)_4$	≥ 0.94	$X'_{1.2\text{-}1.9}$		nH_2O	Common
		Tri	1.2-1.9	Y_6	Z_8	$O_{20}(OH)_4$	≥ 0.94	$X'_{1.2\text{-}1.9}$			
Palygorskite			?	Y_4	Z_8	$O_{20}(OH)_2(OH_2)_4$[c]		X?		$4H_2O$	Uncommon
Sepiolite			?	Y_8	Z_{12}	$O_{30}(OH)_4(OH_2)_4$		X?		$8H_2O$	Uncommon

[a] Negative charge per formula unit is twice that given by Bailey et al. (1971a) because the formula unit used here applies to the contents of a volume defined by the $a \times b$ unit cell base area that is one layer thick. For example, for chlorites this volume is approximately $0.5 \times 0.9 \times 1.4$ nm³, and for palygorskite it is $1.8 \times 0.5 \times 0.65$ nm³.

[b] Y represents cations in octahedral coordination, most commonly Al but many others of similar dimensions can substitute for it isomorphically. Z represents cations in tetrahedral coordination, most commonly Si, but Al frequently replaces some Si in a regular or random manner; e.g., replacement of one in four occurs in the micas and gives rise to the negative charge of two per formula unit; in the brittle micas 2 Al replace two of each four Si.

[c] X' represents a monovalent cation, and X'' a divalent cation.

[d] A''' is an irrelevant cation (usually Al) and A'' is a divalent cation (usually Mg) but extensive isomorphous substitution may occur.

[e] Following Bradley (1940); according to Gard and Follett (1968) the anions are $O_{20}(OH)_2(OH_2)_3$.

The dissolution of CO_2 from plant and microbial respiration processes also has an influence on the formation of clays. The following shows the alteration of calcium feldspar to phyllosilicates kaolinite with the aid of CO_2:

$$CaAl_2Si_2O_8 + 3H_2O + 2CO_2 \leftrightarrow Al_2Si_2O_5(OH)_4 + Ca^{++} + 2HCO_3^-$$

plagioclase   kaolinite

The changes in clay fraction mineralogy during the coverage of soil profile development are especially clear. The clay fraction differences at different weathering stages are evident in the **three Jackson-Sherman weathering stages**, as shown in Table 9-4.

Table 9-4. Jackson- Sherman Weathering Stages

Characteristic minerals in soil clay fraction	Characteristic soil chemical and physical conditions
	Early stage
Gypsum	Very low content of water and organic matter,
Carbonates	Very limited leaching
Olivine/ pyroxene/ amphibole	Reducing environments
Fe(II)-bearing micas	Limited amount of time for weathering
Feldspars	
	Intermediate stage
Quartz	Retention of Na, K, Ca, Mg, Fe(II), and silica:
Dioctahedral mica/ illite	Ineffective leaching and alkalinity
Vermiculite/ chlorite	Igneous rock rich in Ca, Mg, Fe(II), but no
	Fe(II) oxides
Smectites	Silicates easily hydrolyzed
	Flocculation of silica, transport of silica
	into the weathering zone
	Advanced stage
Kaolinite	Removal of Na, K, Ca, Mg, Fe(II), and silica:
Gibbsite	Effective leaching, fresh water
Iron oxides	Oxidation of Fe(II)
(goethite, hematite)	Acidic compounds, low pH
Titanium oxides	Dispersion of silica
(anatase, rutile, ilmenite)	Al-hydroxy polymers

After M.L. Jackson and G.D. Sherman, Adv. Agran 5, 219-318 (1953). M.L. Jackson, Soil Sci. 99, 15-22 (1965)

9.1.3 Agricultural Applications

Many valuable minerals useful for plant life can be obtained in soil due to the complexing property in soil. These minerals are termed **micronutrients**. The oxidation of pyrite in soil causes the formation of acid-sulfate soils or cat clays.

$$FeS_2 + \frac{7}{2}O_2 + H_2O \rightarrow Fe^{2+} + 2H^+ + 2SO_4^{2-}$$

Soils containing marine sediments (such as those from Florida, New Jersey, and North Carolina) are likely to have **acid soils**. A useful test for acid soils is the peroxide test.

$$FeS_2 + \frac{15}{2}H_2O_2 \rightarrow Fe^{2+} + 11^+ + 2SO_4^{2-} + 7H_2O$$

The level of sulfate and pH released will indicate the potential of acid soils. The adjustment is by adding lime to neutralize the soil.

In case of **alkaline soil**, due to the presence of Na_2CO_3, aluminum or iron sulfate can be added:

$$2Fe^{3+} + 3SO_4^{2-} + 6H_2O \rightarrow 2Fe(OH)_3^{3+} + 6H^+ + 3SO_4^{2-}$$

Sulfur also can be added in such a manner that sulfur-oxidizers in soil can convert it into sulfuric acid for neutralization.

Fertilizers consisting of N, P, and K are designated by numbers (e.g., 6-12-8) signifying that the concentration of N is 6%, P_2O_5 is 12%, and K_2O is 8%. Liquid NH_3, obtained by the Haber process, can be directly added to soil, or injected into the water. Ammonium nitrate, NH_4NO_3, is also a common fertilizer but can cause explosion. Super phosphate is more soluble than the parent mineral fluorapatite.

$$2Ca(PO_4)_3F + 14H_3PO_4 + 10H_2O = 2HF + 10Ca H_4(PO_4)_2 \cdot H_2O$$

In general, the fertilizers are all soluble in water; therefore, the runoff from land became a problem for subsequent eutrophication. The best solution is to have the fertilizers either graft or bond to soil humus for prevention of leaching capability by water. Alternatively, organic compounds containing N, P, and K can be made to minimize this potential. These are termed **hard fertilizer** in contrast to the water-soluble fertilizer, the **soft fertilizer**.

An important property of soil is ion exchange. Evidently, the transport of many metal ions is essential for plant life, not to mention the storage and distribution of water. It has been documented that about 4.5 tonnes of calcium ion can be stored in 1 hectare of land. The need for common fertilizer composed of potassium, nitrogen, and phosphorus is based on the slightly poor transport ability of soil due to the bulkiness of potassium ion and the weak coordination bond between soil grains and nitrate or phosphate group. Usually larger ions are not readily ion-exchangeable.

The 16 elements that are essential to plant growth in soil are H, B, C, N, O, Mg, P, S, Cl, K, Ca, Mn, Fe, Cu, Zn, and Mo. Of these, B, Cl, Mn, Fe, Cu, Zn, and Mo are micronutrients (absorbed in trace amount) and Mg, Ca, and S are secondary nutrients. The remaining 6 elements are macronutrients. Usually, the **enrichment factor in biomass** (EF_B) can be correlated by a log-log plot of EF_B versus the ionic potential (IP) for the metals. This type of plot is termed a Banin-Navrot plot. They are similar regardless of plant or animal origin.

An example is illustrated by Figure 9-13. Both Ef_B and IP are defined as follows:

$$EF_B = \frac{\text{concentration of elements in organism}}{\text{concentration of elements in crustal rock}} \qquad [9\text{-}1]$$

$$IP = \frac{\text{valence of free cation of the elements}}{\text{radius of free cation of the elements}} \qquad [9\text{-}2]$$

Other elements can be toxic to plants. The mechanism of phytotoxicity is still not understood. Potential toxic metals can act either by displacement of essential metal and complexation or reacting by modification to the metal. Strong complex forming metals usually have covalent characters. This approach can be qualified with a **Misono softness parameter** (Y) as follows:

$$Y = 10 \frac{I_Z R}{Z^{\frac{1}{2}}} I_{Z+1} \qquad [9\text{-}3]$$

when R is the ionic radius of metal ions whose valence is Z and whose ionization potential is I_Z (e.g. for Cu^{2+}, $I_2 = 1958$ KJ mole^{-1} and $I_3 = 3554$ KJ mole^{-1}). A classification of metals by plotting the ionic potential to Misono softness is seen in Figure 9-14. Therefore, the toxicity sequence in general is in line with the direction, the softest Lewis acid being the most toxic, as shown in Table 9-5. Some commonly used parameters for nutrient uptake for plant are listed in Table 9-6. C_s is the nutrient bulk concentration in soil (mol m^{-3} soil), and C_l is the nutrient soil solution concentration (mol m^{-3} solution).

Table 9-5. Representation Metal Toxicity Sequence

Organisms	Toxicity sequence
Algae	Hg>Cu>Cd>Fe>Cr>Zn>Co>Mn
Flowering plants	Hg>Pb>Cu>Cd>Cr>Ni>Zn
Fungi	Ag>Hg>Cu>Cd>Cr>Ni>Pb>Co>Zn>Fe
Phytoplankton (freshwater)	Hg>Cu>Cd>Zn>Pb

Based on Sposito, Ref 9-8

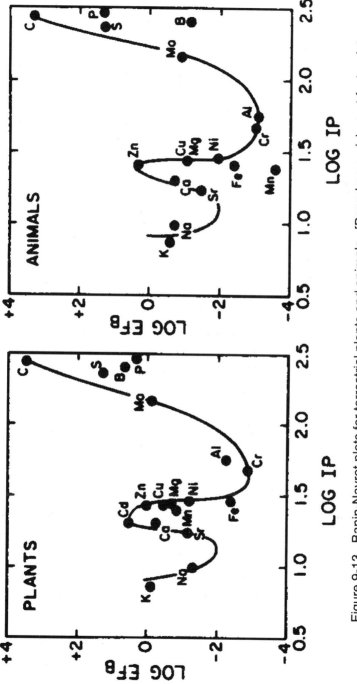

Figure 9-13. Banin-Navrot plots for terrestrial plants and animals. [Based on enrichment factor data compiled by A. Banin and J. Navrot, Origin of life: Clues from relations between chemical compositions of living organisms and natural environments, *Science* 189: 550-551 (1975).]

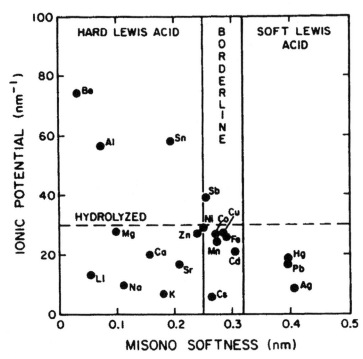

Figure 9-14. Classification of metals according to ionic potential and Misono softness. (From G. Sposito, Ref. 9-8).

Table 9-6. Representative Values of the Soil Parameters

Nutrient species	c_t (mmol m^{-3})	β_e/θ	D_e (10^{-10} m^2 s^{-1})	τ_D^b (days)
NO_3^-	3,000-11,000	1.0	2.5	1
$H_2PO_4^-, HPO_4^{2-}$	0.6-5.0	$10-10^3$	$10^{-4}-10^{-2}$	2×10^4
K^+	100-2,500	1-50	$10^{-2}-10^{-1}$	200
Ca^{2+}	500-5,000	10-100	$10^{-3}-10^{-2}$	2×10^3
Mg^{2+}	500-5,000	1-60	10^{-2}	200
SO_4^{2-}	700-1,500	2-10	1	1
$H_3BO_3^0$	10-3,500	1-3	2-4	1
Fe^{2+}	0.1-1.0	10^3	10^{-4}	2×10^4
Mn^{2+}	0.1-100	1-100	$10^{-3}-10^{-1}$	2×10^3
MoO_4^{2-}	0.01-0.2	10-2,000	$10^{-2}-1.0$	200

[a] Compiled from S.A. Barber, Soil Nutrient Bioavailability, Wiley, New York, 1984, after Sposito, 1989.
[b] $\tau_D=2\delta^2/D_e$, $\delta= 3$ mm, D_e = lower value in column 4.

$$\beta_e = \frac{dC_s}{dC_l} \qquad [9\text{-}4]$$

The ratio of the two quantities is termed the **nutrient buffer power**, β_e, usually measured by the slope of nutrient adsorption isotherm. If D_l is the nutrient diffusion coefficient in soil solution and

$$D_e = \frac{\theta D_l f}{\beta_e} \qquad [9\text{-}5]$$

where θ is the volumetric water content of the soil, f is an impedance factor (tortuosity), and D_e is called the **Nye diffusivity parameter**. Furthermore, if q is the surface excess of the nutrient and ρ_b is the dry bulk density of the soil, then β_e is related to the slope of nutrient adsorption isotherm.

$$\beta_e = \theta \left(1 + \frac{\rho_b dq}{dC_l} \right) \qquad [9\text{-}6]$$

A diffusion time constant, τ_D, is defined as

$$\tau_D = 2 \frac{\delta^2}{D_l} \qquad [9\text{-}7]$$

where δ is rhizosphere distance in mm.

9.1.4 Geotechnical Applications

Soil stabilization and soil erosion will be addressed now. Soil erosion is the removal of surface layers of soil by the agencies of wind, water, and ice. Erosion involves a process of both particles and transport by these agencies and is initiated by drag impact or tractor forces acting on individual particles of soil at the surface. Erosion may also occur along the stream bank where the velocity of the flowing water is high and the resistance of the bank materials is low. Piping or spring sapping is another type of erosion caused by seepage and emergence of water from the face of an unprotected slope.

The susceptibility of a soil to erosion is known as its **erodibility**. Some soils (e.g., silts) are inherently more erodible than others (e.g., well-graded sands and gravel). In general, increasing the organic content and clay size fraction of a soil decreases erodibility. Erodibility also depends on other factors including soil texture, past moisture content, void ratio, pH, and composition or ionic strength of the eroding water.

Other properties of soil include the storage capability of soil, which is actually caused by adsorption. Because soil can supply trace metal nutrient for plants, it is considered as a support for the plant kingdom. Soil is classified as an ion exchanger. It is possible to be used as a waste disposal method for treating radioactive metals. Soil can behave as a catalyst for the denitrification of many organic compounds. The silicate chemistry indeed exhibits a host of channeling and spacial sites in the many wonderful 3-dimensional tessellations.

A suggested hierarchy of erodibility based on the unified soil classification system gray symbols is as follows:

$$\text{most erodible} \rightarrow \text{least erodible}$$

$$ML > SM > SC > MH > OL \gg CL > CH > GM > GP > GW$$

This erodibility hierarchy is simple and is based on graduation and plasticity indices of remolded or disturbed soils. It fails to take into account effects of soil structure, void ratio, and antecedent moisture content.

Erosion of soils is a function of climate, topography, vegetative cover, soil properties, and the activities of animals and humans. A so-called **universal soil loss equation** (USLE) is commonly used to estimate sediment loading from surface erosion. It can be written as follows:

$$Y = \sum_{i=1}^{n} A_i (R, K, L, S, C, P, Sd) \qquad\qquad [9\text{-}8]$$

where Y = sediment loading from erosion (ton/yr)
n = number of subareas
A_i = acreage of subarea (acre)
R = rainfall factor
K = soil erodibility factor
L = slope-length factor
S = slope-gradient factor
C = cover factor
P = erosion control practice factor
Sd = fraction of total erosion which is delivered to stream

Suitable values for these parameters for a specific condition and area are commonly found in monographs and tables. For example, the C factor can be located in Table 9-7. R for a certain location can be found in Figure 9-15. Wischmeier has published a convenient nomograph that can be used to determine erodibility k-values of soils due to water erosion. The nomograph is valid for exposed subsoils at construction sites and farmlands. Five soil parameters are required:

- percent silt and very fine sand (0.002-0.10 mm)
- percent sand (0.10-2.0 mm)
- percent organic matter
- structure
- permeability

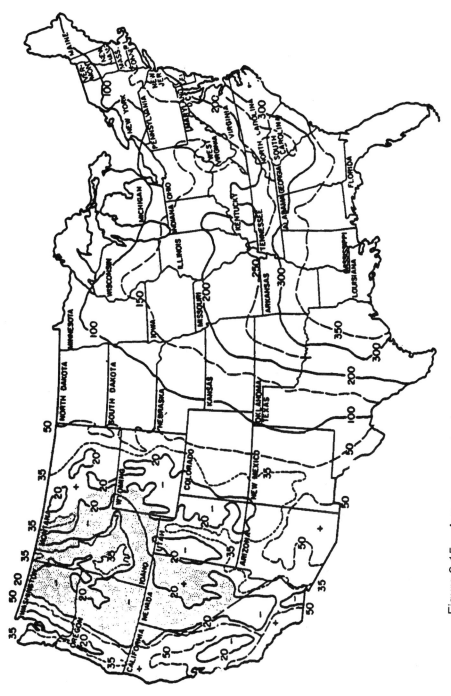

Figure 9-15. Average annual values of rainfall factor, R.

Table 9-7. Reductive Protection of Ground Cover Against Erosion: Values For C Factor

Land-User Groups	Examples	Range of "C" Values
Permanent Vegetation	Protected woodland	0.0001-0.45
	Prairie	
	Permanent pasture	
	Sodded orchard	
	Permanent meadow	
Established Meadows	Alfalfa	0.0004-0.3
	Clover	
	Fescur	
Small Grains	Rye	0.07-0.5
	Wheat	
	Barley	
	Oats	
Large-Seeded Legumes	Soybeans	0.1-0.65
	Cowpeas	
	Peanuts	
	Field peas	
Row Crops	Cotton	0.1-0.70
	Potatoes	
	Tobacco	
	Vegetables	
	Corn	
	Sorghum	
Fallow	Summer fallow	1.0
	Period between plowing and growth of crop	

U.S. EPA Office of Research and Development (ORD): *Loading Functions for Assessment of Water Pollution from Nonpoint Sources*. National Technical Information Service, Springfield, VA (1976), p. 59.

The first three parameters will often suffice to provide a reasonable approximation of the erodibility. This approximation can be refined by including information on permeability and soil structure as indicated on the nomograph. The soil erodibility, K, can be found in Figure 9-16 along with information on soil structure.

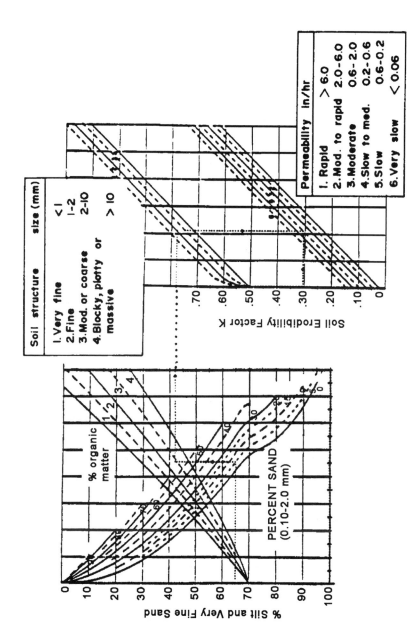

Figure 9-16. Soil erodibility nomograph for calculating values of K. [For example, assume a lab analysis indicates soil with: 65% silt and very fine sand; 5% sand (.10-2.0 mm); 3% organic matter; fine granular soil (1-2 mm); slow to medium permeability (0.2-0.6 in/hr). The sample k value is 31.]

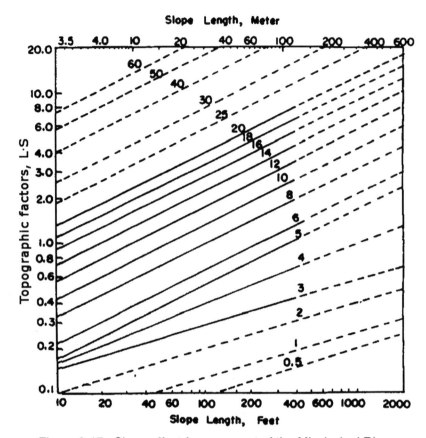

Figure 9-17. Slope effect for area east of the Mississippi River.

The topographic factor (LS) can be found in Figure 9-17. For example, for a plot of land east of the Mississippi River with a slope of 6% and a slope length of 250, the topographic value is 1.0%. The cover factor depends on the type of vegetation on the land. The P factor for a crop land with 6% slope and contour strip-cropping is 0.25 (see Table 9-8). The fraction of total erosion that is delivered to the sediment is available from the U.S. Department of Agriculture.

Various control technologies for erosion are available. The common ones for agriculture are terracing and diversions. In siliculture, good plans for road design and transportation are crucial for erosion control. Biological methods for using biopolymer as cementing agents are in the development stages.

According to **Mohr-Coulomb failure criteria**, rupture along a plane in a material occurs by a critical combination of normal and shear stresses and not by normal and shear stress alone. The relation between normal and shear stress on failure plane can be expressed as

$$\tau = c + \sigma \tan\phi \qquad\qquad [9\text{-}9]$$

where τ is shear stress at failure and σ is the normal stress on the failure plane. Cohesion is shown by c, and ϕ is the angle of friction. Because granular soils have no cohesion ($c=0$), chemical additives are used for increase of cohesion for improving the bonding space, as shown in Figure 9-18. The biopolymers from bacteria may be an alternative for this application in the future; for example, the application of slime-forming bacteria to the earth structures (such as shell and core of earth dams and embankments) can be similar to lime or ash stabilization. The procedure consists of adding a percentage of stabilizer to the soil layers during compaction. To enhance the strength of subsurface soil of a structure's foundation, pressure injection of slime-forming bacteria may be utilized. The same idea can be used for mitigating liquefaction during an earthquake, if the bacteria can be injected below the foundation.

Incidentally, coal mining, if not well designed, has multiple adverse impacts on the environment:

- disturbance of the land by strip mining with its attendant effect on streams
- acid mine drainage problems
- subsidence of the land resulting from less coal being left in place to support the overburden
- mine fires
- refuse banks from mining and coal preparation.

Good examples of these effects of the improper exploitation of natural resources can be illustrated by the whole commonwealth of Pennsylvania, the strip mining of the Dakotas, and the heap mining of copper in Arizona.

Table 9-8. "P" Values for Erosion Control Practices on Cropland

Slope	Type of Control Practice				
	Up and Downhill	Cross-slope Farming Without Strips	Contour Farming	Cross-slope Farming with Strips	Contour Strip Cropping
2.0-7	1.0	0.75	0.50	0.37	0.25
7.1-12	1.0	0.80	0.60	0.45	0.30
12.1-18	1.0	0.90	0.80	0.60	0.40
18.1-24	1.0	0.95	0.90	0.67	0.45

U.S. EPA Office of Research and Development (ORD): *Loading Functions for Assessment of Water Pollution from Nonpoint Sources*. National Technical Information Service, Springfield, VA (1976), p. 64.

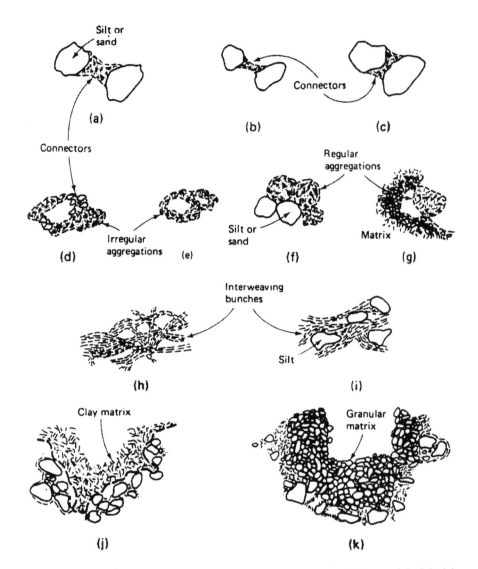

Figure 9-18.   Schematic representations of particle assemblages; (a), (b), (c) connectors; (d) irregular aggregations linked by connector assemblages; (e) irregular aggregations forming a honeycomb arrangement; (f) regular aggregations interacting with silt or sand grains; (g) regular aggregation interacting with particle matrix (h) interweaving bunches of clay; (i) interweaving bunches of clay with silt inclusions; (j) clay particle matrix; (k) granular particle matrix (after Collins and McGown, 1974, *loc. cit.*).

9.1.5 Biodegradation in Soil

Many soil bacteria, fungi, and animals, along with catalytic surfaces provided by clay minerals are capable of removing or detoxifying a variety of harmful substances, as shown in Figure 9-19. Figure 9-20 illustrates the decomposition of the organic matter and the formation of humic substances in soil. In addition, the pathway of the microbial degradation of cotton fiber is given in Figure 9-21. The unique ability of soil to detoxify and adsorb substances that would be considered pollutants if released to the air and water environments has caused it to be used extensively as a treatment medium for such waste materials. Land filling and land treatment techniques have been used in environmental engineering practices. Because the structure of synthetic compounds is getting more complex than the naturally occurring substances, the biodegradation rate of the xenobiotics is becoming much slower. The accumulation of these toxic materials in soils presents a great danger to us.

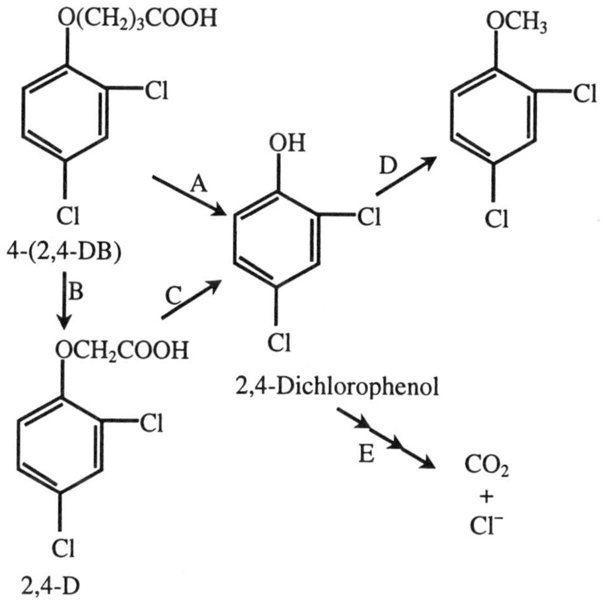

A: Defusing (*Flavobacterium*)
B: Activation (soil)
C: Detoxication (*Arthrobacter*, soil)
D: Addition reaction (*Arthrobacter*)
E: Degradation (*Pseudomonas*, soil)
Initial steps in the metabolism of several phenoxyalkanoate herbicides.

Figure 9-19. Degradation reactions in contrast to other processes.

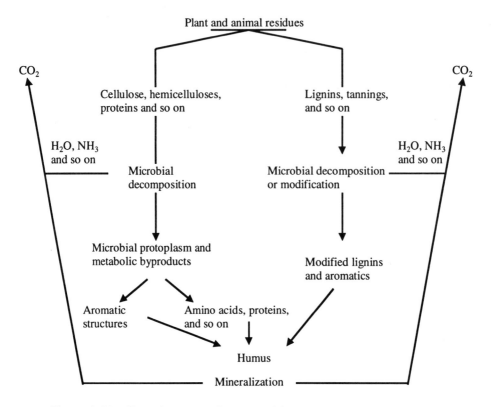

Figure 9-20. Organic matter decomposition and the formation of humic substances in soil (from Stevenson, 1964, with permission of the editor).

After soil is formed, it can undergo compaction, consolidation, pore pressure, lateral pressure and settlement. Most important is that soil is host to a great variety of microorganisms. There is an overall presence of enzymes in soil as detected by enzyme tests. In ecological language, the organisms are as syntrophism (e.g., commensalism, predation, and competition); they are not neutralism or amensalism (see the chapter on biological processes in this series of books). The final bioproduct is humus, which is refectoric.

9.2 GLOBAL SOIL POLLUTION

Sodium cyanide has long been known to be a very poisonous substance. But as shown in Table 9-9, it is not as toxic as some manmade organic materials. Dioxin (2,3,7,8-TCDD) has the lowest LD50 content based on mass. Humans also think that incineration is the safest way to destroy toxic materials. It is astonishing to find that there are still many toxic substances produced among which dioxin is one created from the incineration of wastes at high temperature as shown in Table 9-10. PCBs have been widely used as plasticizers, transformer fluids, lubricants, and hydraulic fluids. Although their application is now restricted, large quantities have been widely spread in our environment. They are widely present in a wide variety of marine creatures (refer back to Figs 3-4, 3-5, and 3-6) and inextricably entwined in the food chain.

A case study of dioxin contamination in the United States is presented here. Dioxin (2,3,7,8-TCDD), is a byproduct of herbicides, some of them used during the Vietnam War as shown in Table 9-11. It was later found to be an extremely toxic material that causes death, birth defects, etc. Many people have been exposed to it, as demonstrated by Table 9-12.

Table 9-9. Toxicities of Selected Poisons

Substance	Molecular Weight	Minimum Lethal Dose (mol/ kg)
Botulinum Toxin A	9.0×10^5	3.3×10^{-17}
Tetanus Toxin	1.0×10^5	1.0×10^{-17}
Diptheria Toxin	7.2×10^4	4.2×10^{-12}
2,3,7,8-TCDD[a]	322	3.1×10^{-9}
Saxitoxin	372	2.4×10^{-8}
Tetrodotoxin	319	2.5×10^{-8}
Bufotoxinc[b]	757	5.2×10^{-7}
Curare	696	7.2×10^{-7}
Strychnine	334	1.5×10^{-6}
Muscarinc[c]	210	5.2×10^{-6}
Dtisopropylfluorophosphate	184	1.6×10^{-5}
Sodium Cyanide	49	2.0×10^{-4}

[a] Source: Pland and Kende. These data were compiled by Mosher et al. and the values indicate only relative toxicity. It should be noted that the values deal with different species, routes of administration, survival times and, in one case, mean lethal dose rather than minimum lethal dose. Except where noted, administration was by the intraperitoneal route in mice.

[b] LD_{50} on oral administration in the guinea pig.

[c] Intravenous injection in the cat.

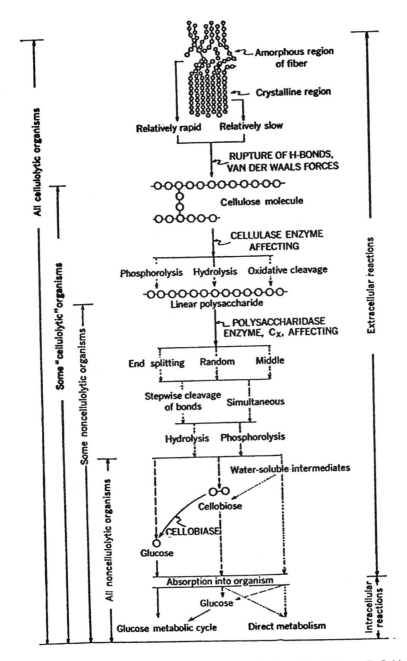

Figure 9-21. The microbial degradation of cotton fiber (From R.G.H. Siu, "Microbiological Decomposition of Cellulose," Reinhold Publishing Corporation, 1951).

Table 9-10. HRGC/LRMC Analytical Results for CDD/CDF in Ash Sample from Incineration of Waste Containing PCP[a]

CDD/ CDF	Total Number of Apparent Isomers	Total Detected (ng/ g)	Minimum Detectable Concentrated (ng/ g)
MCDF	3	75	0.1
DCDF	8	25	0.3
TrCDF	8	15	0.6
TCDF	7	7	0.5
PCDF	5	8	1
HxCDF	5	5	1
HpCDF	2	6	1
OCDF	1	2	1
MCDD	1	1	0.1
DCDD	4	5	0.3
TrCDD	5	2	0.6
TCDD	4	4	0.2
PCDD	5	32	1
HxCDD	5	81	1
HpCDD	2	117	1
OCDD	1	198	1

[a] Recent analysis by Wright State University. The essential points of analysis are:

- Extraction of CDD and CDF from the sample matrix using organic solvents.

- Preliminary separation of CDD and CDF from other constituents of the matrix, which also have been extracted (including other chlorinated materials) using acid-base treatments and liquid chromatography with alumina, silica gel, and /or other suitable columns.

- Further fractionation of the CDD and CDF using normal and reverse-phase high-performance liquid chromatography (HPLC).

- Analysis of the prepared abstracts containing CDD and CDF using capillary-column gas chromatography, mass spectrometry (GC-MS). Both low- and high- resolution mass spectrometry may be used in this analysis, depending on the sample.

Table 9-11. Agent Orange Had Far Less Dioxin Than Earlier 2,4,5-T herbicides Used in the Vietnam War

Code Name	Herbicide	Quantity, Gal	Period of Use	2,3,7,D-TCDD, PPM
Orange	2,4-D; 2,4,5-T	10,646,000	1965-1970	1.98
White	2,4-D; pilioram	5,633,000	1965-1971	—
Blue	Cacodylic acid	1,150,000	1962-1971	—
Purple	2,4-D; 2,4,5-T	145,000	1962-1965	32.8[a]
Pink	2,4,5-T	123,000	1962-1965	65.6
Green	2,4,5-T	8,200	1962-1965	65.6
Total:		17,705,200		

[a] Assumed level from one known and four probable samples of purple. Note: pink and green levels are twice that of purple because they were full-strength 2,4,5-T. Sources: Proceedings from 2nd Coninuing Education Conference on Herbicide Orange, May 1980; and Air Force OEHL Technical Report on Toxicology, Fate and Risk from Agent Orange and Dioxin, October 1978.

Sources: Proceedings from 2nd Continuing Education Conference on Herbicide Orange, May 1980; and Air Force OEHL Technical Report on Toxicology, Fate and Risk from Agent Orange and Dioxin, October 1978.

Table 9-12. More Than 800 Workers Have Been Exposed to Dioxin in Industrial Accidents

Date	Exposed	Location of Accident	Remarks
1948	250	Monsanto's 2,4,5-trichlorophenol plant in Nitro, West Virginia.	122 cases of chloracne being studied; so far, 32 deaths vs. 48.4 expected; no excess deaths from malignant neoplasms or circulatory disease; studies continue.
1963	75	BASF's 2,4,5-trichlorophenol plant at Ludwigshafen, West Germany	55 cases of chloracne, 42 severe; 17 deaths so far vs. 11 to 25 expected (4 gastrointestinal cancers and 2 oatcell lung cancers); most common injuries were impaired senses and sliver damage; studies continue.
1956	?	Rhone-Poulenc's 2,4,5-trichlorophenol plant in Grenoble, France	17 cases of chloracne, also elevated lipid and cholesterol levels in blood.
1963	106	NV Philips' 2,4,5-T plant in Amsterdam, The Netherlands	44 chloracne cases (42 severe), of whom 21 also had internal disturbances; 8 deaths so far (6 possible myocardial infarctions); some symptoms of fatigue; full report planned.
1964	61	Dow Chemical's 2,4,5-tricholorphenol plant at Midland, Michigan	49 cases of chloracne; deaths so far 4 vs. 7.8 expected, 3 cancer deaths vs. 1.5 expected, one a soft tissue sarcoma; studies continue

Table 9-13.   2,3,7,8-TCDD Is One Compound in a Family

All dibenzo- -dioxins have a three-ring structure consisting of two benzene rings connected by oxygen atoms:

And 2,3,7,8-tetrachlorodibenzo- -dioxin is one of the 75 possible chlorinated dioxins:

Related are chlorinated dibenzofurans:

Dioxin precursors combine to form dioxin in the general reaction:

For example, 2,3,7,8-TCDD is the most likely result from the reaction of 2,4,5-trichlorophenol:

Chlorophenols can be formed from phenols (by combustion) and chloride by oxidation to chlorine.

Table 9-13 shows the pathway of dioxin production. Two major types of chlorinated compounds are formed: the chlorinated dibenzo dioxins (CDD) and the chlorinated dibenzo furans (CDF). Dioxin is very refractory and very resistant to biodegradation. Referring to Figure 9-22 for persistence of pesticides in soil, the half-life for dioxins in soil was thought to be less than a year, but in the case of dioxin concentration in Missouri, it was found that high residual dioxin concentration persists even after 15 years. The detailed mechanism of dioxin's effect on cellular biology is not yet fully understood, as shown in Figure 9-23.

The dioxin's precursor comes from chlorinated aromatics. Most of these are the remains of pesticides and insecticides. Most chlorinated insecticides left in the soil will yield these precursors for CDD or CDF (e.g., see Table 9-10). Some common chlorinated insecticides are shown in Figure 9-24.

An interesting study was made to determine the sperm density in the United States as a function of time, as shown in Table 9-14. The results are frightening, for the sperm density is decreasing sharply. The decrease may be a result of the exposure to the toxic substances shown in Table 9-15. Here we have another hint that if pollution cannot be properly controlled, the human species may be extinct in the near future.

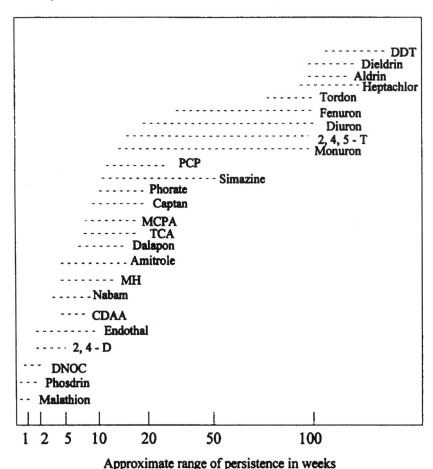

Figure 9-22. Persistence of pesticides in soil.

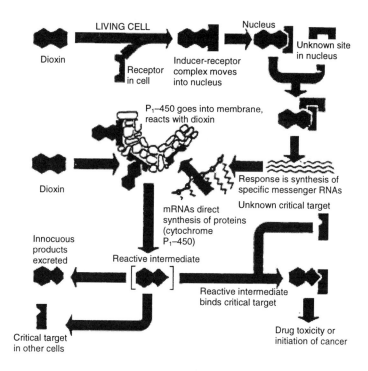

Figure 9-23. Dioxin's effects on cellular biology.

Table 9-14. Sperm Density in United States Males As a Function of Time

Year	Number of Cases	Mean Sperm Density (10^6 cell/ ml)	Comments	Reference
1929	271	100[a]		7
1938	200	120[a]	0.5% had $<20 \times 10^6$/ ml	8
			25% had $<60 \times 10^6$/ ml	
1949	49	145	Range = $50\text{-}321 \times 10^6$/ ml	9
1950	100	101		10
1951	1,000	107	29% had $<60 \times 10^6$/ ml	9
			44% had $>100 \times 10^6$/ ml	
1974	386	48		
1975	1,300	79	Median = 65×10^6/ ml	12
1975	100	81	Median = 61×10^6/ ml	5
1977	4,122	70	Median = 50×10^6/ ml	6
			23% had $<20 \times 10^6$/ ml	
1979	22	62		13
	24	22[b]		

[a] Wives were pregnant at the time.

[b] Occupational dibromochloropropane exposure for 60 days or more.

Table 9-15. Statistical Parameters Relating Sperm Density to Toxic Substances

Precursor Structures	m/z	Number of Chlorines	Slope (SD/ Intensity)	R Squared Change	Signific- ance
	427	6	0.22	0.048	0.040
	241	3	-0.11	0.054	0.010
Heptachlorobiphenyl	373	6	-0.15	0.037	0.005
	343	4	-0.16	0.021	0.005
	230	5	-0.01	0.017	0.006
Trichlorophenol	231	4	0.08	0.012	0.007
Hexachloronaphthalene	313	5	0.01	0.011	0.009
Pentachlorophenol	263	5	0.00	0.013	0.010
Hexachlorobiphenyl	339	5	-0.10	0.005	0.015
	463	7	-0.04	0.008	0.020
	318	3	-0.09	0.015	0.021
Tetrachlorodiphenylether (?)	287	3	-0.09	0.008	0.026
Hexachlorobenzene	282	6	0.04	0.005	0.035
	212	3	0.20	0.003	0.049
	239	4	0.03	0.003	0.065
	241	3	0.02	0.002	0.087
	224	3	-0.14	0.002	0.115
	251	5	-0.12	0.001	0.152
Pentachlorobiphenyl	305	4	-0.09	0.001	0.198
Octachlorobiphenyl	407	7	-0.03	0.001	0.251

9.2.1   DDT and PCB

We have noticed the global pollution of various organochloro-compounds even in pristine portions of land such as an isolated island without human activities, as shown in Table 9-16. Chemically DDT stands for 2,2-di-p-chlorodiphenyl-1,1,1-trichloroethane (see Fig. 9-24 for structure). It is readily synthesized with high yield in the laboratory (it used to be a standard organic chemistry preparation for students taking their first course in organic chemistry) by a Friedel-Crafts reaction of trichloroacetaldehyde and chlorobenzene.

Table 9-16. DDT and PCB Residues in Oceanic Birds and Fishes

Species	Locality	DDT (Wet Wt., ppm)	PCB (Wet Wt., ppm)
Shearwater	Mexico	3.0	0.4
	California	11.3	1.1
	New Brunswick	40.9	52.6
	California	32.0	2.1
	New Brunswick	70.9	104.3
Petrel	Bermuda	6.4	—
	California	66.0	24.0
	Mexico	9.2	1.0
	Mexico	3.2	0.35
	Baja California	953	351
	New Brunswick	164	192
	New Brunswick	199	697
Anchovy	San Francisco Bay	0.33-0.59	—
	Monterey	0.90	—
	Morro Bay	0.74	—
	Port Hueneme	3.04	—
	Los Angeles	14.0	1.0
English sole	San Francisco Bay	0.19-0.55	0.05-0.11
	Monterey	0.76	0.04
Shiner perch	San Francisco Bay	1-1.4	0.4-1.2
Jack Mackerel	Channel Islands	0.56	0.02
Hake	Puget Sound	0.18	0.16
	Channel Islands	1.8	0.12
Bluefin tuna	Mexico	0.22-0.56	0.04
Yellowfin tuna	Galapagos Islands	0.07	—
	Central America	0.62	0.04
Herring	Baltic Sea	0.68	0.27
Plaice	Baltic Sea	0.018	0.017
Cod	Baltic Sea	0.063	0.033
Salmon	Baltic Sea	3.4	0.30

Source: Robert M. Garrels et al., *Chemical Cycles and the Global Environment*, p. 139; adapted from Robert W. Risebrough, "Chlorinated Hydrocarbons," in D. H. Hood edition, ed., *Impingement of Man on the Oceans* (New York: Wiley-Interscience, 1971), pp. 259–286.

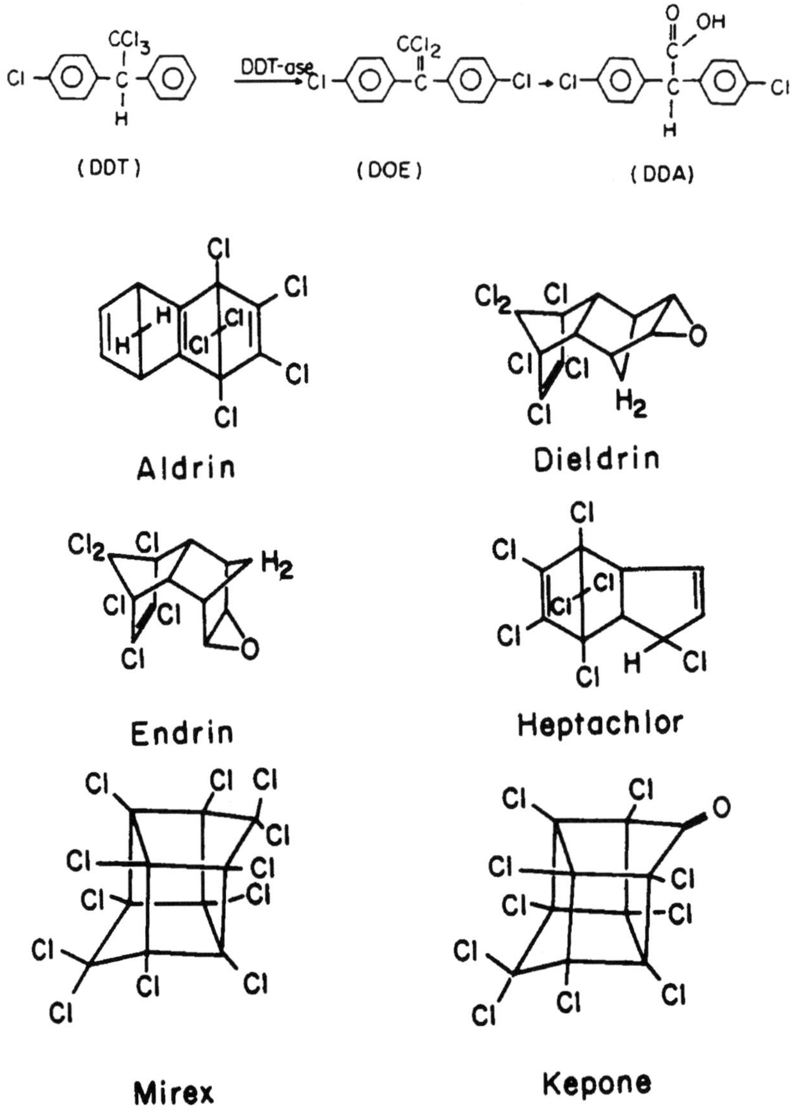

Figure 9-24. Organochlorine insectides. The top row shows the transformation of DDT.

In World War II, massive amounts of DDT were used to combat insect-borne diseases such as malaria, typhus, and diarrhoea. It was so effective against mosquitoes, lice, and flies that in a short time these insects were totally wiped out, saving many lives. Similar to the miracle sulfa drugs that were in vogue during World War I, the deaths due to infectious disease were greatly reduced.

The metabolic product of DDT is DDE, 2,2-di-p-chlorodiphenyl-1,1-dichloroethylene. This is a simple dehydrochlorination reaction. DDT can be transformed to DDE, which is nontoxic, by an insect enzyme called DDT-*ase*, and further into DDA, as demonstrated in Figure 9-24. For industry production of DDT, there are frequently isomers in which the o-positions, rather than the p-positions, are substituted by chlorine as by-products. These isomers can act as estrogen mimics and can interfere with normal sex hormones. Development of breast and ovarian cancers in females and demasculinization in males is the consequence.

Because a total ban or severe restrictions have been imposed on DDT use for the past quarter of a century, the persistent nature is well known. Almost daily we still have intake of DDT in some quantity without exceptions. Table 9-17 indicates the DDT exposure of an average meal, whether taken from restaurants or elsewhere.

PCBs in general are a mixture of the isomers of polychlorinated biphenyls originating from the chlorination of biphenyl under Lewis acid conditions. The number of chlorine atoms introduced into the biphenyl system can range from 1 to 10 (10 being perchlorobiphenyl) and can also be in any position. There are 209 chlorinated biphenyls possible, and these are termed **congeners**. The trade name of Aroclor by Monsanto has been commonly adopted for the formulations of dielectric fluids. Usually, there is a four-digit code associated with each Aroclor product, for example, Aroclor 1242. The first two digits (12) indicate that 12 carbons are in the system (the system includes naphthalene, e.g., Aroclors 1016). The last two digits indicate weight percentage of chlorine in the system. In the case of Aroclor1242 there is 42% chlorine in sample. Analytically, GC-MS can pin down most congeners. If one of the ortho-positions is unsubstituted by chlorine for PCBs, it can be consequently oxidized followed by the formation of PCDFs (polychlorinated dibenzofurans) (refer to Table 9-13).

Incidents of human poisoning from PCBs are well documented. The Yusho poisoning in Japan in 1968 affected 2000 people. There were similar incidents involving cooking with PCB-containing rice oil in Yu-Cheng, Taiwan in 1978. Most patients develop an acne-like rash, feelings of fatigue, joint pains, and some liver cancer.

Table 9-17. Estimated Daily Content of DDT and DDE in Complete Meals in the U.S.

Year Location	Source	Number	DDT	DDE[a] as DDT	Total[a] as DDT	DDE as DDT (% of total)
1953-1954						
Wenaichee, Washington	Restaurant	18	0.178	0.102	0.280	37
Tacoma, Washington	Prison	7	0.116	0.063	0.179	55
1954-1955						
Talahassee, Florida	Prison	12	0.202	0.056	0.258	21
1956-1957						
Walla Walla, Washington	College dining room for meat abstainers	11	0.041	0.027	0.068	39
1959-1960						
Anchorage, Alaska	Hospital	3	0.184	0.029	0.213	14
1961-1962						
Washington, D.C.						
Baltimore, Maryland						
Atlanta, Georgia	Market basket survey	36[b]	0.026[c]	0.017[c]	0.045[c]	40[c]
Minneapolis, Minnesota						
San Francisco, California						
1962-1964						
Wenaichee, Washington	Restaurant	12	0.038	0.049	0.087	56
Wenaichee, Washington	Household	17	0.514	0.193	0.507	40

Table 9-17. (Continued)

Year Location	Source	Number	DDT	DDE[a] as DDT	Total[a] as DDT	DDE as DDT (% of total)
1962-1964						
Atlanta, Georgia						
Baltimore, Maryland						
Minneapolis, Minnesota	Market basket survey	25[b]	0.023[c]	0.013[c]	0.036[c]	36[c]
St. Louis, Missouri						
San Francisco, California						
1964						
Baltimore, Maryland	Market basket survey	1[b]	0.025[c]	0.017[c]	0.040[c]	43[c]

[a] Total daily content in milligrams

[b] This figure refers to the number of diet samples, each consisting of the total normal 14-day food intake for males 16 to 19 years old, which were tested. In some instances, additional diet samples were taken and aliquots were analyzed for pesticide content of various classes of foodstuffs, but no compensate value was given.

[c] The author did not calculate the daily DDT or DDE intake. However, using the author's mean dietary concentrations of DDT and DDE and the mean daily food intake of 3.78 kg from the market basket survey, the reviewer has calculated the values shown.

(Source: from R.A. Horne, *Chemistry of Our Environment*, Wiley-Interscience, 1978.)

$$2R-Cl + 2\,Na \longrightarrow R-R + 2NaCl$$

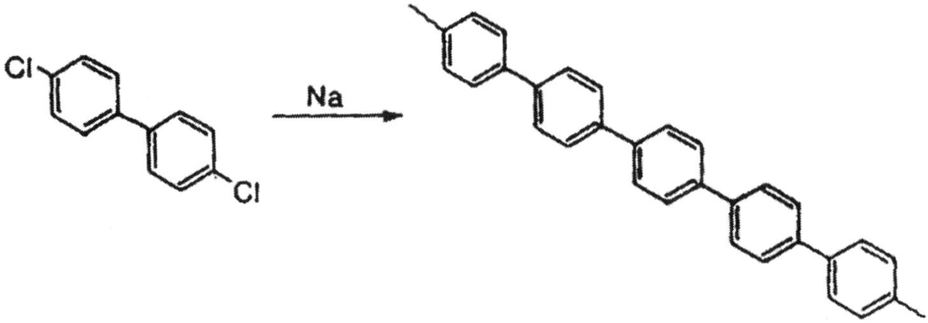

Figure 9-25. Wurtz reaction.

Polybrominated biphenyls (PBBs) behave similarly to PCBs. PBBs have been used as fire retardants. When PBBs are mistakenly added to cattle feed, as in the 1973 Michigan incident, the cattle have to be destroyed as well as the dairy products exceeding $100 million.

The total destruction of PCBs is often to the "six nines" level — that is, at 99.9999% level. A high temperature incineration method including plasma techniques is needed. Other chemical dechlorination methods, including the Wurtz reaction with sodium metals, are also possibilities, as shown in Figure 9-25. A photochemical method, ultrasound, as well as the use of anion radicals, is another way.

$$ArCl \xrightarrow{h\nu} Ar + Cl$$

$$ArCl + amine \rightarrow amine^{+}\bullet + ArCl^{-}\bullet$$

$$ArCl^{-}\bullet \rightarrow Ar\bullet + Cl^{-}$$

$$ArCl^{-}\bullet + H^{+} \rightarrow ArH + Cl$$

$$Na + ArH \rightarrow Na^{+} + ArH^{-}\bullet$$

$$C_{10}H_8{}^{-}\bullet \,(\text{naphthalenide, radical ion}) + ArCl \rightarrow C_{10}H_8 + ArCl^{-}\bullet \rightarrow Ar + Cl^{-}$$

Electrochemically generated anion radicals in micellar form at alkaline conditions in isopropyl alcohol also are used.

$$ArCl^-• \rightarrow Ar + Cl^-$$

$$Ar + (CH_3)_2CHOH \rightarrow ArH + (CH_3)_2C(OH) •$$

$$(CH_3)_2C(OH)• + OH^- \rightarrow (CH_3)_2CO^-• + H_2O$$

$$(CH_3)_2CO^-• + ArCl \rightarrow (CH_3)_2 = O + ArCl^-•$$

Another type of dechlorination includes nucleophilic displacement; for example, using polyethylene glycol with KOH,

$$ArCl + 2KOH \rightarrow ArOK + KCl + H_2O$$

In this case, stabilization of OH^- is done by placing the reaction as in a crown ether cavity in the aprotic medium. Finally nucleophilic displacement for aryl chloride dechlorination may involve a benzene intermediate with triple bonds in the aryl rings and the subsequent addition of water to form phenyls. In this regard, $R_3Si - PR_2$ can also be used.

$$R_3Si - PR_2 + ArCl \rightarrow ArPR_2 + R_3SiCl$$

9.3 PESTICIDES AND STOCKPILE WASTES

Soil pollution could be typified as the malfunctioning of soil as an environmental component following its contamination with certain compounds, particularly as a result of human activities. Pesticides are a major source of soil pollution. Hence, the discussion in this section will focus on them. Insects share plant foods with us, but some of them are carriers of devastating diseases. Prior to 1940, the majority of pesticides were organic materials such as arsenate, lead, and mercury. They are often referred to as **first generation pesticides,** which are highly poisonous to insects. They are, however, highly poisonous to humans and animals as well. The discovery of DDT, which possesses insecticidal properties and relatively low human toxicity, marks the era of **second generation pesticides**. Tremendous efforts were made by large numbers of chemists to synthesize various organic pesticides. DDT was banned in the

United States in 1973 because (1) it is highly **persistent**, meaning that it does not degrade in a reasonable period of time, and (2) it is oil soluble and hence accumulates in fat tissues, consequently building up in the food chain. A detailed discussion of this subject has been discussed previously.

TCDD (2,3,7,8- tetrachlorodibenzo-p-dioxin), a by-product in the manufacture of the herbicide 2,4,5-trichlorophenoxyacetate (2,4,5T) from 2,4,5 trichlorophenol, causes birth defects, skin disorders, liver damage, and other adverse effects (refer to Table 9-13).

Several nonpersistent classes of pesticides have been developed. Organophosphates and carbamates are the two major representative classes. The organophosphates all have a phosphorous atom connected by a double bond to either a sulfur or an oxygen atom and also by single bonds to oxygen or sulfur atoms with attached organic groups, as shown in Figure 9-26. The structure of carbamates is characterized by their having a carbon atom connected by a double bond to an oxygen atom and by single bonds to oxygen on one side and nitrogen on the other, each with attached organic groups. Both organophosphates and carbamates can react with water and oxygen to be decomposed within a few days once exposed in the environment. The LD_{50} and mode of action of these insecticides are listed in Table 9-18.

It is also interesting to note that the herbicide paraquat and its metabolite are related to 1-methyl-4-phenyl-1,2,3-tetrahydropyridine (MPTP), and MPTP can induce a kinsonian-like state in humans. Thus, pesticides may be causative agents for Parkinson's disease, as shown in Figure 9-27.

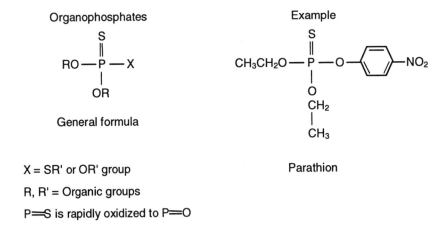

Figure 9-26. Organophosphates; for example, parathion.

Table 9-18. Lethal Dosage (LD_{50} Oral) for Pesticides in Test Rats

Pesticides	LD_{50} oral (mg/ kg)
Insecticides	
Aldrin	54-56
Arochlor (PCB)	250
Carbaryl	540
DDT	420-800 (60-75 dog)
Dichlorvos	56-80
Dieldrin	50-55
Endrin	5-43
Heptachlor	90
Lead arsenate	825 (192 sheep)
Lindane	125-200
Malathion	480-1500
Methoxychlor	5000-6000
Mirex	300-600
Nicotine	50-60
Parathion	4-30
Paris Green	22
Pyrethrins	820-1870
Herbicides	
2,4-D	666
Dioxin (impurity in 2,4,5-T)	0.03
Pentachlorophenol	27-80
Sodium arsenite	10-50
2,4,5-T	300 (100 dog)
Fungicides	
Copper sulfate	produces jaundice in sheep and chickens upon several months' exposure
Mercurials (ethylmercury-p-toluenesulfonanilide)	100

CARBAMATE PESTICIDES

$[C (= O) - NH -]$ or $[C (C\text{-}OH) = N\text{-}]$

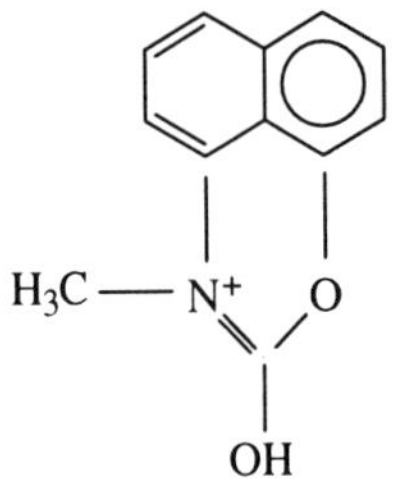

CARBARYL
(SEVIN]

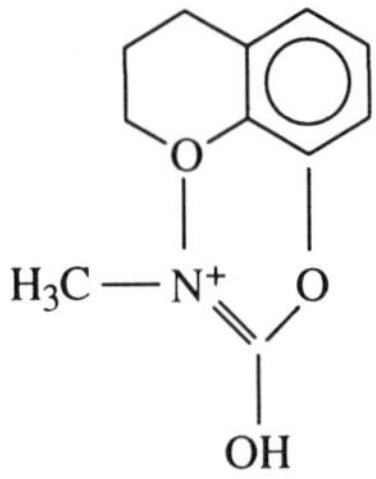

CARBOFURAN
[FURA DAN)

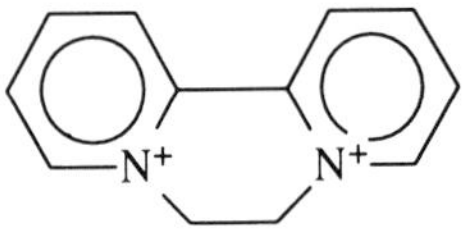

DIQUAT

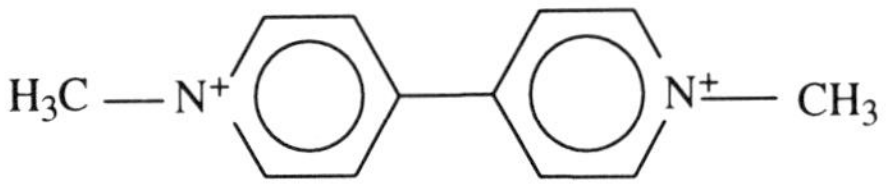

PARAGUAT

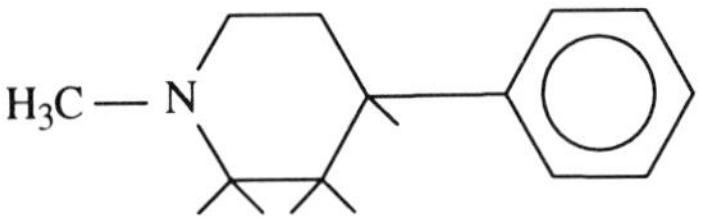

MPTP (1 - methyl - 4 - pheyl - 1, 2, 3, 4 - Tetra hydropyridine)

(CAUSE OF PARKINSON'S DISEASE)

Figure 9-27. Carbonate [-C(=O)-NH-] pesticides and MPTP.

9.3.1 The Third Generation Insecticides

In addition to the persistence of pesticides creating environmental problems, insect resistance has plagued producers of pesticides, resulting in the development of a new class of pesticides. The aims of third generation pesticides are to operate more selectively and to be composed of as natural compounds as possible. It was found that juvenile hormones can regulate growth and, if applied externally, prevent insects from developing. On the other hand, the molting hormone will hasten the insects' aging process, hence, shortening their life span. Pheromones are molecules that act as messengers between insects, guiding them to each other, to food supplies or to avoidance of dangerous places. By applying sex pheromones, sex attractants, male insects can be excited to such a frenzy that they cannot find mates. Table 9-19 lists some insect sex pheromones while Table 9-20 tabulates some insect attractants.

Another way to control the insect population is sterilization. Male insects can be sterilized by **chemosterilants** or radioactivity. When these sterile insects are released to mate with females, they often displace potent ones, greatly reducing the population. Aziridine is one of the commonly used chemosterilants. Another group of chemicals that are useful in controlling insect damage to crops is the **antifeeding compounds**. The pests will die of starvation because they are prevented from eating the protected crops. Plants usually have some natural antifeeding compounds; for example, drismane was always found in the analysis of petroleum and coal compounds. Another pest control technology is the introduction of predators or diseases that may eliminate pests. Extensive care should be taken to ensure that they will not turn to other prey or hosts when the species to be controlled is gone. Another biological means is to introduce genetic defects into a species. In some cases, winter hibernation has been eliminated, insuring that the insects will perish during the cold winter.

Plants and animals are usually able to generate various chemicals for different purposes. For example, Figure 9-28 shows volatile substances found in *Acanthomyops Claviger*. **Interorganismic chemical effects** can be categorized into allelochemic effects (**interspecific** interactions or allelopathy) and **intraspecific** allomones (**intraspecific** interactions). Allelochemic effects (such as repellents, escape substances, venoms, and suppressants) have an adaptive advantage to the producing organism. Interspecific allomones include kairomones such as attractants, inductants, signals, and stimulants, and give an adaptive advantage to the receiving organism.

Table 9-19. Examples of Insect Sex Pheremones

Insect	Pheremone
Gypsy moth	$CH_2CH(CH_2)_4CH$—$CH(CH_2)_9CH_3$ with CH_3 and O (epoxide) substituents cis (disparture)
Fall armyworm moth	$CH_3COCH_2(CH_2)_7CH{=}{=}CH(CH_2)_8CH_3$ cis
Cabbage looper moth	$CH_3COCH_2(CH_2)_8CH{=}{=}CH(CH_2)_9CH_3$ cis
European corn borer moth[a] Red-handed leaf roller moth[b] Smartweed borer[c]	$CH_3COCH_2(CH_2)_9CH{=}{=}CHCH_2CH_3$ cis
Oriental fruit moth[d]	$CH_3COCH_2(CH_2)_4CH{=}{=}CH(CH_2)CH_3$ cis
Pine beetle	(structures: two isopropenyl/diol structures labeled with CH₂, HO, H₃C, CH₃; and a bicyclic terpene with CH₃, CH₁₀, CH₁, OH)
Western pine beetle	(bicyclic oxane structures with O, —CH₂CH₃, CH₃ and CH₃) cis and trans

[a] The Iowa variety requires 4% trans-isomer while the New York variety requires 97% trans-isomer for maximal activity.

[b] 6-7% trans-isomer is required for maximal activity.

[c] 50% trans-isomer is required for activity.

[d] 8% trans-isomer is required for maximal activity.

Table 9-20. Insect Attractants

Insect attracted	Attractant	
Mediterranean fruit fly		(Siglure)
Mediterranean fruit fly		(Medlure)
Melon fly		
Ants		
Sugar beet wireworm	$CH_3CH_2CH_2CH_2CO_2H$	
June beetle		

Table 9-21 lists some allelopathic interactions. Pheromones, which are signals for reproductive behavior, social regulation, control of caste differentiation, alarm and defense, territory and trail marking, and food location, are examples of intraspecific chemical effects. A good hunting dog rolls itself in mud to disguise its scent before hunting. Understanding these interorganismic chemical effects is helpful when designing a well-integrated pest management plan.

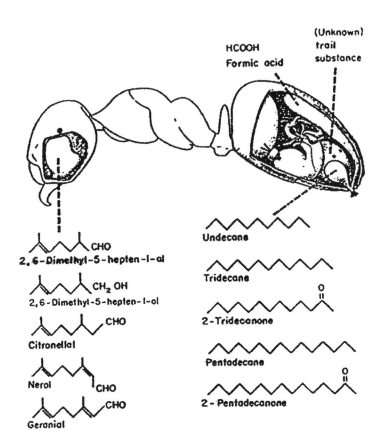

Figure 9-28. The structural formulas of volatile substances found in various exocrine glands of *Acanthomyops claviger*. Terpenes are located in the mandibular gland in the head, and alkenes and keytones in Dufour's gland of the abdomen. All but pentadecane and 2-pentadecanone are efficient alarm substances. (From Wilson, 1970, with permission of Academic Press, Inc., and the author.)

A summary of the third-generation insecticides are listed here:
(1) Attractants — sex pheromones e.g., hexalure (for pink moth)
   (a) to trap male
   (b) for male confusion
   (c) for mating of sterilized female
(2) Juvenile hormones:
     dormancy for linden bug — keep it becoming mature
(3) Molting hormone:
     accelerate to become over-mature antijuvenile hormones e.g., precocenes (for controlling the larva stage)

(4) Insect predators
     bacteria and viruses:
         Bacillus propillial against Japanese beetle
(5) Antifeeding compounds:
     drismane

Table 9-21. Some Allelopathic Interactions

Source	Target	Chemical Nature
Plant versus Plant		
Hard chaparral	Competing herbs	Phenols
Soft chaparral	Competing herbs	Terpenes (camphor, cineole, etc.)
Walnut tree	Competing plants	Jugione
Grass (aristida eligantha)	N-fixing bacteria and green algae in soil; lack of N discourages other plants	Phenolic acids
Eucalyptus	An example of self-toxicity	
Plant versus Animal		
Buttercup	Grazing animals	Protoanemonia
Larkspur	Grazing animals	Neurotoxic alkaloids such as delphinine
Tobacco	Aphids	Nicotine
Foxglove	Vertebrates	Steriod cardiac glycomides
Oleander	One leaf can be fatal to humans	
Oak tree	Vertebrates, moth larvae and also fungi and viruses	Protein-binding tannins
Balsam fir	Insects	Hormones and hormone-analogue
Hypericum	Herbivores (some beetles [Crysolina] detoxify hypericia)	Hypericia (causes intense photosensitivity and blindness)
Cruciferae	Cabbage butterflies, moths, weevils, beetles and aphids	Mustard oils such as allyl isothiocyanane
Animal versus Animal		
Skunk	Anything	Butyl mercaptan
Bombardier beetle	Predators	Quinone spray
Monarch caterpillars	Birds	Cardiac glycomides
Grasshopper	Birds	obtained by feeding on milkweed
Animal versus Bacteria		
Man	Bacteria	Lynoxyme
Plant versus Fungus		
Orchid	Fungus	Orchinol
Potato	Fungus	Chlorogenic and caffeic acids

9.3.2 Neurotoxin and Stockpile Waste

The transmission of nerve impulse in the body between one nerve and another involves the release of acetylcholine (ACh) from the first nerve that diffuses across the interneuronal space to trigger a different nerve. To prevent continued stimulation, the ACh is hydrolyzed by the enzyme **acetylcholinesterase** (AChE) present in the space. A serine molecule in the active site of AChE is acetylated, and the acetyl group is spontaneously hydrolyzed to regenerate the AChE. The organophosphorus compounds are potent inhibitors of AChE. In this case, we say that the organophosphate insecticides are acetylcholinesterase (AChE) inhibitors, as shown in Figure 9-29.

The same mechanism operates on humans when the body is confronted with a **chemical warfare** (CW) agent. For example, such nerve gases as G-agent or V-agent (please compare their structure with organophosphate, for example, parathion, shown in Figure 9-26.

$$
\begin{array}{cc}
\text{O} & \text{O} \\
\parallel & \parallel \\
\text{R'O}-\text{P}-\text{F} & \text{R'O}-\text{P}-\text{SR} \\
\mid & \mid \\
\text{R} & \text{R} \\
\text{G-agent} & \text{V-agent}
\end{array}
$$

In the case of sarin (G-agent), where R is C_3H_7 and R' is CH_3, the LCt_{50} is 50-100 mg-min m^{-3}. This lethal dosage is expressed both in amount and contact time. For example, the CS agent, shown in Figure 9-30, is used for riot control by police in most cities, and has LCt_{50} of 70,000 mg-min/m^3 for which 1-5 ng/m^3 is sufficient for control.

Currently, some of these toxic chemical agents, including GB, VX,

$$(R = (iPr)_2\, N(CH_2)^{2-},\ R' = Et)$$

and the blister agents (e.g., mustard, 2, 2'-dichlorodiethyl sulfide) are stacked in eight contingenous United States sites, as shown in Figure 9-31, and also elsewhere in Russia. These must be destroyed by the end of 2004 due to the decomposition rate of the stabilizer of the munitions with explosives, including M55 rockets and M23 land mines. Destruction of these chemicals is done by incinerators, followed by pollution abatement. There are four types in general use: the hearth type, rotary kiln, conveyor, and the hopper type as shown in Figure 9-32. The efficiency is more than 99.99%. A full scale facility is operational at Johnston Atoll, an island in the Pacific Ocean.

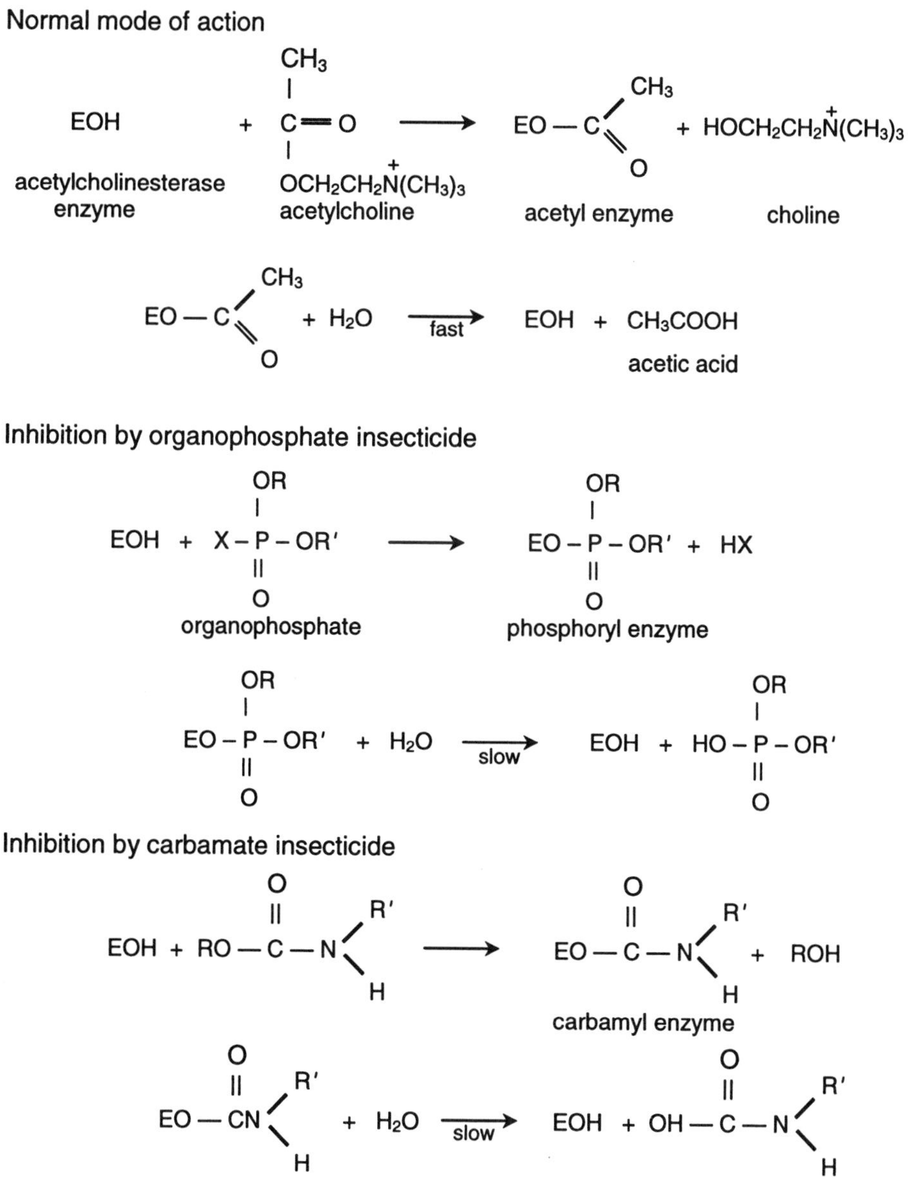

Normal mode of action

$$EOH + \underset{\substack{\text{acetylcholinesterase} \\ \text{enzyme}}}{} \quad \underset{\substack{\text{acetylcholine} \\ \overset{|}{OCH_2CH_2\overset{+}{N}(CH_3)_3}}}{\overset{\overset{CH_3}{|}}{\underset{|}{C}}{=}O} \longrightarrow \underset{\text{acetyl enzyme}}{EO-C\overset{CH_3}{\underset{O}{\diagdown}}} + \underset{\text{choline}}{HOCH_2CH_2\overset{+}{N}(CH_3)_3}$$

$$EO-C\overset{CH_3}{\underset{O}{\diagdown}} + H_2O \xrightarrow{\text{fast}} EOH + \underset{\text{acetic acid}}{CH_3COOH}$$

Inhibition by organophosphate insecticide

$$EOH + \underset{\substack{\text{organophosphate}}}{X-\overset{\overset{\displaystyle OR}{|}}{\underset{\underset{\displaystyle O}{||}}{P}}-OR'} \longrightarrow \underset{\substack{\text{phosphoryl enzyme}}}{EO-\overset{\overset{\displaystyle OR}{|}}{\underset{\underset{\displaystyle O}{||}}{P}}-OR'} + HX$$

$$EO-\overset{\overset{\displaystyle OR}{|}}{\underset{\underset{\displaystyle O}{||}}{P}}-OR' + H_2O \xrightarrow{\text{slow}} EOH + HO-\overset{\overset{\displaystyle OR}{|}}{\underset{\underset{\displaystyle O}{||}}{P}}-OR'$$

Inhibition by carbamate insecticide

$$EOH + RO-\overset{\overset{\displaystyle O}{||}}{C}-N\overset{R'}{\underset{H}{\diagdown}} \longrightarrow \underset{\text{carbamyl enzyme}}{EO-\overset{\overset{\displaystyle O}{||}}{C}-N\overset{R'}{\underset{H}{\diagdown}}} + ROH$$

$$EO-\overset{\overset{\displaystyle O}{||}}{C}N\overset{R'}{\underset{H}{\diagdown}} + H_2O \xrightarrow{\text{slow}} EOH + OH-\overset{\overset{\displaystyle O}{||}}{C}-N\overset{R'}{\underset{H}{\diagdown}}$$

Figure 9-29. Mechanism of acetylcholinesterase inhibition.

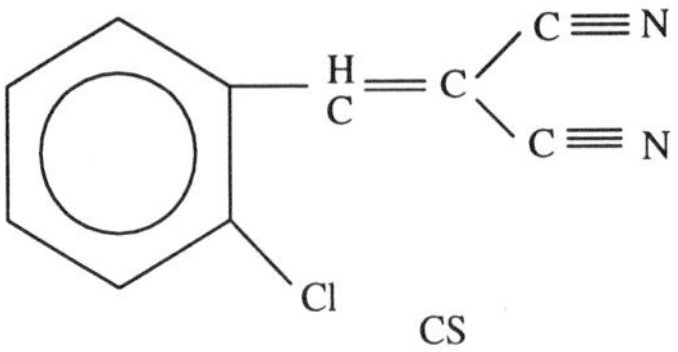

CS

$LC_{t_{50}}$

$70,000$ mg-min m^{-3}

Figure 9-30. Structure of CS agent and its lethal dosage.

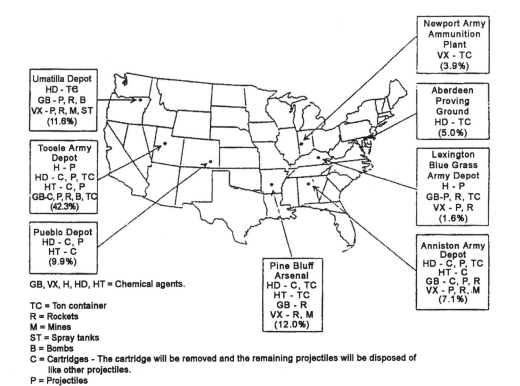

Newport Army Ammunition Plant
VX - TC
(3.9%)

Umatilla Depot
HD - TC
GB - P, R, B
VX - P, R, M, ST
(11.6%)

Aberdeen Proving Ground
HD - TC
(5.0%)

Tooele Army Depot
H - P
HD - C, P, TC
HT - C, P
GB-C, P, R, B, TC
(42.3%)

Lexington Blue Grass Army Depot
H - P
GB-P, R, TC
VX - P, R
(1.6%)

Pueblo Depot
HD - C, P
HT - C
(9.9%)

Anniston Army Depot
HD - C, P, TC
HT - C
GB - C, P, R
VX - P, R, M
(7.1%)

GB, VX, H, HD, HT = Chemical agents.

Pine Bluff Arsenal
HD - C, TC
HT - TC
GB - R
VX - R, M
(12.0%)

TC = Ton container
R = Rockets
M = Mines
ST = Spray tanks
B = Bombs
C = Cartridges - The cartridge will be removed and the remaining projectiles will be disposed of like other projectiles.
P = Projectiles

Figure 9-31. United States stockpiles of nerve agents and munitions.

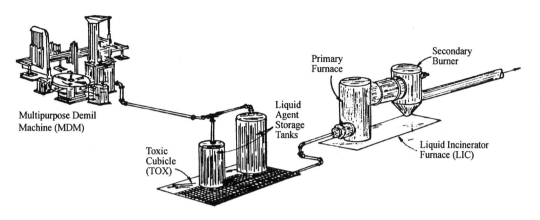

(A) LIC Hearth Type

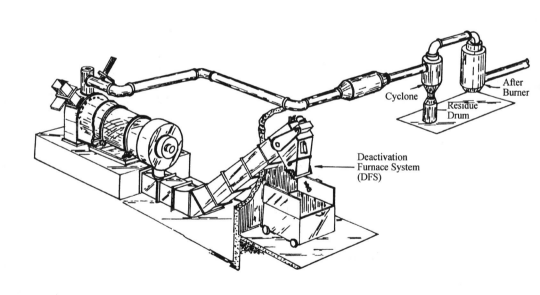

(B) DFS Rotary Kiln

Figure 9-32. Incinerator types. (A)ZIC herthtype, (B) DFS rotary kiln, (C) MPF conveyor, (D) DUN hopper type.

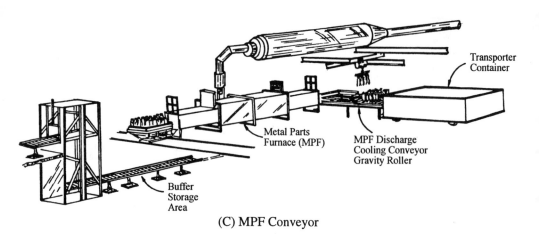

(C) MPF Conveyor

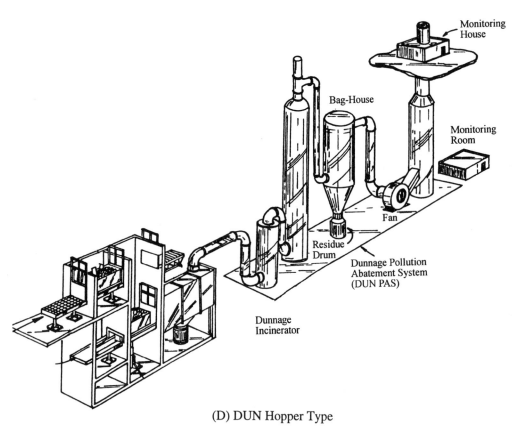

(D) DUN Hopper Type

Figure 9-32 (Continued). Incinerator types. (A)ZIC herthtype, (B) DFS rotary kiln, (C) MPF conveyor, (D) DUN hopper type.

REFERENCES

9-1. D. J. Greenland and M. H. B. Hayes, The Chemistry of Soil Constituents, Wiley, New York, 1978.

9-2. G. H. Bolt and M. G. M. Bruggenwert, Soil Chemistry, Elsevier, New York, 1978.

9-3. K. W. Brown, G. B. Evans, Jr., and B. D. Frentrup, Hazardous Waste Land Treatment, Butterworth, Boston, 1983.

9-4. A. D. McLaren and G. H. Peterson, Soil Biochemistry, Marcel Dekker, New York, 1967.

9-5. R. D. Holtz and W. D. Kovacs, An Introduction to Geotechnical Engineering, Prentice-Hall, Englewood Cliffs, New Jersey, 1981.

9-6. R. Lal and B. A. Stewart, Soil Degradation, Springer-Verlag, New York, 1990.

9-7. F. J. Stevenson, Humus Chemistry: Genesis, Composition, Reactions, Wiley, New York, 1982.

9-8. G. Sposito, The Chemistry of Soils, Oxford University Press, New York, 1989.

9-9. T. F. Yen, Chemistry of Marine Sediments, Ann Arbor Science, Ann Arbor, Michigan, 1977.

9-10. N. F. Janes, Recent Advances in the Chemistry of Insect Control, Special Publication No. 53., Royal Society of Chemistry, London, 1985.

9-11. H.A. Horne, The Chemistry of Our Environment, Wiley, New York, 1978.

9-12. G.V. Chilingarian, E.C. Donaldson, and T.F. Yen, Subsidence Due to Fluid Withdrawal, Elsevier, Amsterdam, 1995.

9-13. M.D. Erickson, Analytical Chemistry of PCBs, 2nd ed., Lewis, Boca Raton, Florida, 1997.

9-14. R.W. Miller and R.L. Donahue, Soils in Our Environment, 7th ed., Prentice-Hall, Englewood Cliff, New Jersey, 1995.

9-15. W. Salomons, Biogeodynamic of Pollutants in Soils and Sediments: Risk Assessment, Springer-Verlag, Berlin, 1995.

9-16. Geopolymerization of Biopolymers: A Preliminary Inquiry, D. Kim, I. G. Petrisor and T. F. Yen, Carbohydrate Polymers, 56(2), 213–217 (2004).

PROBLEM SET

1. An 830-acre watershed is located 5 miles south of Indianapolis in central Indiana. Given the information presented below, compute the projected sediment from sheet and rill erosion in terms of average daily loading.

    180 acres of cropland

    Corn; conventional tillage, average annual yield of 40 to 45 bushels/acre; cornstalks remain in fields after harvest; contour strip-cropped planting method; soil is Fayette silt loam; slope is 6%; slope length is 250 feet.

    220 acres of pasture

    No canopy of tree or brush; cover at surface is grass and grasslike plants; 80% ground cover; soil is Fayette silt loam; slope is 6%; slope length is 200 feet.

    430 acres of woodland

    50% of area covered with tree canopy; 80% of area covered with leaf and limb litter; undergrowth is managed and therefore minimal; slope is 12%, slope length is 150 feet.

HAZARDOUS WASTE AND REMEDIATION

Soils are good media for receiving a great variety of wastes. Due to excessive loading, some specific soil sites are exceeding the saturation limit. In this chapter, the solid wastes (including municipal and agricultural types) and their associated problems of disposal and leachate control are first evaluated. The concept of recycling and reuse is discussed, especially with regards to the plastic wastes. Current developments of degradable plastics are emphasized, and the problem concerning heavy metals that add a burden to the soil is addressed.

Basics related to hazardous wastes and their nature are presented. The widespread amounts of harmful toxic substances occurring in soil are mentioned, especially with reference to the heavy metals. Finally, the remediation technology of soil is discussed, including the principles of bioaccumulation, soil vapor extraction, immobilization and encapsulation, and some approaches to in-situ and on-site remediation techniques.

10.1 SOLID WASTE CHEMISTRY

Solid waste and **refuse** have the same meaning and are used synonymously. In highly populated areas, the magnitude of the problem resulting from the generation, collection, and disposal of solid waste is unsurmountable. In the United States it was estimated that the national average of solid waste generation amounts to 3.1 kg/capita-day, as shown in Table 10-1. Even in of 1968 survey the US collected and disposed of 1.27 Pg of solid waste in urban areas, and New York City alone used 60 hectares of Staten Island for disposal and incineration of solid waste. The cost is three times higher than that of the total coal mining of West Virginia and the subsequent delivery to New York City. To state how grave the problem is, one can simply say that each year the United States population throws out 48 billion cans, 28 billion bottles, 4 million tons of plastic, 30 million tons of paper and 100 million tires. Most of this debris has to be disposed by towns and cities at a cost that is only exceeded by local expenditures on education and highway construction.

Based on the waste classification of the Incineration Institute of America, solid wastes include trash, rubbish, garbage, refuse, animal and agriculture wastes, and even some forms of liquid wastes. Table 10-2 shows sample municipal waste composition from the east coast of the United States. From the table, the majority of the organic portion of the wastes can be equivalent to cellulose except ca. 5% of leather, rubber, and plastics.

The organic portion of solid waste can be treated as **cellulose**, especially if one wishes to convert it into useful products. Table 10-3 shows the chemical analyses of a great variety of municipal wastes. Variation of the energy content ranges from 2,000 to19,000 Btu/lb.

Table 10-1. Annual and Daily Production of Solid Wastes in the United States

Type of waste	Total mass (10^{10} tonnes/ yr)	Per capita (kg/ day)
Household, commercial, municipal	226	3.1
Collected	172	2.4
Uncollected	54	0.7
Industrial	100	1.4
Mineral	1000	14
Agricultural	1860	25
Farm animal wastes	1360	19
Crop residues	500	7
Total (rounded)	3200	44

Data from R.J. Black et al., "The National Solid Wastes Survey: An Interim Report," U.S. Dept. of Health, Education and Welfare. Washington, D.C. 1968.

Table 10-2. Sample Municipal Composition - East Coast U.S.

WEIGHT PERCENT

Physical		Rough Chemical	
Cardboard	7%	Moisture	28.0%
Newspaper	14	Carbon	25.0
Miscellaneous paper	25	Hydrogen	3.3
Plastic film	2	Oxygen	21.1
Leather, molded plastics, rubber	2	Nitrogen	0.5
garbage	12	Sulfur	0.1
Grass and dirt	10	Glass, Ceramics, etc.	9.3
Textiles	3	Metals	7.2
Wood	7	Ash, other inserts	5.5
Glass, ceramics, stones	10	Total	100.0
Metallics	8		
Total	100		

Conversion of wastes by pyrolysis, yields of products.

Table 10-3. Ultimate Analysis of Typical Municipal Refuse Components

Refuse Component	C (%)	H (%)	O (%)	N (%)	S (%)	Inerts [a]	Btu/lb	% Moisture	% as delivered
Newspapers	49.14	6.10	43.03	0.05	0.16	1.43	7.974	5.97	10.33
Brown paper	44.90	6.08	47.84	0	0.11	1.01	7.256	5.83	6.12
Magazine paper	32.91	4.95	38.55	0.07	0.09	22.47	5.254	4.11	7.48
Corrugated boxes	43.73	5.70	44.93	0.09	0.21	5.06	7.043	5.20	25.68
Paper food cartons	44.74	6.10	41.92	0.15	0.16	6.50	7.258	6.11	2.27
Vegetable and food waste	49.06	6.62	37.55	1.68	0.20	1.06	1.795	78.29	2.52
Plastics (average)	78.00	9.00	13.00				15.910		0.84
Evergreen trimmings	48.51	6.54	40.44	1.71	0.19	0.81	2.708	69.00	1.68
Lawn grass, green	46.18	5.96	36.43	4.46	0.42	1.62	2.058	75.24	1.68
Ripe tree leaves	52.15	6.11	30.34	6.99	0.16	3.82	7.984	9.97	2.52
Wood	49.00	6.00	42.00			2.28	6.840	24.00	2.52
Glass, ash, ceramics						100.00			8.50
Metals						100.00	2.66		7.53

[a] Inerts – ash, glass, metal, stone, ceramics
Source: compiled from R.G. Bond and C.P. Straub, handbook of Environmental Control, Vol.2. Soild waste, CRC, W. Palm Beach, Florida, 1993.

Chemistry and biochemistry of cellulose are essential to the understanding of thermal decomposition and enzyme hydrolysis related to conversion and landfill technology. For example, the biodegradation scheme of cellulose is illustrated in Figure 10-1, and the pyrolysis and combustion schemes are given in Figure 10-2.

A great variety of unit process equipment and devices for material recovery in the solid waste processing plant are illustrated in Table 10-4. These common and simple devices are also illustrated for reference.

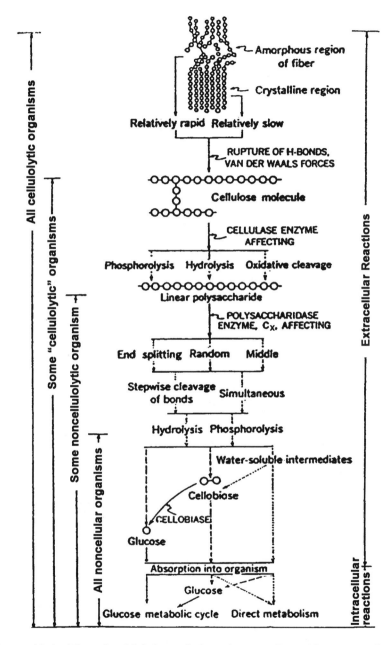

Figure 10-1. The microbial degradation of cotton fiber. (Source: R.G.H. Siu, "Microbiological Decomposition of Cellulose", Reinhold Publishing Corporation, 1951.)

Table 10-4a. Unit Process Equipment, Trommel, Shredder, Air Classifier, Magnet

TROMMEL – a perforated, rotating horizontal cylinder used to break open trash bags, remove glass in large enough pieces for easy recovery, and remove small abrasive items such as stones and dirt. Though trommels are used extensively in gravel and ore processing and for screening incinerator residue, there has been little experience in using a trommel before a shredder, and much must still be learned about its design and operation with raw refuse.

SHREDDER – a size-reduction machine which grinds mixed refuse to a more uniform particle size for further processing and breaks up composite items into their individual materials. Solid waste shredders are generally horizontal shafts with swinging hammers (as shown in the sketch) or vertical shaft ring grinders. The shredder must consistently produce the specified particle size range and must not permanently deform any of the materials in a way which interferes with subsequent processing steps.

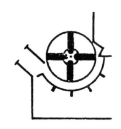

AIR CLASSIFIER – a controlled system used to separate light objects from the heavy fraction of waste. One form incorporates a column of air moving vertically upward as shredded refuse falls downward through it, as in the "zig-zag" type shown (A). Another type is a vibrating table with blowers arranged so that air is passed through the refuse and the light fraction is carried on. A third type employs a tilted, rotating drum (B). Each type must be investigated to determine performance relative to others, for example, in preparing the light fraction of refuse as a fuel.

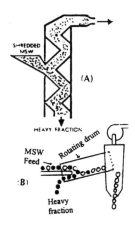

MAGNET – a standard industrial tool, used to remove iron and steel from the refuse stream. The magnet may be place after the air classifier, where it recovers the steel (principally as a "cleaner" fraction than if it were located directly after the shredder). However, alternative locations should be investigated because the steel must be free of organic materials to be suitable for reuse. One possible location is following the rising current separator so that the metal is washed prior to removal.

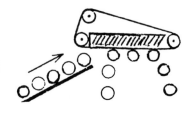

Table 10-4b. Unit Process Equipment, Screens, Air Knife, Rising Current Separator, Heavy-Media separator

SCREENS – devices used to separate materials by size and prepare fractions within the narrow size ranges required by some separation equipment. Although screening is a simple operation, in many instances the exact size of screen openings and the type of screens can be chosen only by experience.

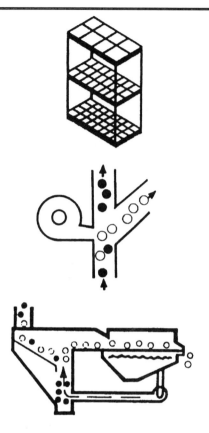

AIR KNIFE – a jargon term for a blower device intended here to separate steel cans from more massive pieces of iron and steel, a necessary step because the two require different processes for reuse. Experimentation is required to discover the best design for this application as well as the best location in the flow chart. An air knife may also dry the metal if placed after the rising current separator.

RISING CURRENT SEPARATOR – a unit housing a flowing current of water used to carry off or wash away organic materials such as food wastes, heavy plastics, and wood from the air classified heavy fraction. Because the water is pumped upwards through the refuse, many materials which normally would sink in water are able to float and be removed.

HEAVY-MEDIA SEPARATOR – a tank of dense fluid, actually a suspension of a mineral in water. When the mixture of glass, aluminum, and other non-ferrous metals is immersed in the liquid, the fluid density can be controlled so that the aluminum and glass float while the other metals sink. The NCRR ETEF is capable of producing appropriate fractions of municipal solid waste, otherwise unavailable for investigation of such factors as separating efficiency, throughout rates, particle size range, fluid density range, and ease of removal of the suspension from the product.

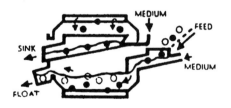

Table 10-4c. Unit Process Equipment, Electrostatic Separator, Electrodynamic Separator, Electronic-Optical Sorter

ELECTROSTATIC SEPARATOR – a device utilizing the principle that electrical conductors lose an induced static charge faster than insulators. In this way, an electrostatic sorter can separate conducting materials (e.g. glass) after the particles are charged in a high voltage electrical field. This technique for sorting solid waste fractions is under investigation in several places; further investigation and refinement is necessary.

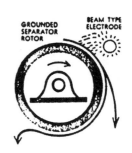

ELECTRODYNAMIC SENSOR – sometimes termed an "aluminum magnet" or eddy current separator, this separator uses the electrodynamic induction of a magnetic field as a method of sorting. If an alternating current is passed through a piece of metal, and properly applied, the metal temporarily becomes magnetic and can be deflected and separated; this principle has been used to separate aluminum and other non-ferrous metals. In pilot trials, aluminum can be separated clean through enough for reuse, but much must still be learned about larger scale processing.

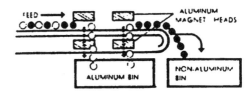

ELECTRONIC-OPTICAL SORTER – a unit which separates glass from stones and pieces of ceramics and sorts the glass according to color. A photo-electric detector determines the color of the pieces into the proper containers. This process is used in the food industry to separate good beans or rice from bad.

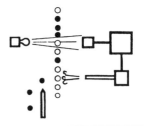

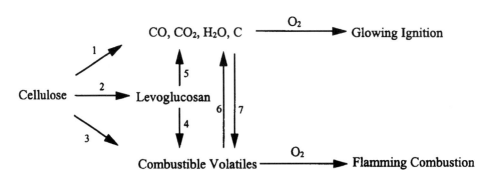

Figure 10-2. The general reactions involved in pyrolysis and combustion of cellulose.

10.1.1 Recycling and Conversion

It should be mentioned that wastes and resources are related. For example, through biodegradation, biodecomposition, or biodisintegration, the wastes can be converted into useful energy, raw material, chemical feedstocks, and so on. Also, many wastes are the end products of biological deteriorations or breakdowns from past useful product or manufactured goods. This process can be illustrated as follows:

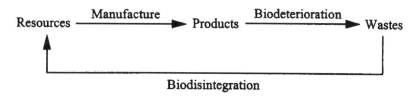

Of course, the conversion is not only limited through biological processes; a variety of physical-chemical methods, so-called recycling, are available.

The buzzwords **waste minimization** and **waste reduction** have to do with the manufacturing of the products from the resources. Waste minimization involves the selection of alternative treatment processes that reduce the quantity or quality of the wastes requiring ultimate disposal. Waste reduction means the cutting down of the amount of wastes from their sources. The most efficient approach is source reduction and control of the manufacturing processes. Recent initiation of the many environmentally-benign or environmentally-friendly chemical processes is just a beginning of the effort.

The direct use of solid wastes to generate power is perhaps one of the most common traditional incineration methods. Other methods will include the partial oxidation or thermal treatment under limited air or inert atmosphere — that is, **pyrolysis** techniques. There are many versions of pyrolysis: flash pyrolysis, fluidized bed pyrolysis, hydropyrolysis etc.; yet they are based on the same principle, thermal decomposition. Figure 10-3 illustrates cellulose molecules under thermal decomposition conditions. Actually, the thermal decomposition of cellulose as depicted by Figure 10-2 is quite complex. Most organic materials under pyrolysis conditions will yield the following fractions:

- gas
- char — solid phases
- tar and oil — liquid phase
- residue (could include some inorganics)

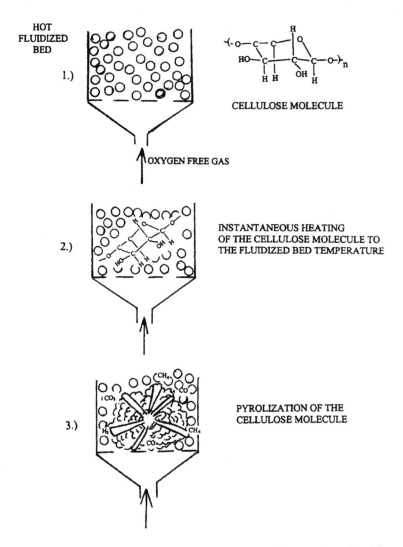

Figure 10-3. Chemical reduction of the cellulose molecule. (1) and (2) show the cellulose molecule being introduced into the hot fluidized bed. (3) shows the cellulose molecule literally being rearranged by exploding into its pyrolysis gas products.

The chemical composition of these functions are illustrated in Table 10-5. Of course, the type of raw waste will greatly affect the distribution of the four fractions, shown in Table 10-6, as well as the energy distribution obtained, shown in Table 10-7.

The presence of metals and glasses with organics during pyrolysis will also be possible. In this case, the glass, nonferrous and ferrous metals in waste will be recycled. Actually Owen-Illinois has recorded a green glass as shown in Table 10-8. A common Garrett glass recovery process is shown in Figure 10-4.

Table 10-5. Typical Products of Pyrolysis

Char fraction, 20 wt.%; Heating value 9000 Btu/ lb.	
48.8 wt.%	Carbon
3.9	Hydrogen
1.1	Nitrogen
0.3	Sulfur
31.8	Ash
0.2	Chlorine
13.9	Oxygen (by difference)

Oil fraction, 40 wt.%; Heating value 4.8 MM Btu/ bbl. (10,500 Btu/ lb)	
57.5 wt.%	Carbon
7.6	Hydrogen
0.9	Nitrogen
0.1	Sulfur
0.2	Ash
0.3	Chlorine
33.4	Oxygen (by difference)

Gas fraction, 27 wt.%; Heating value 550 Btu/ cu. ft.	
0.1 mol%	Water
42.0	Carbon monoxide
27.0	Carbon dioxide
10.5	Hydrogen
<0.1	Methyl chloride
5.9	Methane
4.5	Ethane
8.9	C_3 to C_7 hydrocarbons

Water fraction, 13 wt.%	
Contains:	Acetaldehyde, Acetone, Formic acid, Furfural Methanol, Methylfurfural, Phenol, Etc.

Table 10-6. Average Yields of Pyrolysis Products from Douglas Fir Bark, Rice Hulls, Grass Straw, and Cow Manure

	Weight Percentage of Dry Feed			
	Oil	Char	Gas	Water
Douglas Fir Bark	35-50	50-25	5-15	10
Rice Hulls	40	35	10	15
Grass Straw	50	20	15	5
Cow Manure	30	45	15	10

Table 10-7. Energy Distribution - Pyrolysis of Wastes

Materials Pyrolyzed	Energy in Raw Refuse, Million Btu/ ton	Energy in Pyrolysis Products, Million Btu/ ton of Raw Refuse[a]			
		Gas	Char	Tar and oil	Total[a]
Household Refuse	17.8	7.1	5.9	2.3	15.3
Industrial Refuse	9.1	6.1	1.7	0.6	8.4
Scrap Tires	33.5	8.7	15.9	6.8	31.4
Battery Cases - Hard Rubber	26.9	6.0	19.0	1.6	26.6
Waste Bark and Sulfite Liquor	13.6	6.6	4.7	1.7	13.0
Battery Cases - Plastic	37.7	8.3	0.3	30.4	39.0
Rice Hulls	13.2	3.6	7.0	1.6	12.2
Rice Straw	12.2	4.0	5.9	1.5	11.4
Cattle Manure	14.2	6.3	5.2	1.2	12.7
Paper Mill Sludge	10.7	4.6	4.4	0.2	9.2
Raw Sewage	14.2	8.3	3.2	2.3	13.8

[a] Does not account for energy required for pyrolysis nor for heat of drying wastes.

Hydrogenation or hydropyrolysis is an important process to increase the liquid portion of the fractions. So-called waste-to-oil or **refuse derived fuel (RDF)** is based on the principle that cellulose or carbohydrate can be converted into hydrocarbons either under or upon the base catalyzed condition or hydrogen atmosphere.

Table 10-8. Chemical Compositions of Green Glass

Oxide	Typical Production Green (%)	Recovered from San Francisco Refuse (%)
SiO_2	71.6	72.34
Al_2O_3	1.5	2.20
MnO	0.01	0.008
CaO	10.8	9.76
MgO	0.7	0.79
Na_2O	14.5	13.77
K_2O	0.3	0.84
Fe_2O_3	0.19	0.106
PbO	0.003	0.031
Cr_2O_3	0.23	0.048
TiO_2		0.048

Source: Data supplied by Owen, Illinois

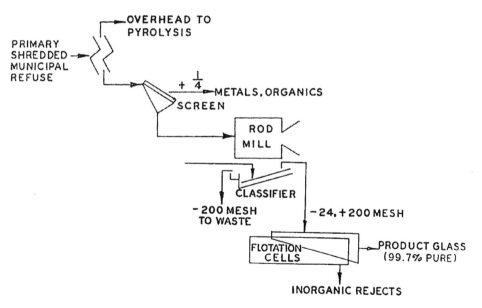

Figure 10-4. Garrett glass recovery process.

The alkaline catalyzed mechanism is as follows:

1. at 160°C $Na_2CO_3 + 2CO + H_2O \rightarrow 2HCOONa + CO_2$

2.
$$\begin{array}{c} H\ \ H \\ |\ \ \ | \\ -C-C- \\ |\ \ \ | \\ OH\ OH \end{array} \xrightarrow{H_2O} \begin{array}{c} H \\ | \\ -C{=}C- \\ | \\ OH \end{array} \longrightarrow \begin{array}{c} H \\ | \\ -C-C-+HCOO^- \\ |\ \ \| \\ H\ \ O \end{array} \rightarrow \begin{array}{c} -CH_2-CH-+CO_2 \\ | \\ O{\cdot} \end{array}$$

3. at 250°C
$$\begin{array}{c} -CH_2-CH- \\ | \\ O{\cdot} \end{array} + H_2O \longrightarrow \begin{array}{c} -CH_2-CH- \\ | \\ OH \end{array} + OH^-$$

4. $OH^- + CO \rightarrow HCOO^-$

5. $(2y + 2)HCOO^- + C_x(H_2O)_y \rightarrow C_xH_{2y+2} + yH_2O + (2y + 2)CO_2$

Here, $C_x(H_2O)_y$ is a carbohydrate, C_xH_{2y+2} is a hydrocarbon, and x = y. If x ≠ y, water molecules may be used for adjustment. For hydrogenation, more hydrogen has to be introduced to the system:

$$C_x(H_2O)_y \xrightarrow{900°C} xCO + yH_2$$

$$xCO + 2(x + 1) H_2 \xrightarrow{< 300°C} C_xH_{2x+2} + xH_2O$$

Here, $(x + 2)H_2$ mole hydrogen has to be introduced in order to have a liquid hydrocarbon (refer to Fischer-Tropsch synthesis in Chapter 1). For fuels, a rule of thumb is that the more the hydrogen atoms are accommodated to carbon atoms, the more the solid form (C or C_nH_m, m<<n) will transform to a liquid form (C_nH_{2n+2}) and finally to a gas as methane (CH_4).

A typical flow chart of resource recovery from wastes is shown in Figure 10-5. An actual process at work in San Diego is illustrated in Figure 10-6.

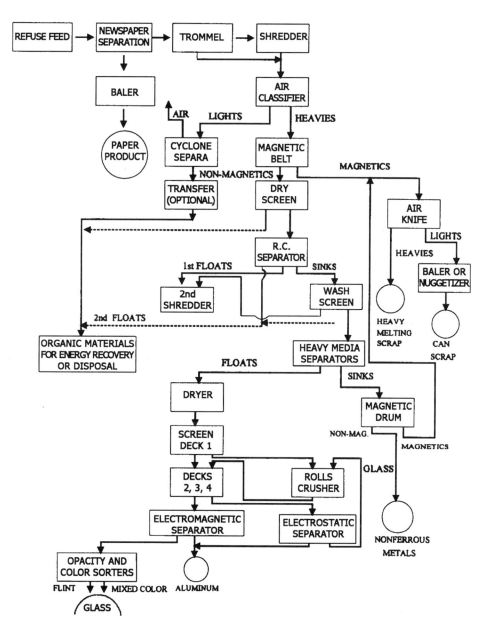

Figure 10-5. Resource recovery processing.

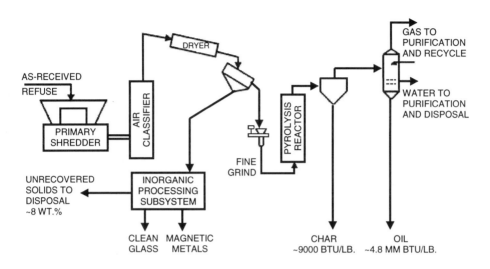

Figure 10-6. Recycling of solid wastes via pyrolysis.

10.1.2   Disposal: Landfill and Leachate

Bioconversion of solid wastes is also a traditional method of resource recovery, as shown in Figure 10-7 by means of microorganisms. During bioconversion, humus also will be formed. A general scheme is:

$$C_aH_bO_cN_dS_e + (a - b/4 - c/2 + 3d/4 + e/2)H_2O \rightarrow (a/2 + b/8 - c/4 - 3d/8 - e/4)CH_4$$

$$+ (a/2 - b/8 + c/4 + 3d/8 + e/4)CO_2$$

$$+ dNH_3 + e\,H_2S$$

Sometimes the equation can be written with the wastes expressed by $CH_aO_bN_c$, for example,

Cardboard	$CH_{1.604}$	$O_{0.734}$	$N_{0.003}$
Newspaper	$CH_{1.521}$	$O_{0.655}$	$N_{0.002}$
Garbage	$CH_{1.652}$	$O_{0.497}$	$N_{0.057}$

The products in this manner can be

	H_2O	CO_2	CH_4
Cardboard	0.234	0.484	0.561
Newspaper	0.294	0.474	0.526
Garbage	0.381	0.438	0.561

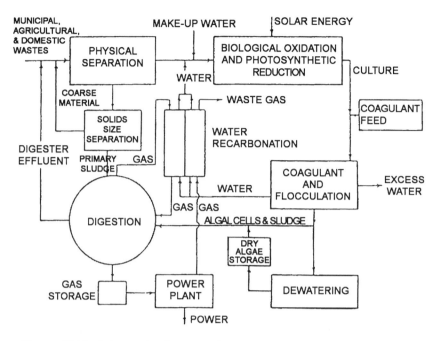

Figure 10-7. Schematic diagram of system for the biological conversion of solar energy to electrical power.

The gas obtained from these experiments often has a CO_2/CH_4 volume ratio of 0.75, which is close to the value anticipated. To purify the bioproduced gas into useful fuel (such as even low-Btu gas), often the removal of water by absorption, adsorption, and scrubbing of impurities (such as H_2O, H_2S, and NH_3) is needed. The gas purification process is illustrated in Table 10-9.

A typical sanitary waste **landfill** is sketched in Figure 10-8 with compact clay liners from the Michigan Department of Natural Resources. The importance is that the design has to be some distance away from residue and water table, and in the bottom there must be room for leachate collection. Usually, the top is covered by an impermeable cover and may be equipped with gas producing wells. At the cap of the cover the soil is unstable; it is better to have its use as a recreation site—e.g., the Rose Bowl in Pasadena, CA is a landfill as designed by F.W. Bowerman of the University of Southern California.

The volume of a landfill can be estimated from the following:

$$V = \frac{PEC}{D_c} \qquad [10\text{-}1]$$

where V = volume of land in m³, P = population,

$$E = \text{ratio of cover (soil) to compact fill} = \frac{V_{sw} + V_c}{V_{sw}} \qquad [10\text{-}2]$$

(where V_{sw} = volume of solid waste in m³, V_c = volume of cover in m³), C = average mass of solid waste per capita per year in kg/person, and D_c = density of compacted fill in kg/m³.

Table 10-9. Gas Purification Process

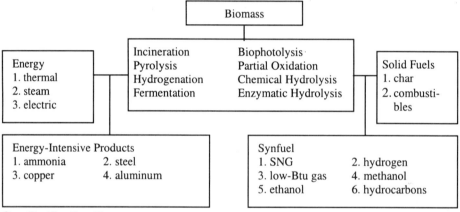

Gas Purification Processes

Objective	Processes
Removal of water vapor	Absorption by hydroscopic liquids diethylene or triethylene glycol salt brine
	Adsorption in activated solid dessicents
	Molecular sieve
	Activated carbon
	Silica gel
	Activated alumina
	Condensation by compression/cooling
Removal of sour gas	Molecular sieve method
	Scrubbing with alkaline solution
	Membrane separation

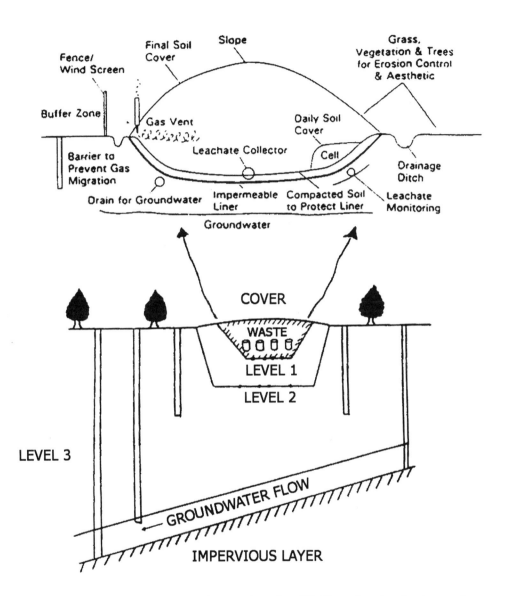

Figure 10-8. Cross section through a sanitary landfill. Three levels of safeguard in hazardous waste landfills: Level 1. Linear plus leachate collection/treatment. Level 2. Backup liner plus leachate collection/treatment. Level 3. Wells to monitor and if needed, control leachate plume.

There are three stages of biological activities of landfills, as shown in Figure 10-9:

1. aerobic decomposition stage
    - usually short since high BOD with little oxygen
    - increase in landfill temperature
    - dissolution of highly soluble salts
    - few organics produced

2. anaerobic decomposition stage (2 stages)
    stage 1
    - facultative anaerobes produce large amounts of low molecular weight aliphatic acids and carbon dioxide
    - reduction in leachate pH and redox environment
    - dissolution of sparingly soluble inorganic salts produce leachate of high conductivity

    stage 2
    - increase in leachate pH (decrease in conductivity) due to degradation of low molecular weight acids - methane produced from CO_2 and organic acids

3. Final aerobic stage
    - incoming oxygenated water and cessation of biological activity

The preceding list is the basis for the production of landfill gas.

A serious concern about landfills is the generation of **leachate** as a result of infiltration of surface water passing through the waste disposal site. In modern landfills there are dual leachate collection systems; one is located between the two impermeable liners for the bottom and sides, and another above the top liner of the double-liner system. The flexible liner is made of geofabric or geomembrane material such as chlorosulfonated rubber or chlorinated polyethylene, which is resistant to biodegradation. Leachate is collected in perforated pipes that are imbedded in granular material.

Usually, the leachate contains constituents from the wastes which are water soluble and also the products of chemical and biochemical transformation of the wastes. A typical composition is shown in Table 10-10. The pH values vary but are always in the lower range, as shown in Table 10-11. One problem is that the lower volume state of Mn^{2+} and Fe^{2+} may become insoluble Mn^{4+} and Fe^{3+} hydrated oxides upon exposure to air and thus clog the leachate collection system.

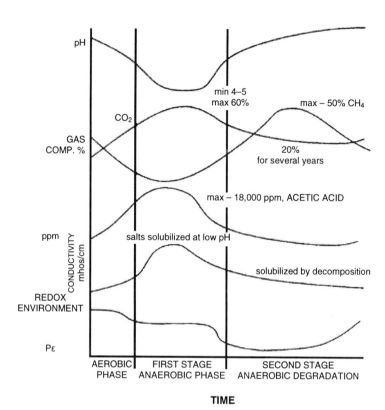

Figure 10-9. The theoretical degradation of a landfill. Ref: Pojasek, R.B., Ed. Toxic and hazardous waste disposal, volume 2 (Ann Arbor Science, Ann Arbor, Michigan, 1977).

Table 10-10. Characteristics of Leachate Sanitary Landfills (mg/L)

Constituent	Range*	Typical value
Organic strength, COD	1,000-30,000	10,000
BOD$_5$	200-20,000	6,000
Total solids	2,000-5,000	3,000
Total nitrogen	20-1,000	200
Alkalinity (as CaCO$_3$)	200-5,000	400
Soluble salts (Cl, SO$_4$)	200-3,000	500
Iron	50-800	100
Lead	1-10	2
Zinc	25-250	50
pH	5-8	6

Adapted from Chian and DeWalle, 1977, Tchobanoglous et. al., 1977, and Vesilind and Rimer, 1981.
*Except for pH

Other biochemical processes used for refuse conversion include anaerobic and aerobic decomposition. In the former method, solid waste is mixed with sewage sludge and the mixture is digested. Although operational problems made this process impractical on a massive scale, single household units admixed with human excreta have been used. The aerobic decomposition is well known as **composting**, usually utilizing long rows of shredded refuse as windows and allowing sufficient oxygen to penetrate the compost pile. This is usually referred to as **static pile composting** and can produce excellent soil conditioners.

10.2 PLASTIC WASTES

From domestic wastes, the percentage of waste plastics has risen from 2–3% to 3–4% steadily since 1970. Most plastics found in solid wastes are the **thermoplastic** types. In contrast, the **thermosetting** ones only account for a small fraction, as shown in Figure 10-10. The thermoplastic types usually consist of the following:

- polyolefins, including high- and low-density polyethylene (HDPE and LDPE), and polypropylene (PP)
- styrenes, including polystyrene (PS) and acrylonitrile-butadiene-styrene (ABS)
- vinyls, primarily polyvinyl chloride (PVC) and polyvinylidene chloride (PVDC)

Furthermore, the majority of the thermoplastics found in wastes are related to packaging in one way or another; for example, adhesives, coatings, containers, films, sheets, etc., as indicated by Table 10-12. Of course in waste streams, items such as hardware, toys, and furniture of the nonpackaging portion are also found, but it is only a small portion.

Because plastics are resistant to biodegradation by nature, the alternative is thermal decomposition. We will attempt to explain some problems regarding this approach. In general, the bulk of a given type of plastic is from a given type of polymer. Polymers decompose into fragment molecules (radicals) or monomers depending on their structure, via heat, radiation, and mechanical means. Under pyrolysis or incineration, the polymer decomposition is equivalent to depolymerization. It is essential to know that depolymerization is different from polymerization, because it can rarely achieve 100% monomer recovery, especially for a polymer mixture. In many cases, the monomer formation is interrupted by the activity of the free radical as a result of cross-linking. Similar to the general free radical reactions, the presence of hydrogen atoms tends to stabilize the reactive radicals, whereas oxygen atoms enhance cross-linking. Table 10-13 represents the depolymerization mechanisms. In pyrolysis, saturated hydrocarbons (such as polyethylene or polypropylene) are expected to dehydrogenate to create either internal or vinyl unsaturation. Both sites can initiate polymerization or polycondensation reactions. As a result, this will lead to char formation, as shown in Figure 10-11. In case of heat treatment under deficient oxygen conditions, various active functional groups on the polymer intermediates will result, as shown in Table 10-14.

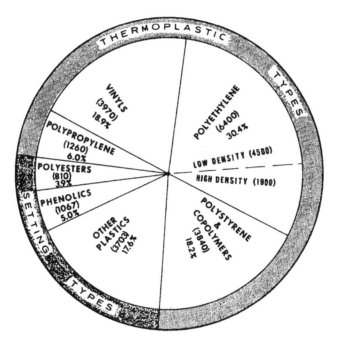

Figure 10-10. Plastics production in the U.S.A. in 1971 reached 21,050 millions pounds. Graph shows distribution by types, millions of pounds in parentheses, and percent of total production.[1] (Source: J.W. Jenson, J.L. Holman and J.B. Stephenson in T.F. Yen, Recycling and Disposal of Solid Wastes, P260, Ann Arbor Science, 1974.)

Analytical tools such as differential thermal analysis (DTA), which gives the temperature at which heat is liberated or absorbed, and thermogravimetric analysis (TGA), which provides a record of weight changes as the temperature is increased, simulate the actual thermal decomposition of the polymers. In general, polymers containing carbon and hydrogen will yield carbon monoxide and aldehydes under recycling operations. Polymers containing nitrogen, halogens, and sulfur will yield toxic and corrosive gas such as hydrogen cyanide, hydrogen chloride, and sulfur dioxide. A typical TGA curve of polyvinyl chloride is shown in Figure 10-12. The combustion products based on TGA studies of polyurethane are listed in Table 10-15.

Table 10-11. The pH Ranges of Representative Leachates

Landfill Site	Age of refuse (yr)	pH
Hughes (1971)	17	7.0
Pohland (1975)	1	5.33
Pohland (1975)	2	5.3
Pohland (1975)	3	5.3
Merz (1954)	1.5	5.6-7.5
Emcon (1974)	2	4.7-5.4
Chain & Dewalle (1977)	0.25	5.63
Reinnart & Ham (1971)	0.33	5.97
Fungaroli (1971)	2	3.7-8.5
Qasim & Burchinal	0.33	5.88-6.48
Zenone (1974)	15	5.7-6.9
Zenone (1974)	13	7.1-7.6
Johansen & Carlson (1976)	3.5	5.9
Johansen & Carlson (1976)	2.5	5.2
Wigh (1979)	3	5.4
Wigh (1979)	5	5.5
Fungaroli & Steiner (1979)	2	5.53
Fungaroli & Steiner (1979)	4.5	5.36
Fuller (1978)	6 mo.	5.5-6.3
Rovers & Farquhar (1973)	1	5.3-11.5
SCS Engineers Site A (1976)	Active	5.37-6.11
SCS Engineers Site D (1976)	2	5.3-6.65
Apgar & Langmuir (1971)	<1	6.6
Apgar & Langmuir (1971)	5.3	6.46
Meichtry (1971)	>3	5.75
Meichtry (1971)	>6	7.4
Summary range values		3.7-11.5 (Norm = 5.3)

Table 10-12. Plastic Resins Consumed by Packaging Industries in 1971

| Products | Consumption by Plastic Types, Millions of Pounds | | | | | |
	Polyethylene	Vinyls	Styrenes	Polypropylene	Other	Total
Adhesives		43			11	54
Coatings	424	110		4	75	613
Closures	50	15	21	35	28	149
Containers and lids	1,030	145	810	65	80	2,130
Film and sheet	1,400	110	45	100	65	1,720
Total	2,904	423	876	204	259	4,666

Table 10-13. Depolymerization (Source: G.A. Zerlant and A.M. Stake in T.F. Yen, *Recycling and Disposal of Solid Wastes*, Ann Arbor Science, 1974, p. 177.)

		Monomer Yield

PMMA

$$-CH_2-\underset{\underset{COOCH_3}{|}}{\overset{\overset{CH_3}{|}}{C}}- \longrightarrow -\underset{\underset{COOCH_3}{|}}{\overset{\overset{CH_3}{|}}{CH}}-\overset{\bullet}{C}- \longrightarrow 100\%$$

Stable
Nonreactive

PMS

$$-CH_2-\overset{\overset{CH_3}{|}}{\underset{\text{(phenyl)}}{\overset{\bullet}{C}}}- \longrightarrow -CH-\overset{\overset{CH_3}{|}}{\underset{\text{(phenyl)}}{C}}- \longrightarrow 100\%$$

PS

$$-CH_2-\underset{\text{(phenyl)}}{CH}- \longrightarrow -CH_2-\overset{\bullet}{\underset{\text{(phenyl)}}{C}}-$$

Unstable
Reactive

40% at 300–400°C. More monomer with higher nitrogen pressure.

Fragments at <500°C. More fragments with higher nitrogen pressure.

PE

$$-CH_2-CH_2- \longrightarrow -CH_2-\overset{\bullet}{CH}- \longrightarrow \text{Fragments} - 400°C$$

TFE

$$-CF_2-CF_2- \longrightarrow -CF_2-\overset{\bullet}{CF}_2-$$

>95% – 500°C, low pressure 16% – 600°C, atmospheric
Fragments – 1200°C, low pressure

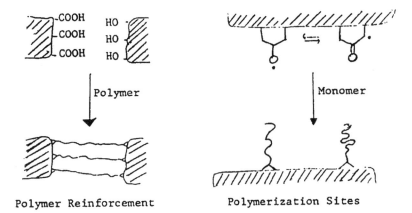

Polymer Reinforcement   Polymerization Sites

Figure 10-11. Reactions involving charred residue. (Completely charred polymeric materials have chemically reactive groups that can be sited for covalent-bonding reactions).

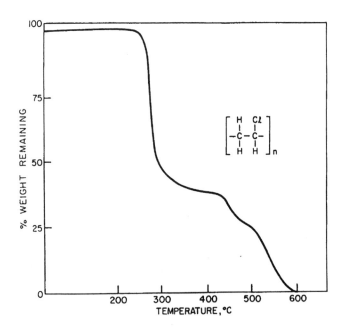

Figure 10-12. TGA of polyvinyl chloride.

Table 10-14. Pyrolysis and the Effect of Oxygen

$$CH_3-CH_2-CH_2-CH_2- \longrightarrow CH_3-\overset{\bullet}{C}H-CH_2-CH_2-$$

$$CH_3-CH=CH-CH_2- \quad \text{Internal}$$

or

$$CH_2=CH-CH_2-CH_2- \quad \text{Vinyl}$$

Both arrows indicate active site for
polymerization or crosslinking

Organic Materials	Polymers
Excess $\quad 2O_2 + CH_4 \rightarrow \quad CO_2 + 2H_2O$ Sulfur Oxides Nitrogen Oxides Deficient $O_2 + 2CH_4 \rightarrow \quad 2CO + H_2$ $CO + H_2 \rightarrow$ Formaldehyde $CO + 2H_2 \rightarrow$ Methanol Acids Ketones Paraffins Aromatics	CO_2 Water CO Aldehydes Alcohols Acids Ketones Paraffins Aromatic Polymers $\rightarrow$ Carbon Black Amide Polymers $\rightarrow$ Nitrogen Containing Gases (Nylon, Melamine) Halogen Polymers $\rightarrow$ Halogen Containing Gases (PVC, TFE)

For the finished plastic products, a great variety of chemical additives have to be introduced. In some formulations the contents of these additives could exceed 20% by weight, and many of them are highly toxic — for instance, fillers and fire retardants used for plastics, including antimony oxide, barium metaborate, talc, mica, zinc borate, titania, molybdenum disulfide, sulfur and carbon. Reinforcements such as glass fibers, dusts, and pure metal powders of which the particles are small (ca. 1µm), are often used. Furthermore, many of the raw additives are not pure material; for example, zinc oxide contains 0.08% lead oxide and 0.05% cadmium oxide. In some plastic recycling (for example, spent propellant and projectile binders) the inorganics become the desirable recoverables (for example, aluminum or magnesium powder).

Table 10-15. Combustion Products of Polyurethane (from TGA)

Compound	Quantity (mg/ g)
CO_2	425-800
CO	175-300
Cyanide ion (as HCN)	5-50
Methane	2-5
Ethylene	2-5
Ethane	1
Propylene	2-4
Propane	<1
Methanol	0.01-0.03
Acetaldehyde	0.03-0.05
1-Butene	<1
Butane	<1
Propionaldehyde	1-5
Acetone	1-5

Antioxidants (such as stable free radicals) are introduced to stabilize the polymer from ozone and ultraviolet attack. In recent years, **photodegradation** for polymers has become important. Photosensitizers are added to aid polymer degradation by the transfer of electronic energy from a donor molecule to a polymer. Examples of these are nitroso compounds, quinones, benzophenones, and diketones, which can be photoexcited to triplet states, or can be formed as adduct biradicals. These biradicals are charge-transfer complexes that weaken the polymer backbone, causing chain scission. This principle is the basis of **biodegradable polymers**, as shown in Figure 10-13. The high altitude weather balloon is made in this manner to be able to self-destruct in a given period in the future.

Separation schemes based on the differences in density of refractive indexes have been able to provide methods for sorting different classes of polymers if the feed stream is a mixture of plastics. Based on liquid media separation, it is possible to sort out five major types of plastics from a mix feed as follows:

	Density	Media
Polypropylene	0.90	water-alcohol 0.91
Low density polyethylene	0.92	water-alcohol 0.93
High density polyethylene	0.94-0.96	salt water 1.20
Polystyrene	1.05-1.06	
Polyvinyl chloride	1.22-1.38	

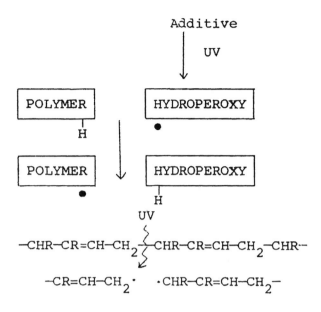

Figure 10-13. Photodegradation.

Finally, a potential application of recycling technology has been developed for waste plastics, as shown in Figure 10-14. We also would like to emphasize the importance of scrap tire recycling. More than two billion used tires have been stockpiled in the United States; and of the additional 285 M tires discarded each year, only 100 M are recycled, leaving the remaining 185 M for illegal dumps or heap piling across the country. In many tropical regions in the world, the abandoned tires may provide breeding for disease-carrying mosquitoes or rats. Stockpiling of tires often cause tire fires just as coal piles; their self-ignition fire is difficult to extinguish. Burning will induce smoke that is often toxic. The popular method for tire recycling is the crumb rubber process, which is a mechanical size reduction of rubber into granular form for fillers of playgrounds and roads. A new process involving a multistage of solvent swelling and ultrasound assisted chemical degradation can remove a high level of the sulfur linkages (devulcanization). Virgin polymers as well as carbon blacks can be recovered. Also recently devulcanization of rubber can be accomplished by milling through an ultrasound-equipped press.

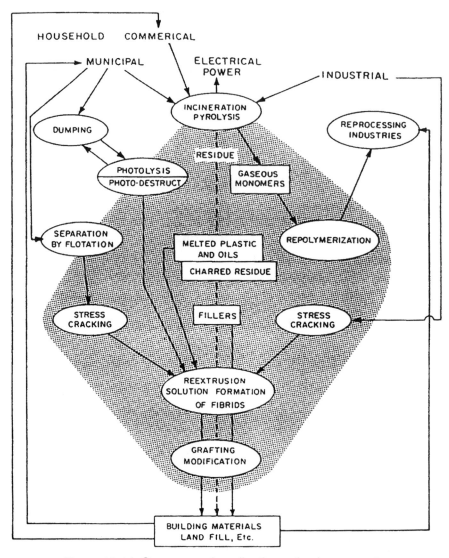

Figure 10-14. Sources and applications of polymer waste.

10.3 Heavy Metals

In addition to the organic carcinogenic chemicals discussed in the previous chapter, another environmental concern is that metallic elements that are toxic to human beings are being spread around increasingly as a result of industrial production and use.

Table 10-16 lists **trace metals** that may pose health hazards in the environment. These elements occur naturally in the Earth's crust, and human beings have been living with them throughout evolutionary history. However, the amounts of trace metals that are being stirred up and spread around have been increasing greatly to a toxic level with the advance of industrialization. Many metallic elements are essential to life. But excess amounts will be very toxic to humans. It seems that our maximum amount of nutrient requirement is different for different metals. For example, there are only 5 g of iron and 80 mg of copper in our body. Many metal elements are deposited in the lithosphere (as sulfide ores) and finally are stored in the pedosphere from the hydrosphere via biosphere cycle.

Table 10-16. Trace Metals That May Pose Health Hazards in the Environment

Element	Sources	Health Effects
Nickel	Diesel oil, residual oil, coal, tobacco smoke, chemicals and catalysts, steel and nonferrous alloys	Lung cancer
Beryllium	Coal industry (new uses proposed in nuclear power industry, as rocket fuel)	Acute and chronic system poison, cancer
Boron	Coal, cleaning agents, medicinals, glass making, other industrial	Nontoxic except as borane
Germanium	Coal	Little innate toxicity
Arsenic	Coal, petroleum, detergents, pesticides, mine tailings	Hazard disputed, may cause cancer
Selenium	Coal, sulfur	May cause dental caries, carcinogenic in rats, essential for mammals in low doses
Yttrium	Coal, petroleum	Carcinogenic in mice over long-term exposure
Mercury	Coal, electric batteries, other industrial	Nerve damage and death
Vanadium	Petroleum (Venezuela, Iran), chemicals and catalysts, steel and nonferrous alloys	Probably no hazard at current levels
Cadmium	Coal, zinc mining, water mains and pipes, tobacco smoke	Cardiovascular disease, suspected hypertension in humans, interferes with zinc and copper metabolism
Antimony	Industry	Shortened life span in rats
Lead	Auto exhaust (from gasoline), paints (prior to about 1948)	Brain damage, convulsions, behavioral disorders, death

With permission from table in *Chemical Engineering News*, 19:30, July 19, 1971.)

Heavy metals, are strictly those beyond Rb (At. wt. is 37); yet commonly they refer to Cd, Cr, Co, Cu, Fe, Pb, Mn, Hg, Ni, Ag, and Zn (At. wt. is higher than 20, and their density is higher than water). The nonmetals, such as arsenic and selsnium, being also quite toxic are also included in as trace metals. They in particular readily form arsenate and selenate, the forms belonging to the bulk of the first generation insecticides; for example, lead arsenate $Pb_3(AsO_3)_2$, etc. A wealth of information regarding trace metals can be found in the annual conferences of "Trace Metals in the Environment" for the last two decades, which are currently held at the University of Missouri at Rolla. Their proceedings are excellent references.

There are two criteria for heavy metals appearing in the literature:

- **Interference Factor** (IF) $= \dfrac{\text{total anthropogenic emissions}}{\text{total natural emissions}}$

- **Technophility index** (TP) $= \dfrac{\text{Annual mining activity}}{\text{mean concentration of element in crust}}$ [10-3]

10.3.1 Mercury

Mercury has the potential to cause nerve and brain damage. Table 10-17 lists the industrial consumption of mercury in the United States. The largest consumption of mercury is in chlor-alkali production. Figure 10-15 gives a schematic diagram of a mercury cell for chlor-alkali production. For each ton of chlorine produced, there are about 0.1 to 0.2 kilograms of mercury lost to the atmosphere. It is estimated that 25,000 tons of chlorine are produced each year. Table 10-18 lists the major sources of mercury in the environment. It has been said that the production of chlorine is proportional to the release of mercury in the environment. The chlorination of water is for environmental protection; yet we trade off with another evil. The Reed Paper Company controversy in 1970, where fish in the adjacent river were found to contain 0.5 ppm of mercury, is a good case to illustrate this point. In recent years, a new method for the chlor-alkali process that does not involve mercury has been developed.

The mercury-containing effluents from various industries often leave substantial quantities of mercury in the sediment of nearby lakes and bays. The sediments often contain the methane-producing bacteria that slowly convert the mercury deposits into methyl mercury, which quickly enters the food chain, as shown in Figure 10-16. The methylation of Hg is done through vitamin B_{12} in fish.

$$L_5\text{-Co-CH}_3 + Hg^{2+} \rightarrow L_5Co^+ + CH_3Hg^+$$

$$2CH_3Hg^+ \rightarrow Hg(CH_3)_2 + Hg^{2+}$$

Table 10-17. Industrial Consumption of Mercury in the United States (1969)

Industry	Consumption (tonne)
Chlor-alkali	1575
Electrical apparatus	1417
Paints	739
Scientific instruments (thermometers, barometers, etc.)	531
Dentistry	232
Catalysts	225
Agricultural (seed dressings, etc.)	204
Laboratory use	155
Pharmaceuticals	55
Pulp and paper	42
Other	736
Total	5911

Table 10-18. Sources of Mercury in the Environment

Source	Quantity of mercury (tonnes)	
	Production	Estimated release[a]
Total world production of Hg from ores (1900-1979)	361,000	120,000
World production of Hg from ores (1970)	10,000	3,300
Total world release from fossil fuels (1900-1970)	--	150,000
World release from fossil fuels (1970)	--	4,800
Total anthropogenic release (1900-1970)	--	270,000
Annual release by rock weathering	--	800
Total release by rock weathering (1900-1970)	--	5,700
Total quantity of Hg in oceans	--	45,000,000
(approximately 60,000 year half-life before sedimentation)		
Total Hg released by weathering during earth's lifetime	--	1,600,000,000

Data from J. Gavis and J.F. Ferguson, *Water Res.*, 6, 989–1008 (1972)

[a] Estimated on the basis of one-third of production since this fraction was unaccounted for in the United States from 1945-1958.

Through bioaccumulation (a detailed discussion regarding bioaccumulation will be given later in this chapter) or bioamplification, the mercury contaminants propagate in the food chain, as shown in Figure 10-17. Although new technology has been introduced to almost all mercury-using plants, and this has eliminated most of the mercury in their effluents, relatively large amounts of mercury already have been dumped in sediments and will continue to be a source of methyl mercury for many years to come.

Both aryl and alkyl mercurials are efficient fungicides; for example, phenyl mercuric dimethyldicthiocarbonate, PhHg Se (=S)—N(CH₃)₂, is used in papermills as a slimicide and a mold-retardant for paper. Ethylmercuric chloride, C_2H_5HgCl, is used as a seed fungicide. Often, seed grains are coated with alkyl mercury halides for protection. This practice becomes a major source for mercury poisoning.

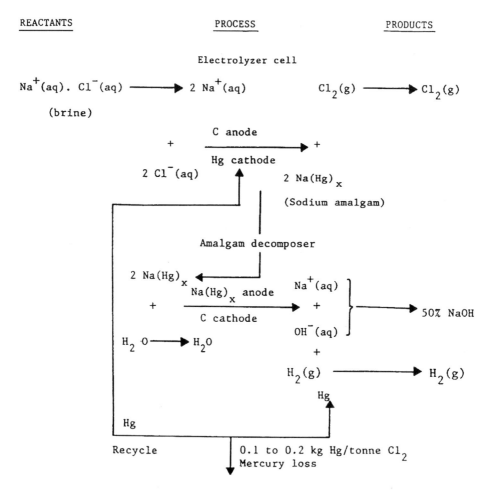

Figure 10-15. Schematic diagram of mercury cell for chlor-alkali production.

An infamous case of mercury poisoning occurred in Minamata, Japan, between 1953 and 1960. A total of 111 cases of poisoning and 43 deaths were reported in the fishing village. The so-called "Minamata diseases" gave symptoms such as numbness of lips and limbs; impaired vision, hearing, and speech; and difficulties in walking and coordination. Later on, it was found that the mercury in fish averaged 5–20 ppm and people were actually poisoned by CH_3Hg^+ in fish.

Consumption of mercury in food can range from no ill effect (0.1 mg/day) to a toxic or a lethal dose (20 mg/day), as shown in Figure 10-18. The expression "mad as a hatter" derives from exposure of hat-makers to mercury in the form of $Hg(NO_3)_2$ in the felt. Other occupations that have this potential hazard include gilding and mirror-making as well as thermometer manufacturing.

(i)

Mechanism of Propagation

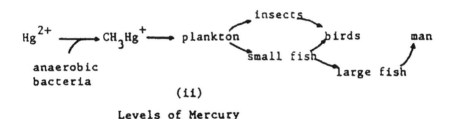

(ii)

Levels of Mercury

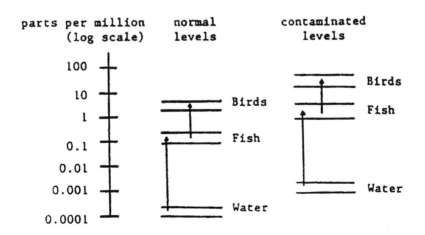

Figure 10-16. Propagation of mercury in food chain.

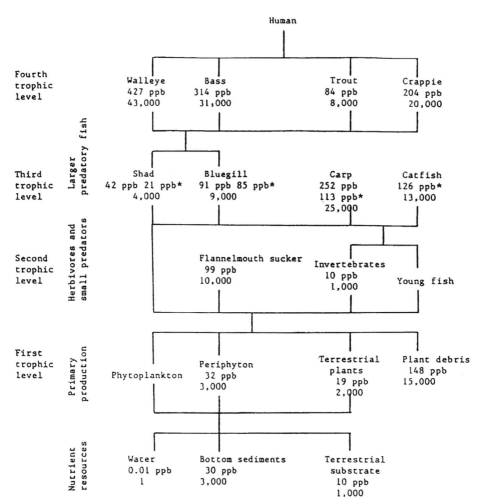

Figure 10-17. Bioamplification of mercury in Lake Powell, Utah and Arizona. The mean level of mercury in parts per billion (ppb) and the amplification factor above the surrounding water are given for each species. Thus, the Walleye contains a concentration of mercury 43,000 times that of the water. Data from L. Potter, D. Kidd, and D Standiford. *Environ. Sci. Technol.* 9 (1), 41–46 (1975).

They all involve neurological disorders. Even today we still practice mercury almagum by dentists in a patient's mouth. The antidote for acute mercury poisoning is the British anti-Lewiside (BAL). The complex BAL·Hg is not a simple monomeric chelate; instead, it forms a polymer of (BAL·Hg)$_n$, as shown in Figure 10-19.

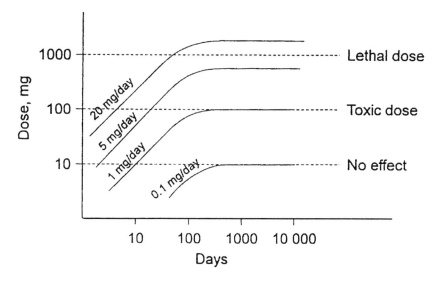

Figure 10-18. Accumulation curves for different levels of mercury in the diet. (N.J. Bunce, 1994)

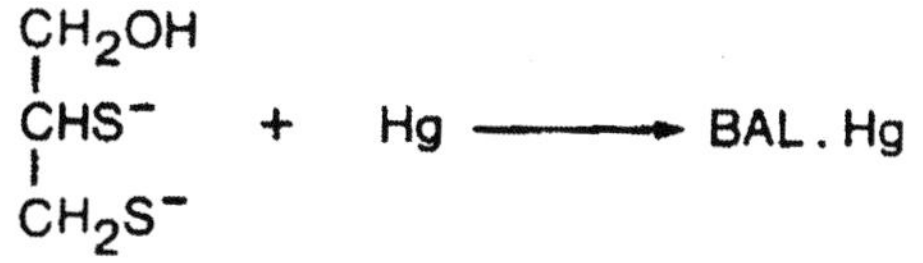

$$\begin{array}{c} CH_2OH \\ | \\ CHS^- \\ | \\ CH_2S^- \end{array} \quad + \quad Hg \quad \longrightarrow \quad BAL \cdot Hg \qquad \log K = 25.7$$

BAL

$$\cdots S-Hg-S \underset{CH_2CH}{\diagdown} \quad \overset{S-Hg-S}{\diagdown} \underset{CH_2CH}{\diagdown} \quad \overset{S-Hg-S}{\diagdown} \cdots$$
$$\qquad\qquad \underset{CH_2OH}{|} \qquad\qquad \underset{CH_2OH}{|}$$

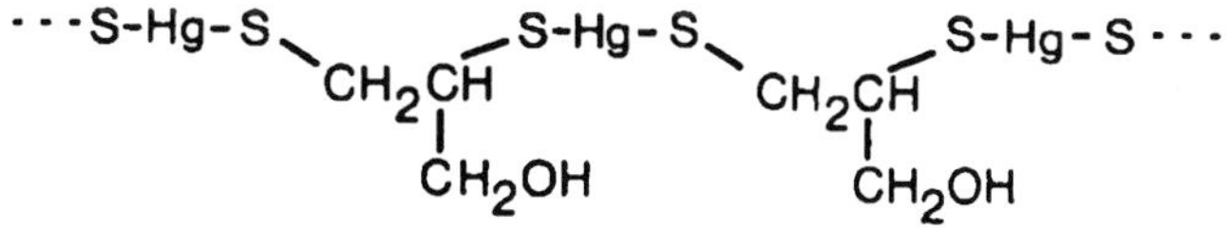

$(BAL \cdot Hg)_n$

Figure 10-19. Polymer form of BAL.

10.3.2   Lead

Due to tetraethyl lead as an important additive to gasoline, the roadway pollution of lead in highways are worldwide. It is not unusual to have 1000 ppm of lead concentration in the soil of an urban area. Some airborne lead concentrations are illustrated in Table 10-19. Auto-exhaust includes $PbCl_2$, $PbBrCl$, and $PbBr_2$, because dichloroethane and dibromoethane are added for avoidance of valve-sticking problem of Pb deposit. Other sources of lead include the spent lead-acid battery. Physiological effects of lead poisoning for the most part include damage to heme synthesis or kidney function or permanent nerve dysfunction if the blood lead level reaches 0.33–0.8 ppm. The mean blood level for traffic police and automobile-tunnel employees is already 0.3 ppm, as shown in Table 10-20. It has been argued with some cogency that lead poisoning contributed to the decline of the Roman Empire. The ruling aristocracy had lead plumbing and drank wine from lead-lined casks. With the prohibition and decreased use of lead plumbing, lead-based paints, lead ceramic glazes, and leaded gasoline, it is hopeful that our pollution and consequently the poisoning will be diminished.

Table 10-19. Atmospheric lead concentrations at different sites

Location	Type of site	Lead concn. ($\mu g/m^3$)
North Central Pacific Ocean		0.0010
Greenland		0.005
California	White and Laguna Mountains	0.008
California	Remote mountains	0.12
Berlin	Quiet streets	0.4-0.5
Philadelphia		1.6
Berlin	Busy street	3.8
New York	2-75 m from traffic	4.1
Los Angeles	Central city	4.3-6.6
Detroit	5-150 m from traffic	4.8
Los Angeles	4-20 m from traffic	7.6

Source: H.A. Waldron and D. Stöfen, "Sub-clinical Lead Poisoning," Table 4, pp. 10–11. Academic Press, New York, 1974.

Table 10-20. Lead content of human blood

Lead (ppm)	Significance
0.01	"Natural" blood lead level before man began using lead
0.10	Lower limit of "normal" blood level in the United States
0.25	Mean blood lead level in the United States
0.25	Suggested "danger" blood lead level for children
0.30	Mean blood lead level in Glasgow children
0.30	Lowest lead level found in industrially exposed adults having mild symptoms of lead poisoning
0.31	Mean blood lead level in Manchester children
0.40	Upper limit of "normal" blood lead level in the United States
0.40	Lower average blood lead level in children showing lead poisoning symptoms
0.40	Lowest approximate level found in industrially exposed adults having severe symptoms of lead poisoning
0.70	European "danger" threshold for occupational poisoning
0.80	United States "danger" threshold for occupational poisoning

Source: T.J. Chow, Chem. Brit. 9 (6), 260 (1973).

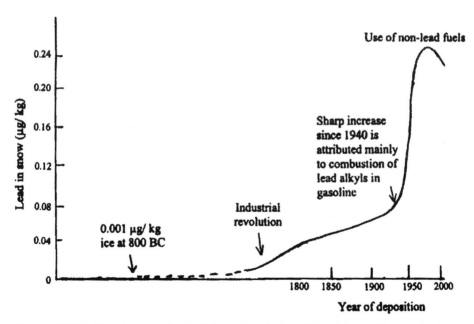

Figure 10-20. The lead content of Greenland snow (Reprinted with permission from S. K. Hall, *Environ. Sci. Tech.*, 6, 31 (1972), © The American Chemical Society).

For children, the high body burden of Pb will cause mental retardation and hyperactivity. According to the United States Centers for Disease Control, the blood level of each child should be kept below 10µg/100mL ideally, and the avoidance of levels up to 25µg/100mL will save society approximately $4,600 in health and special education costs per child. Data from the Greenland ice sheet show that the lead increase was initiated by the Industrial Revolution. However, recent findings verify that there was an increase in 500 b.c. and 400 a.d. as well, as shown in Figures 10-20 and 10-21. More recently, besides used automobile batteries, the cathode ray tubes (CRT) used in televisions and computer monitors contribute significantly to the lead pollution. Thus the e-wastes are an important source of leachable lead waste.

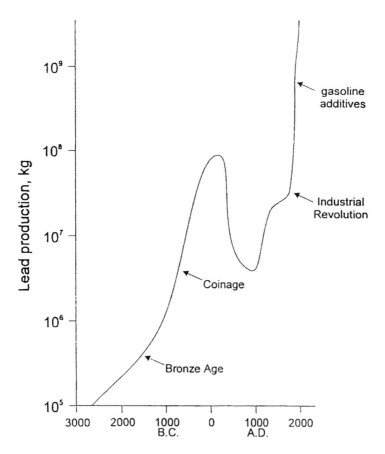

Figure 10-21. Historical production and consumption of lead (J.O. Nriagu, The Biochemistry of Lead, Elsevier/North Holland Biomedical Press, Amsterdam, 1978, and from Pathways, cycling and transformation of lead in the environment, Ed. P.M. Stokes, Royal Society of Canada, 1986).

10.3.3 Cadmium

The chemical nature of **cadmium** is similar to zinc. They both become insoluble in marine sediments as sulfides in the reducing condition especially when the sediment is anaerobic; they become soluble in aerobic condition, for example, the soluble ion pair of $CdCl^+$, due to the oxidative release of CdS. In seawater, most Cd appears as chloride complexes, such as $CdCl^+(29\%)$, $CdCl_2(37\%)$, $CdCl_3^-(31\%)$, and $Cd^{2+}(2\%)$.

Cadmium is known to cause hypertension and kidney damage. Studies have shown that there is correlation between the death of cronical uremia and the drinking water quality — especially that of well water. Figure 10-22 is a schematic of the metabolism of cadmium.

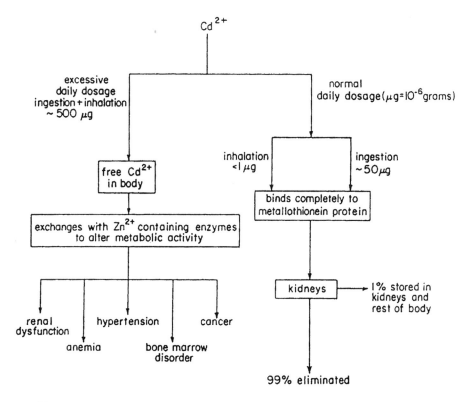

Figure 10-22. Metabolism of cadmium. (Source: Dose date abstracted from L. Friberg, M. Piscator, and G. Nordberg, Cadmium in the Environment. cleveland: CRC Press, 1971), pp. 25, 26.

Outbreaks of cadmium poisoning also have occurred in Japan in the Sasu River Basin. Many people developed a painful condition of their bones in which they were subjected to numerous fractures. This disease is called "itai itai" (ouch ouch). The cadmium content of paper is quite high due to the sizing additives. In this manner, a development of hypertension also can occur due to cigarette smoking.

10.4 HAZARDOUS WASTE

The production and use of large quantities of organic and inorganic chemicals has resulted in the output of a great deal of by-products and waste material, much of which is hazardous. In addition, there are numerous solid wastes generated at the same time (see Table 10-1). It is important to classify the hazardous waste material. Some of the hazardous chemicals will be very persistent if no abatement method is imposed. For example, the half-life of vinylidene chloride is longer than one hundred thousand years by precipitation (Table 10-21).

Table 10-21. Estimated Atmospheric Half-Lives of Five Hazardous Chemicals, Assuming Removal Only by Dissolution in Raindrops

Compound	P_{is}, in torr	x_{is}	α_i	$\tau_{\frac{1}{2}}$, in year
Acrylonitrile	1.14×10^2	1.0	9.05×10^3	0.8
Benzo(a)pyrene	5.46×10^{-9}	2.8×10^{-11}	5.4×10^3	1.4
Ethylene dichloride	8.4×10^1	1.6×10^{-3}	1.9×10^1	3.9×10^2
Tetrachloroethylene	1.85×10^1	1.1×10^{-5}	6.1×10^{-1}	1.2×10^4
Vinylidene chloride	6.17×10^2	3.9×10^{-5}	6.5×10^{-2}	1.1×10^5

Hazardous waste consists of individual waste materials or combinations of wastes that are presently or potentially dangerous to humans or other living organisms. For regulatory purposes, hazardous wastes are defined in terms of flammability (F), corrosivity (C), reactivity (R), and toxicity (T). **Flammable hazardous wastes** include: (1) liquids having a flash point below 60°C, (2) nonliquids liable to cause fires through friction, absorption of moisture, or spontaneous chemical change and liable, when ignited, to burn so vigorously or persistently as to create a hazard, (3) ignitable compressed gases, and (4) oxidizers. **Corrosive hazardous wastes** include: (1) aqueous wastes with a pH of less than or equal to 2.0 or greater than or equal to 12.5, and (2) liquid wastes capable of corroding steel at a rate greater than or equal to 0.250 inches per year.

Reactive hazardous wastes are those including explosiveness (E) that (1) readily undergo violent chemical changes, (2) react violently or form potentially explosive mixtures with water, (3) generate toxic fumes in a quantity sufficient to present a danger to human health, (4) explode when subjected to a strong initiating force, or (5) explode at normal temperatures and pressures. **Toxicity of hazardous wastes** is defined in terms of the level of contaminants in a leachate from hazardous wastes obtained by means of a specified extraction procedure. Basically, a waste is classified as toxic if its leachate, extracted by an acidic medium with a pH of 5.0, contains 100 times or more the approved levels of any of the toxic contaminants identified in the National Interim Primary Drinking Water Standards.

	Type	Action
PE	Production of Evil	Plan for Emergency
R	Reactivity	Recognition
F	Flammability	Forecast
E	Explosiveness	Evaluation
C	Corrosiveness	Control
T	Toxicity	Training

For the above, the buzzword "**perfect**" helps us to remember the major types of hazardous materials. In the case of emergency response to a major accident, certain action also should be taken.

The US Environmental Protection Agency (EPA) estimated, even as of 1976, that 35 billion kilograms of hazardous waste were being produced annually and that only 10% was disposed of in an environmentally acceptable manner. By August of 1983, the figure had jumped to 150 billion kilograms, and by 2000 this figure is about 260 billion kilograms. There are about 14,000 abandoned hazardous waste sites in the United States. The 10 most dangerous dumps are listed in Table 10-22, and there are also the Superfund sites ready for cleanup.

Table 10-23 lists examples of hazardous wastes from some specific sources, while Table 10-24 lists some wastes from discarded commercial chemical products, off-specification materials, containers, and spill residues.

Table 10-22. Ten Sites Ranked as the Most Dangerous Abandoned Hazardous Waste Dumps in the Nation

Site Name	Hazard	Years of Disposal	Type of Waste	Contaminants
FMC Corp. Fridley, Minnesota	Contamination of drinking water source for cities of Fridley and Brooklyn Center; pollution of Mississippi River, the source of potable water for Minneapolis	Early 1930s to early 1970s	Solvents, paint sludges painting waste	Trichlorometh ylene, methyl chloride, benzene, others
Tyboon Corner Landfill New Castle County, Delaware	Surface water and groundwater contamination	na[a]	Sanitary, industrial waste	na[a]
Brown Lagoon Brown Borough, Pennsylvania	Ground water and surface water contamination. Lagoon lies adjacent to Bear Creek, which joins the Allegheny River, a water source for Pittsburgh	na[a]	Wastes from coal mines, oil fields, and chemical firms	na[a]
Industri-Plex 128 (Mark Phillip Trust) Woburn, Massachusetts	Surface water and ground water contaminations	1953 to 1981	Wastes from manufacture of insecticides, explosives, acids, tanned hides, and residues	Arsenic, lead, chromium, others
Liport Landfill Gibbsoboro and Pittman Townships, New Jersey	Surface water, ground water and air pollution; site is located in area of fruit orchards	1958 to 1971	Domestic and Industrial wastes	Benzene, toluene, bis(2-chloromethyl), ether, beryllium, mercury

Table 10-22. (Continued)

Site Name	Hazard	Years of Disposal	Type of Waste	Contaminants
Sinclair Refinery Wellsville, New York	Surface water and possibly ground water contamination	na[a]	Refinery wastes, including oil sludges and fly ash	Mercury, polychlorinat ed biphenyls, oil components
Price Landfill Pleasantville, New Jersey	Groundwater contamination of potable water source of Pleasantville; plume of contamination threatens Atlantic City	1969 to 1976	Sanitary and industrial wastes	Benzene, chloroform, trichloroethyl ene
Pollution Abatement Services, Oswego, New York	Surface water and groundwater pollution; polluted surface water discharges into Lake Ontario	1970 to 1976	Polymer gas, plating wastes, metal sludges, paint wastes, laboratory chemicals	Large quantities of polychlorinat ed biphenyls, others
Laboratory Site Charles City, Iowa	na[a]	na[a]	na[a]	na[a]
Helen Kramer Landfill Mantua Township, New Jersey	Surface water and groundwater contamination	1970 to 1980	Sanitary, construction and nonchemical industrial wastes	na[a]

[a] na = not available

Data courtesy of the American Chemical Society

Table 10-23. Hazardous Waste from Specific Sources

Hazardous Waste Number	Hazardous Waste	Hazard Code
Petroleum refining		
KD 48	Dissolved air flotation (DAF) float from the petroleum-refining industry	(T)
KD 49	Slop oil emulsion solids from the petroleum refining industry	(T)
Leather tanning and finishing		
KD 53	Chrome (bloc) trimmings generated by the following subcategories of the leather tanning and finishing industry: hair pulp/chrome tan/retan/wet finish; hair save/chrome tan/retan/wet finish; retan/wet finish; no beam-house; through-the-bloc; and shearling	(T)
Iron and Steel		
KD 60	Ammonia lime still sludge from coking operations	(T)
Primary zinc		
KD 67	Electrolytic anode slimes/sludges from primary zinc production	(T)

Table 10-24. Discarded Commercial Chemical Products, Off-Specification Species, Containers, and Spilled Residium

Hazardous Waste Number	Hazardous Waste
PO03	Acroleum $H_2C=CH-CHO$
PO13	Barium cyanide, $Ba(CN)_2$
PO24	p-Chloroaniline, $H_2N-C_6H_4-Cl$
PO50	Endosulfan.
PO63	Hydrocyanic acid, HCN
PO65	Mercury fulminate, $HgC_2N_2O_2$
PO61	Nitroglycerine, $CH_2(ONO_2)CH(ONO_2)CH_2(ONO_2)$

10.5 REMEDIATION TECHNOLOGY

A number of processes are employed for the treatment of hazardous wastes. Generally there are (1) biological treatments, (2) chemical treatments, such as wet air oxidation, dechlorination, precipitation, complexation, electrolysis, oxidation, and reduction, (3) physical treatments, such as thermal treatment by microwave, photolysis, or electrokinetic treatment, (4) immobilization, such as microencapsulation, vitrification, adsorption, chemisorption, passivation, reprecipitation, (5) deep well injection, (6) land treatment, (7) solar evaporation, (8) incineration, and (9) resource recovery.

A **toxicity characteristic leaching procedure** (TCLP) has been developed by the U.S. EPA for measuring the likelihood of toxic material being released into the environment and causing harm to organisms. If the material is solid, the appropriate surface-to-weight ratio is selected to cut the solid into a smaller size, and allowed to extract and set the buffers with several pHs; the extracts are then analyzed for a list of specified VOCs and metals to determine if the wastes exceed certain levels of contaminants.

10.5.1 Bioaccumulation

In Columbia River Valley, the ^{32}P pollution for plants is 7500× of that of the soil; for adult swallows, the value becomes 75,000×, and finally, for young swallows, the value reaches 500,000×.

We would like to briefly review the bioaccumulation and the volatility of hazardous wastes in soil. Bioaccumulation can serve as an indicator of how a toxic substance propagates in the food chain and gets into the final receptor, the human being. The **bioconcentration factor** (BCF) is defined as the ratio of the concentration of chemicals at equilibrium in the organism (dry weight) to the mean concentration of chemicals in solution. There are several empirical formulae used to estimate the BCF. To estimate BCF from the octanol-water partition coefficient (K_{ow}), use

$$\log \text{BCF} = 0.76 \log K_{ow} - 0.23 \qquad [10\text{-}4]$$

To estimate BCF from the solubility in water (S), use

$$\log \text{BCF} = 2.791 - 0.564 \log S \qquad [10\text{-}5]$$

To estimate BCF from the soil adsorption coefficient (K_{oc}), use

$$\log \text{BCF} = 1.119 \log K_{oc} - 1.579 \qquad [10\text{-}6]$$

Table 10-25 gives the comparison of estimated values with laboratory measurements of BCF.

Table 10-25. Comparison of Estimated Values with Laboratory Measurements of BCF

| Compound | Physical/Chemical Parameter for Estimate | | | | Estimated BCF | | | Laboratory Measurement |
	K_{over}	S (ppm)	K_{oc}	From K_{over}	From S	From K_{oc}	BCF
Nitrobenzene	851	1,700	—	99	9.1	—	15.1
Carbon tetrachloride	437	800	—	17	14	—	30
p-Dichlorobenzene	2,400	79	—	220	53	—	215
Atrazine	427	33	149	59	86	—	7.97
1,2,4-Trichlorobenzene	17,000	30	—	970	91	—	2,800
Methoxychlor	20,000	0.003	80,000	1,100	16,000	8,100	8,300
Napthelene	50,100	31.0	1,300	2,200	88	80	427
Pentachlorophenol	126,000	14	900	4,400	140	53	770
Hexachlorobenzene	170,000	0.035	3,910	5,600	4,100	280	18,500
Heptachlor	275,000	0.030	—	8,000	4,500	—	9,500
Biphenyl	5,750	7.5	—	420	71	—	437
DDT	562,000	0.0017	23,800	14,000	23,000	27,000	29,400
Aroclor 1254	2,950,000	0.01	42,500,	49,000	8,300	4,000	100,000
Chlordane	1,000,000	0.056	—	21,000	120	—	37,800

10.5.2 Soil Vapor Extraction

Hazardous wastes or chemicals often cause subsurface contamination due to chemical spills, illegal dumping, corrosion-caused leaks, or other causes. It is instructive to know how to estimate the volatilization of chemicals from soil.

For the volatilization from soil, such as TCE, and ethylene dibromide using the 1-dimensional equation,

$$\frac{\partial^2 C}{\partial z^2} - \frac{1}{D}\frac{\partial C}{\partial t} = 0 \qquad [10\text{-}7]$$

where

C = concentration in soil (M/L^3)

z = distance measured normal to soil surface, surface is zero (L)

D = diffusion coefficient in soil (L^2/T)

t = time (T)

Boundary conditions
$C = C_0$ at $t = 0$, $0 \le z \le L$
$C = 0$ at $z = 0$, $t > 0$

$$\frac{\partial C}{\partial z} = 0 \text{ at } z = L \qquad [10\text{-}8]$$

The solution is

$$C(z,t) = \frac{4C_0}{\pi}\sum_{n=0}^{\infty}\frac{(-1)^n}{(2n+1)}\exp\left[-D(2n+1)\pi^2\frac{t}{4L^2}\right]\left[\cos(2n+1)\pi\frac{(L-z)}{2L}\right] \qquad [10\text{-}9]$$

where

L = soil layer length (L)

f = flux of compound $(M/L^2 T)$

$$f = \frac{DC_0}{(\pi Dt)^{\frac{1}{2}}}\left[1 + 2\sum_{n=1}^{\infty}(-1)^n e^{\frac{-n^2 L^2}{Dt}}\right] \qquad [10\text{-}10]$$

Increasing L or decreasing t or D and if the summation is small, then

$$f = \frac{DC_0}{(\pi Dt)^{\frac{1}{2}}} = C_0\left(\frac{D}{\pi t}\right)^{\frac{1}{2}} \qquad [10\text{-}11]$$

To obtain the concentration of volatiles in the soil column, we will list the following:

Solving the 1-dimensional diffusion equation,

$$C(z,t) = C_0 \mathrm{erf}\left[\frac{z}{2(DT)^{\frac{1}{2}}}\right] \qquad [10\text{-}12]$$

Solving the error function,
(a) for values of x ≥ 2

$$\mathrm{erf}(x) \approx 1 - \left[\frac{e^{-x^2}}{x\pi^{\frac{1}{2}}}\right] \qquad [10\text{-}13]$$

(b) for values of x → 1, set x = 1 + v (v << 1)

$$\mathrm{erf}(x) \approx \mathrm{erf}(1) + \frac{2v}{e\pi^{\frac{1}{2}}} \qquad [10\text{-}14]$$

$$\mathrm{erf}(1) = 0.8427$$

(c) for values of x ≤ 0.1

$$\mathrm{erf}(x) \approx \frac{2x}{\pi^{\frac{1}{2}}} \qquad [10\text{-}15]$$

Volatilization of toxic chemicals from soil is different from that of from water. The basis is the analysis of the heat balance between the evaporating chemicals (or water) and air. Computation of the flux of volatile pollutants can be made if the diffusion coefficient, D_v, of the particular chemical in air is known. If not, one can evaluate from two other chemicals as long as the diffusion coefficients are known; for example,

$$\frac{D_1}{D_2} = \left(\frac{M_1}{M_2}\right)^{\frac{1}{2}} \qquad [10\text{-}16]$$

The flux is expressed as follows:

$$f = \frac{\dfrac{\rho_{max}(1-h)}{\delta}}{\left[\dfrac{1}{D_v} + \dfrac{\lambda_v^2 \rho_{max} M}{kRT^2}\right]} \qquad [10\text{-}17]$$

where

f = flux of compound (M/L^2T)

ρ_{max} = saturated vapor concentration at the temperature of the outer air (M/L^3)

h = humidity of the outer air ($0 \leq h \leq 1$)

δ = thickness of stagnant layer through which the chemical must pass (L)

D_v = diffusion coefficient of vapor in the air (L^2/T)

λ_v = latent heat of vaporization (cal/M)

M = molecular weight (M/mol)

k = thermal conductivity of air (cal/LK)

R = gas constant (cal/mol)

T = temperature (K)

In case of trichloroethylene (TCE), the properties are summarized in Table 10-26. An example is illustrated here to show how the volatilization of some common pollutants can be evaluated.

[Example 10-1]   Find the flux of TCE at 20°C.

From Equation 10-17 and Table 10-26

$$f = \cfrac{\cfrac{(4.3 \times 10^{-4}\,\text{g/cm}^3)\,(0.5)}{0.3\text{cm}}}{\cfrac{1}{0.072\,\text{cm}^2/\text{s}} + \cfrac{(63.2\,\text{cal/g})^2\,(4.3 \times 10^{-4}\,\text{g/cm}^3)\,(131.5\,\text{g/mole})}{(61 \times 10^{-6}\,\text{cal/s} \cdot \text{cm} \cdot \text{K})\,(1.987\,\text{cal/mol} \cdot \text{K})\,(293\text{K})^2}}$$

$$= 5.04 \times 10^{-5}\,\text{g/cm}^2\,\text{s}$$

Table 10-26. Chemical and Environmental data for Estimation of Trichloroethylene Volatilization

Parameter	Symbol	Value
Characteristics of TCE at T=293°K		
Saturated vapor concentration	ρ_{max}	4.3×10^{-4} g/cm^3
Diffusion coefficient in air	D_V	0.072 cm^2/s
Heat of vaporization	λ_V	63.2 cal/g
Thermal conductivity of air	k	61×10^{-6} cal/s-cm-K
Gas constant	R	1.987 cal/mole-K
Diffusion coefficient of vapor through soil	D	0.039 cm^2/s
Vapor pressure of TCE	P_{vp}	60 mm Hg
Vapor pressure of water	P_{H2O}	17.54 mm Hg
Diffusion coefficient of water vapor through air	D_{H2O}	0.239 cm^2/s
Initial concentration in soil (assumed)	C_0	0.05 g/cm^3
	α	1
Adsorption coefficients (assumed)	β	0
	K_{oc}	360
Solubility	S	1100 mg/L
Molecular weight	M	131.5 g/mole
Ratio C_0/C_g (=S/ρ_{max})	K_H	2.56 cm^3 air/cm^3 water
Environmental Characteristics (assumed)		
Humidity	h	0.5 (=50%)
Stagnant air layer thickness	δ	0.3 cm
Temperature	T	293 K
Wind speed	V	100 cm/s
Soil solid density	ρ_{solid}	2.65 g/cm^3
Soil bulk density y=$(1-\eta-\theta)\rho_{solid}$	ρ_b	1.32 g/cm^3
Volumetric soil water content	θ	0.2 cm^3/cm^3
Soil air content	η	0.3 cm^3/cm^3
Depth of soil column	L	20 cm
Water vapor flux per unit area	f_w	6.7×10^{-2} g/cm^2/day

The soil vapor extraction system is established quite well now, as shown in Figure 10-23. The modeling is based on both Darcy's law as well as continuity equations, assuming homogenous soil properties and vapor composition. The formulation for computation can be summarized by the following steps:

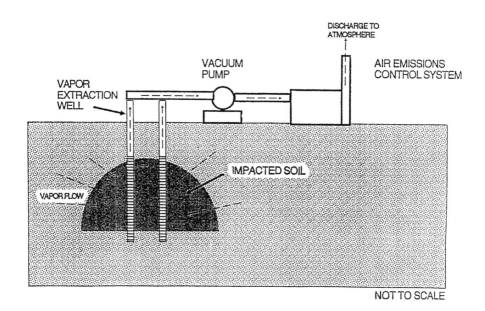

Figure 10-23. Gasoline impacted soil vapor extraction system.
(Source: T.J. Dolan, Environmental Applications, Inc.)

Dalton's law

$$q = \frac{-k\Delta p}{\mu} \qquad [10\text{-}18]$$

Continuity

$$\frac{\partial(\eta, \rho)}{\partial t} = -\nabla(\rho q) \qquad [10\text{-}19]$$

State

$$\rho = \frac{MP}{RT} \qquad [10\text{-}20]$$

Governing equation

$$S \cdot \frac{\partial P}{\partial t} = \nabla(\rho \Delta p)$$

where

$$S = \frac{nM\mu}{RTk} \qquad [10\text{-}21]$$

10.5.3   Encapsulation and Vitrification

Usually the waste block can be coated with polyethylene as a jacket around the waste block. Often low molecular weight polybutadiene can be added to the block during heat under pressure as a binder as well as a cross-linking agent. To ensure that immobilization is permanent, the concept of **microencapsulation** has been adopted. The waste particles are admixed with a binding agent; for example, asphalt water emulsion. After the water is evaporated, the particles are coated with a layer of asphalt. The use of asphalt to encapsulate radioactive waste is worth mentioning because asphalt, especially the asphaltene portion, can tolerate the high energy radiation. No observable decomposition was found for a sample of asphaltene indicated under van der Graff generator for one week.

At a higher temperature range of ca. 1300°C, inorganic wastes, for example the radioactive types, can be encased in a medium of glass; this process is termed **vitrification**. Common glass composition involves different amounts of B_2O_3, Al_2O_3, CaO, MgO, and Na_2O in SiO_2 by fusion under high temperatures. Other vitrification methods include concrete, a mixture of asphalt and concrete, or titanate ceramics.

10.5.4   In-Situ and On Site Soil Remediation Methods

Soils have a great capacity for trapping and storage of a great variety of contaminants. Some of the common practices are listed as follows:

- **Soil Washing** — The contaminated soil is subject to high pressure jets of water to break down the soil structure. The drawback is that the contaminants seem to concentrate in the "fines," and are difficult to treat subsequently.

- **Soil Extraction** — Conventional organic solvents are used even under ultrasound or supercritical fluid extraction conditions.

- **Thermal Desorption** — By heating the contaminated soils in a dry kiln at 200–500°C, the volatile off-gas is further treated. Sometimes other chemicals are added to the soil to enhance the release of contaminants.

- **Soil incineration** — A method of treating at even higher temperatures for releasing all contaminants.

- **Soil Vitrification** — Electrodes are placed vertically in soil and the soil can be melted at 1600-2000°C in the zone that grows downwards from the electrode.

- **Electrokinetic Treatment** — Low-level DC current (mA/cm^2) of which a few volts are sent through the soil. Cations are transported toward the cathode and anions to the anode.

- **In-Situ Bioremediation** — Actually this is referred to as biostimulation, a process designed to degrade contaminants in soil with minimal disturbances, usually aerobically.

Applications are directed towards such hazards as petroleum hydrocarbons from **leaking underground tanks** (UGT). A typical layout scheme of in-situ bioremediation is shown in Figure 10-24. Figure 10-25 depicts applications for liquid and gaseous phase bioremediation. Usually an oxygen release compound is pumped in with the additional nutrients and bacterial seedings. As a parameter for monitoring the concentration of petroleum hydrocarbons, it is preferable to conduct all measurements of BTEX (benzene, toluene, ethyl benzene, and xylenes). In addition, the total chlorinated compound concentrations are also measured, as shown in Figure 10-26. In-situ stimulation is a process whereby a chemical is added. Besides the oxygen release agent such as urea-oxygen adducts, hydrogen peroxide or nitrate solution is quite common. Also, the addition of naturally isolated or indigeneous microorganisms for a particular type of contaminate can be conducted, as shown in Table 10-27.

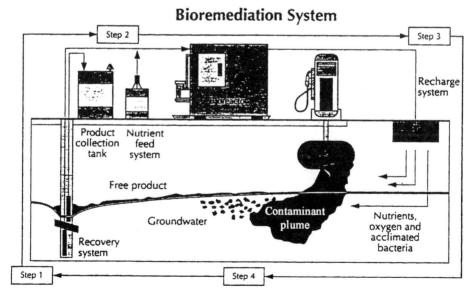

Figure 10-24. A typical in situ bioremediation system (From B.N. Hicks and J.A. Caplan, Pollution Eng. Jan 15, pp 30-33, 1993).

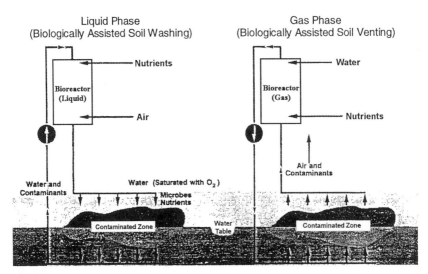

Figure 10-25. Liquid and gaseous phase bioremediation technologies (US DOE, Environmental Restoration of Waste and Management Program DOE/ EM - 0013P, 1991).

Table 10-27. Microorganism Known to Degrade Toxic Organic Pollutants

Organic pollutants	Organism
Phenolic compounds	*Achromobacter, Alcalegenes, Acenitobacter, Arthrobacter, Azotobacter, Bacillus cereus, Flavobacterium, Pseudomonas putida, P. aeruginosa and Nocardia Candida tropicalis, Debaromyces subglobosus, and Trichosporon cutaneoum, Aspergillus, Penicillium, and Neurospora*
Benzoates and related compounds	*Arthrobacter, Bacillus sp., Micrococcus, Moraxella, Mycobacterium, P. putida and P. fluorescence*
Hydrocarbons	*Escherichia coli, P. putida, P. aeruginosa, and Candida*
Surfactants	*Alcaligenes, Achromobacter, Aerobacter aeruginosa, Bacillus, Citrobacter, Clostridium resinae, Corynebacterium, Flavobacterium, Nocardia, Pseudomonas, Candida, and Cladosporium*
Pesticides	
DDT	*P. aeruginosa, 640X*
Linurin	*B. sphaericus*
2,4-D	*Arthrobacter and P. cepacia*
2,4,5-T	*P. cepacia*
Parathion	*Pseumodonas sp. and E.coli; P. stutzeri and P. aeruginosa*

Source: D.L. Wise, Biotreatment Systems, Vol 1, CRC Press, 1988.

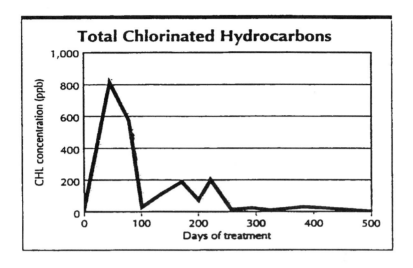

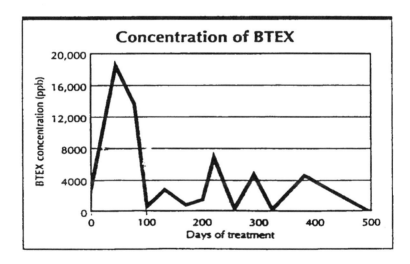

Figure 10-26. Total chlorinated hydrocarbons and BTEX concentrations from one worst-case monitoring well (After B.N. Hicks and J.A. Caplan, loc. cit.)

REFERENCES

10-1. P. A. Veslind and A. E. Rimer, *Unit Operations in Resource Recovery Engineering*, Prentice-Hall, Englewood Cliffs, New Jersey, 1981.

10-2. T. F. Yen, *Recycling and Disposal of Solid Wastes, Industrial, Agricultural, Domestic*, Ann Arbor Science Publishers, Ann Arbor, Michigan, 1975.

10-3. P. J. Knox, *Resource Recovery of Municipal Solid Wastes*, American Institute of Chemical Engineers, New York, New York, 1988.

10-4. K. D. Racke and J. R. Coats, *Enhanced Biodegradation of Pesticides in the Environment*, American Chemical Society, Washington, D.C., 1990.

10-5. D. W. Connell, *Bioaccumulation of Xenobiotic Compounds*, CRC Press, Boca Raton, Florida, 1989.

10-6. E. A. McBean, F. A. Rovers and G. J. Farquhar, *Solid Waste Landfill Engineering and Design*, Prentice-Hall, Englewood Cliffs, New Jersey, 1995.

10-7. C. A. Wentz, *Hazardous Waste Management*, McGraw-Hill, New York, New York, 1989.

10-8. J. T. Pfeffer, *Solid Waste Management Engineering*, Prentice-Hall, Englewood Cliffs, New Jersey, 1992.

10-9. N. J. Bunce, *Environmental Chemistry*, 2nd ed., Wuerz Pub., 1994.

10-10. C. Polprasert, *Organic Waste Recycling*, 2nd ed., Wiley, New York, 1996.

10-11. J.T. Cookson, Jr., *Biremediation: Engineering Design and Application*, McGraw-Hill, New York, 1995.

10-12. E.L. Appleton, "A Nickel-Iron Wall Against Contaminated Groundwater," Env. Sci. Technol. *30* 536A (1996).

10-13. G. D. Andrews and P. M. Subramanian, *Emerging Technolgies in Plastics Recycling*, American Chemical Society, Washington DC, 1992.

10-14. B. Varon, *Soil Pollution: Processes and Dynamics*, Springer-Verlag, Berlin, 1996.

10-15. C. J. Watras and J. W. Huckabee, *Mercury Pollution: Integration and Synthesis*, CRC Press, Boca Raton, Florida, 1994.

10-16. S. Willetts, "Mercury and Arsenic Wastes: Removal, Recovery, Treatment and Disposal", Chemical Ind., *17* 689 (1994).

10-17. C. G. Down and J. Stocks, *Environmental Impact of Mining*, Applied Science, Barking, England, 1977.

10-18. J. Douglas, "Cleaning up Mercury with Genetic Ecology," EPRI Journal, *10*, 20 (1994).

PROBLEM SET

1. A municipal solid waste has the following chemical composition:

   Derive an approximate chemical formula for the organic portion of the waste. You have to normalize the composition on a moisture-free and ash-free basis. Estimate its energy content by the modified Dulong formula

   $$kJ/kg = 337\%C + 1428[\%H - (\%O/8)] + 9\%S$$

2. If a sanitary landfill were set in clay having 50% porosity and if the coefficient of hydraulic conductivity is 1×10^{-7} cm/s, calculate how long it would take the leachate to percolate from the bottom of the landfill through the underlying soil to the groundwater table 1.5 m below, using Darcy's law (assuming that the leachate is not allowed to build up in the landfill and the underlying soil is saturated).

THE BIOSPHERE

This section is devoted to the **biosphere**, the realm of living organisms and their interactions with the lithosphere, the hydrosphere, and the atmosphere. In this section, we will discuss the geochemical and health and risk aspects of the biosphere. The chapter that follows will begin with an introduction to the exosphere, followed by an explanation of the radioactive dating method, which can determine the age of the Earth, followed by a description of some important principles in biochemistry. In the realm of biopolymers, we will address the question of what life is and how it evolves, especially the problem of chemical evolution and autopoiesis. Environmental geochemistry will also be discussed. The exploration and significance of geochemical biomarkers (or the molecular fossils) will be explained. Especially the global geochemical cycles of carbon from the distant past to the future will be examined by the kinetic model.

This section addresses the general introduction of geochemistry, one of the old disciplines in the advancement of modern science. Unfortunately, it has been neglected in the realm of environmental engineering and science except the geotechnical aspect of it. Only until recently, with the emphasis on soil and groundwater remediation, people never paid attention to the geochemical aspect of the minerals and soils. Again, this is a fragmented end-piece of knowledge. We feel the overall underlining principles in geochemistry will have a broad impact to investigators practicing environmental engineering.

Geology is sometimes referred to as earth science and planetary science. Accordingly, geochemistry deals with space chemistry; chemical evolution; transformation and occurrence of minerals and fossilized organic resources; the global geochemical cycles and reservoirs; the origin and diagenesis of organic deposits; the geochemical biomarkers, and so on. To understand the development of evolution, the concept of biochemistry and biopolymer are introduced here.

In a larger scheme, the whole biosphere is stratified and is only limited to certain depths in the atmosphere, lithosphere, and hydrosphere, as shown in Figure E-1. Through biological activity, the plants enrich certain elements such as O, K, and P. The following shows the sequence of abundance of elements in the cosmos in plants:

cosmos: H> He> O> C> K> N> Si> Fe> S> Al> Ca> Na> P> K

plants: O > H> C> K> N> Si > Ca> Mg> P> Na> Fe> S

Figure E-2 shows that the elemental distribution in our blood is very close to that from the granite in the earth. The biosphere is an integral part of the universe and cannot be separated. The lithosphere, hydrosphere, atmosphere, pedosphere, and biosphere are interconnected and unified.

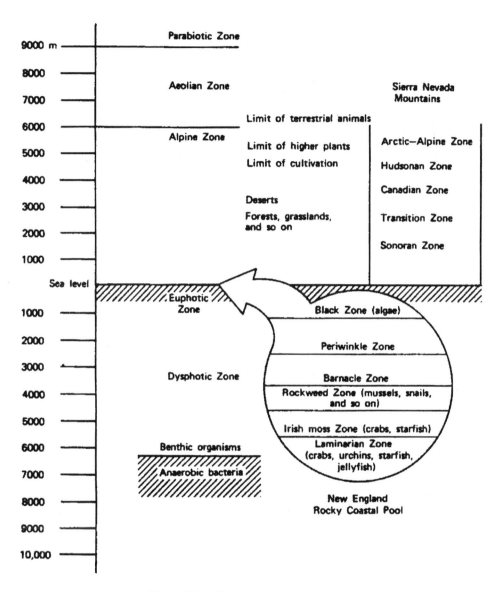

Figure E-1. Biosphere Stratification

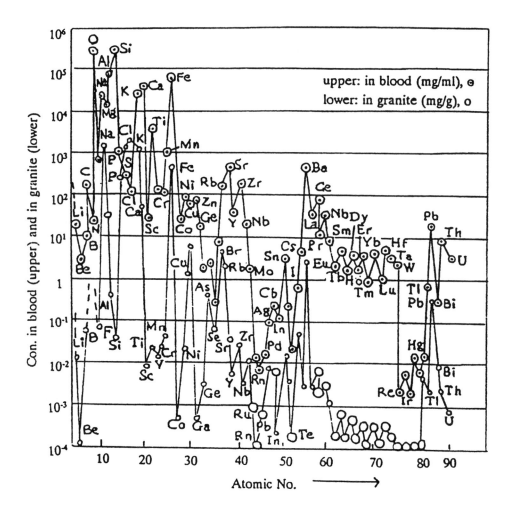

Figure E-2. Elemental distribution in biosphere and lithosphere.
(Source T.F Yen)

CHAPTER 11

GEOCHEMICAL ASPECTS OF BIOSPHERE

11.1 EXOSPHERE

As seen in Figure 11-1, the environment beyond the atmosphere is often called the **exosphere**. It includes the solar system, the galaxy, and the whole universe. It is now believed, according to the "big bang" theory, that the universe originated between 13 and 20 billion years ago from a giant fireball, which emerged from an infinitely dense collection of neutrons. The "big bang" resulted in the formation of hydrogen and some helium nuclei from which the galaxies and stars evolved. The "big bang" initiated a process of evolution that continues now and seems to extend into the future. Figure 11-2 shows the lifecycle of a star.

In astrophysics, the color indexes of galaxies are commonly used as age indicators, old galaxies being redder and brighter than young ones. The spectral type is related to the temperature of the galaxy as follows. When stars evolve, they become redder and brighter as they leave the main sequence for the red

501

giant branch, as shown in Figure 11-3, the **Hertzsprung-Russel diagram**. The location of a star in the diagram can give some information about its age.

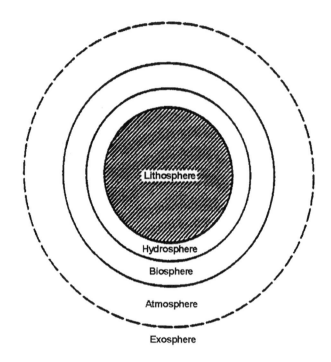

Figure 11-1. Our total environment (After R.A. Horne, 1978).

Now, while the universe is almost entirely made up of hydrogen and helium, it does include a significant amount of heavier elements such as O, C, N, P, S, Fe, and Mg. All these elements are believed to have been produced in the early history of the galaxy, prior to the formation of the solar system. It is also believed that the planets and the sun in the solar system were formed from the same material; that is, they started with the same initial chemical composition. Cosmochemistry mainly deals with the formation and concentration of these major elements and their compounds. Carbon has been found to be a universal element; for example, meteorites contain organic carbon in the range of 0.15% (1.99% by wt.) and this is called **retigen**. Figure 11-4 gives the relative abundance of elements in the universe.

It is now a widely accepted theory that elements heavier than hydrogen and helium are synthesized from these remnants of the "big bang" by nuclear reactions in stars. At high temperatures, the activation barrier to fusion of hydrogen is overcome and a nuclear reaction called "**hydrogen burning**" occurs. As a result, helium is produced, accompanied by a release of energy. If there is a small amount of ^{12}C present, the following catalytic nuclear reaction may be the major pathway for hydrogen burning.

$$^{12}_{6}C + {}^{1}_{1}H \rightarrow {}^{13}_{7}N + \gamma \rightarrow {}^{13}_{6}C + {}_{1}\beta$$

$$^{13}_{6}C + {}^{1}_{1}H \rightarrow {}^{14}_{7}N + \gamma$$

$$^{14}_{7}N + {}^{1}_{1}H \rightarrow {}^{15}_{8}O + \gamma \rightarrow {}^{15}_{7}N + {}_{1}\beta$$

$$^{15}_{7}N + {}^{1}_{1}H \rightarrow {}^{12}_{6}C + {}^{4}_{2}He$$

Summation of the above equations will give the last expression:

$$4{}^{1}_{1}H \rightarrow {}^{4}_{2}He + 3\gamma + 2{}_{1}\beta$$

Note: Notation of nucleide can be either $_{Z}Element^{A}$ or $_{A}Element^{Z}$. Sometimes Z is omitted altogether.

Similar to this hydrogen burning is the process of nuclear fusion process mentioned in Chapter 3. The sun and other stars give off their energy by a similar process: $4\,{}^{1}H \rightarrow {}^{4}He + energy$ at 3.78×10^{23} erg/sec consuming it in 10^{11} years.

G- and cooler star may undergo the following processes:

$$^{1}H + {}^{1}H \rightarrow {}^{2}H + {}_{1}\beta$$

$$^{1}H + {}^{2}H \rightarrow {}^{3}He + \gamma$$

$$^{3}He + {}^{4}He \rightarrow {}^{7}Be + \gamma$$

$$^{7}Be \rightarrow {}^{7}Li + {}_{1}\beta$$

$$^{1}H + {}^{7}Li \rightarrow 2{}^{4}He$$

With the energy from the hydrogen burning, the temperature of the star rises to a level where the activation barrier for hydrogen burning is exceeded. Then helium can be converted to carbon and oxygen as follows:

$$^{3}_{2}He + {}^{4}_{2}He \rightarrow {}^{7}_{4}Be + \gamma$$

$$^{7}_{4}Be + {}^{1}_{1}H \rightarrow {}^{8}_{5}B + \gamma$$

$$^{8}_{5}B \rightarrow {}^{8}_{4}Be + {}_{1}\beta$$

$$^{8}_{4}Be + {}^{4}_{2}He \rightarrow {}^{12}_{6}C$$

$$^{12}_{6}C + {}^{4}_{2}He \rightarrow {}^{16}_{8}O$$

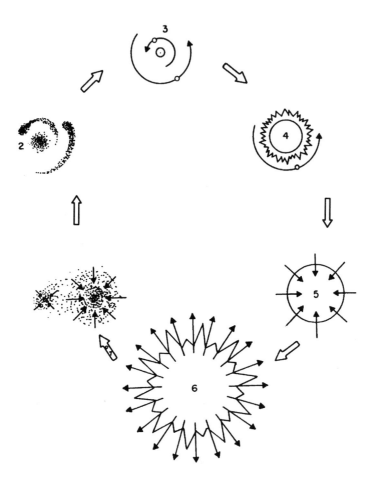

Stage 1, after the big bang the gas cloud starts to contract, the interstellar dust forming a dense mass. Stage 2, proto-star is formed after the shrinking; the impacting of protons begins to start fusion. Stage 3, present day solar system with planets and their satellites. Stage 4, fusion of hydrogen ends. Stage 5, the core shrinks to a red giant. Stage 6, supernova explosion.

Figure 11-2. Life cycle of a star (After R.A. Horne, 1978).

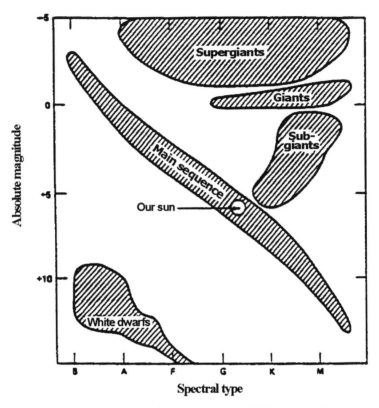

Figure 11-3. Hertzprung-Russell diagram (After R.A. Horne, 1978).

If the temperature becomes even higher, carbon and oxygen atoms can proceed to generate other heavier nuclei such as Ne, Na, Si, P, and S as follows:

$$^{12}_{6}C + {}^{12}_{6}C \rightarrow {}^{20}_{10}Ne + {}^{4}_{2}He$$

$$^{12}_{6}C + {}^{12}_{6}C \rightarrow {}^{23}_{11}Na + {}^{1}_{1}H$$

$$^{16}_{8}O + {}^{16}_{8}O \rightarrow {}^{28}_{14}Si + {}^{4}_{2}He$$

$$^{16}_{8}O + {}^{16}_{8}O \rightarrow {}^{31}_{15}p + {}^{1}_{1}H$$

$$^{16}_{8}O + {}^{16}_{8}O \rightarrow {}^{31}_{16}S + {}^{1}_{0}n$$

Should the temperature reach 3×10^{9} K, the nuclei would have enough energy to overcome the activation barrier of all nuclear reactions. The Ni atom can be generated from the α-**process**, in which the intermediate nuclei photo-disintegrate with the emission of α-particles, which in turn can be captured by the surviving nuclei to build higher mass elements.

$$^{28}\text{Si} \rightarrow 7\,^{4}\text{He}$$

$$^{28}\text{Si} + 7\,^{4}\text{He} \rightarrow\,^{56}\text{Ni}$$

$$2\,^{28}\text{Si} \rightarrow\,^{56}\text{Ni}$$

The ^{56}Ni formed from the α-process is not stable and will convert to ^{56}Fe through an **e-process** (β-decay).

$$^{56}_{28}\text{Ni} \rightarrow\,^{56}_{27}\text{Co} \rightarrow\,^{56}_{26}\text{Fe}$$

A summary of the exosphere and the origin of chemical elements are as follows:

- **Exosphere** 1 light year = 10 T km

- **Solar System** $\rightarrow$ Galaxy $\rightarrow$ Universe
  (10 G km)   (0.1 M light year)   (10 G light year)

- **Cosmochemistry**   He, H_2, OH, CO, CN, CS, SiO, etc.
                        (from microwave resonance)

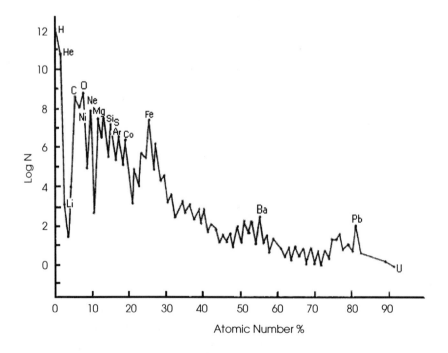

Figure 11-4. The standard abundance distribution—logarithm of the relative number of nuclei as a function of Z. (After Cameron)

- **Stellar Chemistry**

  Hertzsprung-Russell diagram

  $$R \rightarrow N$$
  $$O \rightarrow B \rightarrow A \rightarrow F \rightarrow G \rightarrow K \rightarrow M$$
  (Bluish white) (white) (yellow)   (orange)   (red)
  $$S \rightarrow Se$$
  (red)

  increase in temperature $\leftarrow$

  The main sequence of the color scheme and the examples of some common stars in Hertzsprung-Russell diagram are illustrated as follows:

Sequence type	Color	Examples
O	Blue	Alnitak
B	Blue	Rigel, Regulus, Spica
A	White	Sirius, Vegu, Altair, Denub
F	Yellow-white	Procyon, Canopus
G	Yellow	Sun, Capella, α- Centaurus A
K	Orange	Arcturus, Aldeburan, 61 Cygni
M	Red	Belilgeuse, Antures, Wolf 539

- **M-red giant**

  3-particle collision

  Nucleosynthesis $3\,{}^{4}\text{He} \rightarrow {}^{12}\text{C}$

  $4\,{}^{4}\text{He} \rightarrow {}^{16}\text{O}$

  $({}^{12}\text{C} + {}^{4}\text{He} \rightarrow {}^{16}\text{O})$

  Thus, n numbers of ${}^{4}\text{He}$ can produce stable atoms such as ${}^{20}\text{Ne}$, ${}^{24}\text{Mg}$, ${}^{28}\text{Si}$, ${}^{32}\text{S}$, and so on.

- **α-Process (α-Recombination)**

  photodisintegration at 3×10^{9} K and recombination

  $${}^{28}\text{Si} \rightarrow 7\,{}^{4}\text{He}$$

  $${}^{28}\text{Si} + 7\,{}^{4}\text{He} \rightarrow {}^{56}\text{Ni}$$

- **e-Process (β-Decay)**

  $${}^{56}_{28}\text{Ni} \xrightarrow{\beta} {}^{56}_{27}\text{Co} \xrightarrow{\beta} {}^{56}_{26}\text{Fe}$$

- **s- and r-Process (Neutron Capture)**

$$^{56}_{28}\text{Ni} \xrightarrow{^{1}_{0}n} {}^{57}_{28}\text{Ni} \xrightarrow{^{1}_{0}n} {}^{58}_{28}\text{Ni}$$

s (slow process) — red giant
r (rapid process) — supernova explosion

The preceding processes describe how all the elements in the periodic table are formed.

11.2 RADIOACTIVE DATING

The age of the earth and the solar system can be estimated by the radioactive dating method, based on an accurate determination of the ratio of radioactive isotopes to their stable daughter products. For example, one of the isotopes of potassium is continuously transformed into the elements argon and calcium.

$$^{40}\text{K} \rightarrow {}^{40}\text{C}$$

$$^{40}\text{K} \rightarrow {}^{40}\text{Ar}$$

Let us consider the granitic black mica biotite, a mineral rich in potassium. The predominant potassium isotope is ^{39}K, which is stable, while a small fraction of potassium sites are occupied by the radioactive ^{40}K. Assuming that 12% of the decay results in the formation of ^{40}Ar, there is an accumulation of ^{40}Ar at the expense of ^{40}K with increasing time. Hence, the older a given crystal of biotite, the more ^{40}Ar it should contain. The age of the sample would then be readily estimated by measuring the relative amount of ^{40}K and ^{40}Ar. For a radioactive isotope,

$$P(t) = P(0)e^{-kt} \qquad\qquad [11\text{-}1]$$

where

$P(t)$ = the number of atoms at the geological age, t,
$P(0)$ = the number of atoms at t = 0
k = the decay constant

Assuming a closed system, the mass is conserved. Thus

$$P(0) = P(t) + \Delta D \qquad\qquad [11\text{-}2]$$

where D denotes the total number of daughter atoms produced before time, t. For this special case, 12% of total decays of ^{40}K become ^{40}Ar, and

$$D_A = 0.12 \ [P(0) - P(t)] \qquad [11\text{-}3]$$

Combining the preceding equations, we obtain

$$\frac{D_A}{P(t)} = 0.12\left(e^{kt} - 1\right) \qquad [11\text{-}4]$$

If there are no argon atoms present in the rock at t = 0, then $D_A = \Delta D_A$. The equation can be rewritten and solved for t:

$$t = \frac{1}{k} \ln\left(\frac{D_A(t)}{0.12 \ P(t)} + 1\right) \qquad [11\text{-}5]$$

An isotope of rubidium, which may substitute for potassium in mineral lattice, decays into a single daughter compound, an isotope of strontium as follows:

$$^{87}Rb \rightarrow \ ^{87}Sr$$

The preceding process is also used as a geological clock. To determine the age of the sample, it is necessary to estimate the total amount of ^{87}Rb and ^{87}Sr in the sample and the amount of ^{87}Sr at $t = 0$, $^{87}Sr_0$. Because ^{87}Rb decays only to ^{87}Sr, Equation [11-5] becomes

$$t = \frac{1}{k} \ln\left(\frac{^{87}Sr^*}{^{87}Rb} + 1\right) \qquad [11\text{-}6]$$

where $^{87}Sr^* = \ ^{87}Sr - \ ^{87}Sr_0$.

Estimation of $^{87}Sr^*$ poses a great experimental difficulty, while ^{87}Sr and ^{87}Rb are readily measurable quantities. To solve the problem, $[^{87}Sr^*/^{87}Rb]$ was determined graphically from isotopic data on a series of rocks from some formations exhibiting different ratios of $^{87}Sr/^{87}Rb$. The following relationship could be easily derived:

$$\frac{^{87}Sr}{^{86}Sr} = \left(\frac{^{87}Sr^*}{^{87}Rb}\right)\frac{^{87}Rb}{^{86}Sr} + \frac{^{87}Sr_0}{^{86}Sr} \qquad [11\text{-}7]$$

By plotting $^{87}Sr/^{86}Sr$ against $^{87}Rb/^{86}Sr$, the slope will be $^{87}Sr^*/^{87}Rb$. For stony meteorites, the slope was found to be 0.0615 , as shown in Figure 11-5. Inserting this value back into Equation [11-6], and using $k = 1.39 \times 10^{-11}$ years^{-1}, we obtain the age of the stony meteorites:

$$t = \frac{1}{1.39 \times 10^{-11}} \ln(0.0615 + 1) = 4.3 \text{ billion years}$$

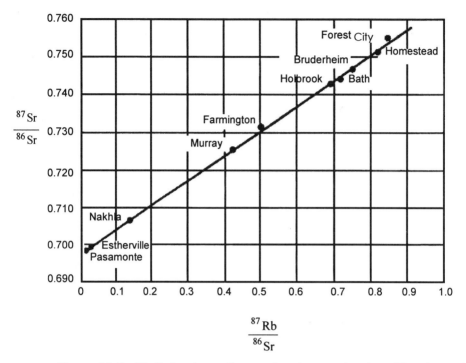

Figure 11-5. Rb-Sr isochron diagram for stony meteorites. The slope of the line gives an age of about 4.3 billion years.

Using this method on various samples, the age of earth was determined to be in the range of 3.2 and 5.7 billion years. Figure 11-6 gives a common classification of geological age.

11.3 BIOCHEMISTRY

The three most important chemical components in organisms are proteins, cellulose (or polysaccharides) and fats. The individual compounds from which these components are made are amino acids, simple sugars (such as glucose), and carboxylic acids. The twenty-some amino acids are the most important compounds found in proteins. Their structures are shown in Table 11-1 and their three-letter and one-letter abbreviations are also listed in Table 11-2. There are five purine and pyrimidine bases generally found in nucleic acids, shown in Table 11-3.

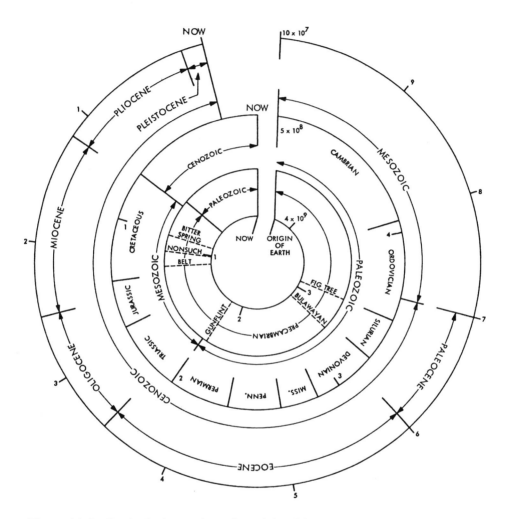

Figure 11-6. Geological ages from the origin of the earth to present (T.F. Yen).

We will now address some of the important characteristics of the amino acids and the derived compounds that make them so significant to our life:

(1) They have amphoteric behavior by which they can adjust themselves to any pH condition.

$$R - CH - COOH \leftrightarrow R - CH - COO^-$$
$$\quad\quad | \quad\quad\quad\quad\quad\quad\quad |$$
$$\quad\quad NH_2 \quad\quad\quad\quad\quad NH_3^+$$

Table 11-1. Some Important Building Blocks of Biochemistry

A. Amino acids $\underset{\underset{\text{NH}_2}{|}}{\text{RCHCOOH}}$

   1. Aliphatic amino acids (hydrophobic R groups)

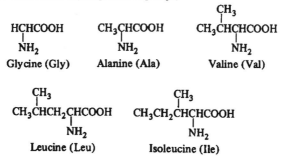

$$\underset{\underset{\text{NH}_2}{|}}{\text{HCHCOOH}} \qquad \underset{\underset{\text{NH}_2}{|}}{\text{CH}_3\text{CHCOOH}} \qquad \overset{\overset{\text{CH}_3}{|}}{\underset{\underset{\text{NH}_2}{|}}{\text{CH}_3\text{CHCHCOOH}}}$$

   Glycine (Gly)   Alanine (Ala)   Valine (Val)

$$\overset{\overset{\text{CH}_3}{|}}{\underset{\underset{\text{NH}_2}{|}}{\text{CH}_3\text{CHCH}_2\text{CHCOOH}}} \qquad \overset{\overset{\text{CH}_3}{|}}{\underset{\underset{\text{NH}_2}{|}}{\text{CH}_3\text{CH}_2\text{CHCHCOOH}}}$$

   Leucine (Leu)   Isoleucine (Ile)

   2. Hydroxyamino acids (hydrophilic R groups)

$$\underset{\underset{\text{NH}_2}{|}}{\text{HOCH}_2\text{CHCOOH}} \qquad \overset{\overset{\text{OH}}{|}}{\underset{\underset{\text{NH}_2}{|}}{\text{CH}_3\text{CHCHCOOH}}}$$

   Serine (Ser)   Threonine (Thr)

   3. Dicarboxylic amino acids and their amides (hydrophilic R groups)

$$\underset{\underset{\text{NH}_2}{|}}{\text{HOOCCH}_2\text{CHCOOH}} \quad \overset{-\text{H}^+}{\underset{+\text{H}^+}{\rightleftharpoons}} \quad \underset{\underset{\text{NH}_2}{|}}{{}^-\text{OOCCH}_2\text{CHCOOH}}$$

   Aspartic acid (Asp)

$$\underset{\underset{\text{NH}_2}{|}}{\text{HOOCCH}_2\text{CH}_2\text{CHCOOH}} \quad \overset{-\text{H}^+}{\underset{+\text{H}^+}{\rightleftharpoons}} \quad \underset{\underset{\text{NH}_2}{|}}{{}^-\text{OOCCH}_2\text{CH}_2\text{CHCOOH}}$$

   Glutamic acid (Glu)

$$\underset{\underset{\text{NH}_2}{|}}{\text{H}_2\text{NCOCH}_2\text{CHCOOH}} \qquad \underset{\underset{\text{NH}_2}{|}}{\text{H}_2\text{NCOCH}_2\text{CH}_2\text{CHCOOH}}$$

   Asparagine (AspNH$_2$ or Asn)   Glutamine (GluNH$_2$ or Gln)

   4. Amino acids having basic functions (hydrophilic R groups)

$$\underset{\underset{\text{NH}_2}{|}}{\text{H}_2\text{NCH}_2\text{CH}_2\text{CH}_2\text{CH}_2\text{CHCOOH}} \quad \overset{+\text{H}^+}{\underset{-\text{H}^+}{\rightleftharpoons}} \quad \overset{+}{\underset{\underset{\text{NH}_2}{|}}{\text{H}_3\text{NCH}_2\text{CH}_2\text{CH}_2\text{CH}_2\text{CHCOOH}}}$$

   Lysine (Lys)

Table 11-1. (Continued)

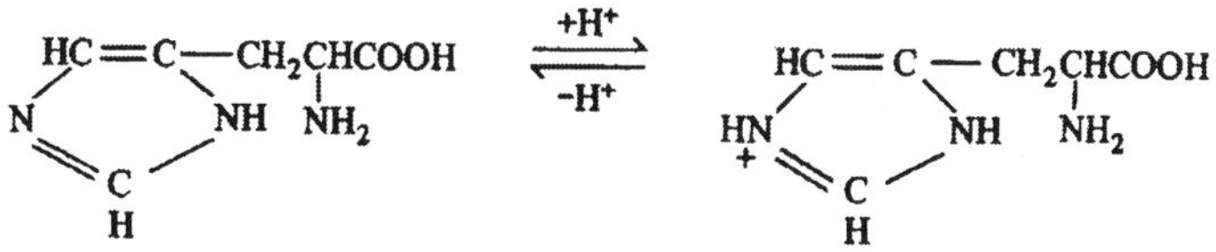

$$HC=C-CH_2CHCOOH \quad \underset{-H^+}{\overset{+H^+}{\rightleftharpoons}} \quad HC=C-CH_2CHCOOH$$

Histidine (His)

$$H_2NCNHCH_2CH_2CH_2CHCOOH \quad \underset{-H^+}{\overset{+H^+}{\rightleftharpoons}} \quad H_2NCNHCH_2CH_2CH_2CHCOOH$$

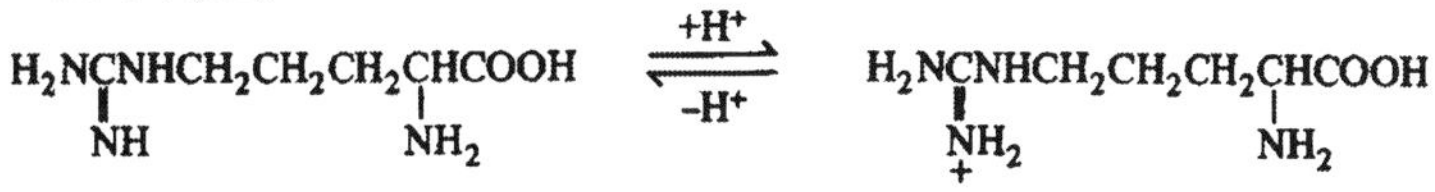

Arginine (Arg)

5. Aromatic amino acids (hydrophobic R groups)

$$\langle O \rangle - CH_2CHCOOH \qquad\qquad HO-\langle O \rangle - CH_2CHCOOH$$

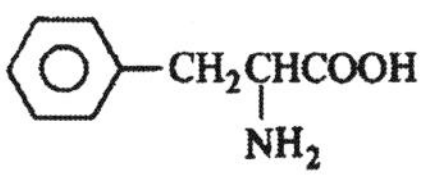

Phenylalanine (Phe) 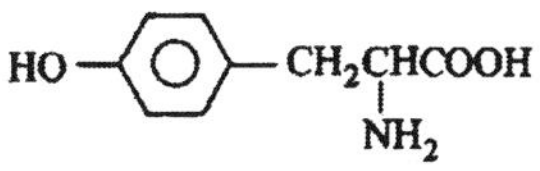   Tryosine (Tyr)

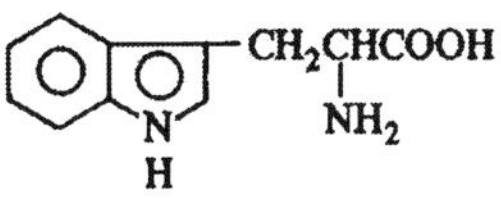

$$- CH_2CHCOOH$$

(histidine has been included
in the preceding category)

Tryptophan (Trp)

6. Sulfur-containing amino acids (hydrophobic R groups except cysteine)

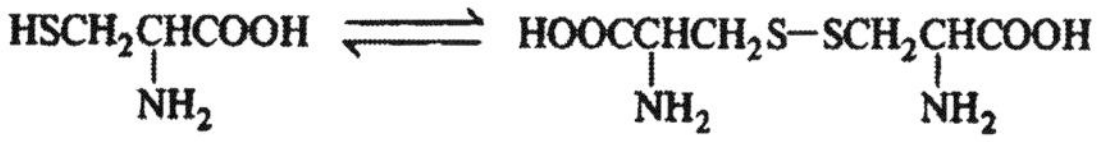

$$HSCH_2CHCOOH \quad \rightleftharpoons \quad HOOCCHCH_2S-SCH_2CHCOOH$$

Cysteine (CySH)   Cystine (CyS-SCy)

$$CH_3SCH_2CH_2CHCOOH$$

Methionine (Met)

7. Imino acid (hydrophobic R group)

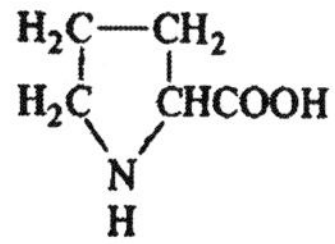

$$H_2C-CH_2$$
$$H_2C \quad CHCOOH$$

Proline (Pro)

Table 11-2. One-and Three-Letter Symbols for the Amino Acids

A	Ala	Alanine	M	Met	Methionine
B	Asx	Asparagine or aspartic acid	N	Asn	Asparagine
C	Cys	Cysteine	P	Pro	Proline
D	Asp	Aspartic acid	Q	Gln	Glutamine
E	Glu	Glutamic acid	R	Arg	Arginine
F	Phe	Phenylalanine	S	Ser	Serine
G	Gly	Glycine	T	Thr	Threonine
H	His	Histidine	V	Val	Valine
I	Ile	Isoleucine	W	Trp	Tryptophan
K	Lys	Lysine	Y*	Tyr	Tyrosine
L	Leu	Leucine	Z	Glx	Glutamine or glutamic acid

* The one letter symbol for an undetermined or nonstandard amino acid is X

Table 11-3. Pyrimidine and Purine Bases

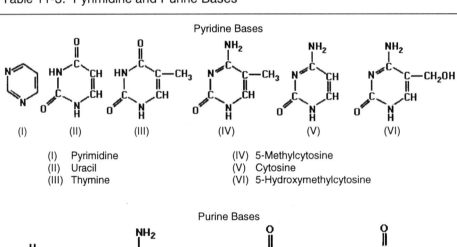

Pyridine Bases

(I) Pyrimidine
(II) Uracil
(III) Thymine

(IV) 5-Methylcytosine
(V) Cytosine
(VI) 5-Hydroxymethylcytosine

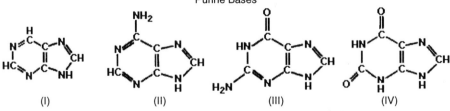

Purine Bases

(I) Purine
(II) Adenine (6-Aminopurine)

(III) Guanine (2-Amino-6-Oxypurine)
(IV) Xanthine (2,6-Dioxypurine)

(2) The amino acid can be either hydrophobic or hydrophilic, depending on the chain length of the R-group. Thus, it can form a biological membrane that can control the solute and solvent transport between the organism and its surroundings.

(3) The amino acid can form polyamino acid (polypeptide) in various molecular weights, as shown in Table 11-4. There are further unique properties, (4) to (7), of the polypeptides (proteins), which will be discussed in the next section.

The metabolic process common in life is biogenesis or anabolism, which is referred to as the building process of macromolecules from simple molecules; and the so-called ergbolism is a light conversion process involving ATP (adenosine triphosphate); lastly, the metabolic breakdown is termed catabolism. The energy flow through living systems is essential for both plants and animals, as shown in Figure 11-7. As basic information, the autotrophic metabolism has a feedback to heterotrophic metabolism, as shown in Figure 11-8.

The biomolecules universally involved in energy conversion are as follows:

Name	Symbol
Adenosine triphosphate	ATP
Nicotinamide adenine dinucleotide	NAD
Nicotinamide adenine dinucleotide phosphate	NADP
Flavin mononucleotide	FMN
Flavin adenine dinucleotide	FAD
Quinones	CoQ
Heme containing molecules, cytochromes	Cyt
Ferredoxin, iron containing, nonheme molecules	Fd

The high-energy compounds always contain the energy-rich phosphate bonds that can easily be cleaved (see Table 11-5). For example,

$$ADP + phosphate + energy \leftrightarrow ATP + H_2O$$

Table 11-4. Linking of Biomonomers to form Active Biomolecules (After J.W. Moore and E.A. Moore, 1976.)

Monomer(s)	Reaction	Biomolecule
1. Purine or Pyrimidine + sugar	⇄	nucleoside

2. Phosphoric acid + nucleoside ⇄ nucleotide

3. Nucleotide + nucleotide ⇄ polynucleotide

Table 11-4. (Continued)

Monomer(s)	Reaction	Biomolecule
4. Fatty acid + alcohol		lipid

$$3\ CH_3(CH_2)_{14}\ C\overset{\displaystyle O}{\underset{OH}{\Big|}} \quad + \quad \begin{array}{c} CH_2-CH-CH_2 \\ | \quad\ | \quad\ | \\ HO \quad OH\ OH \end{array} \rightleftarrows$$

lipid structure $+ 3\ H_2O$

| 5. Amino acid + amino acid | | polypeptide |

amino acid + amino acid structure $\rightleftarrows$ dipeptide structure $+ H_2O$

| 6. Pyrrole + formaldehyde | | Porphyrin-type skeleton |

porphyrin structures $\rightleftarrows$ $+ 4\ H_2O$

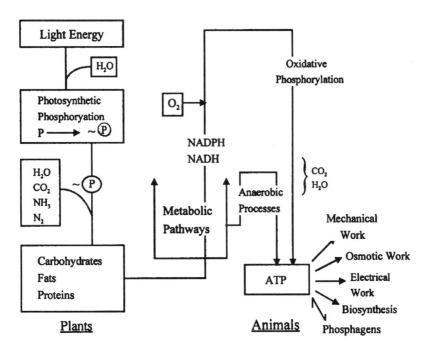

Figure 11-7. Energy flow through living systems.

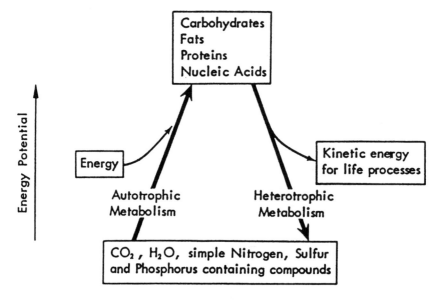

Figure 11-8. Feedback system of autotrophs to heterotrophus.

Table 11-5. Energy Rich Phosphate Bonds

Characteristic Linkage	General Formula	General Designation	Biochemical Example	$\Delta G°$ (kcal/mole)
$\overset{\parallel}{-}\text{C}\overset{}{-}\text{N}\sim$ with H	$\overset{\text{NH}}{\overset{\parallel}{\text{RC}-\text{N}}}\sim \text{(P)}$ with H	Guanidinium phosphate	Creatine phosphate Arginine phosphate	-10.5 -9.0
$\overset{\parallel}{-}\text{C}-\text{O}\sim$	$\overset{\text{CH}_2}{\overset{\parallel}{\text{RC}-\text{O}}}\sim \text{(P)}$	Enolphosphate	Phosphoenol pyruvate	-12.8
$\overset{\parallel}{-}\text{C}-\text{O}\sim$	$\overset{\text{O}}{\overset{\parallel}{\text{RC}-\text{O}}}\sim \text{(P)}$	Acylphosphate	Acetyl phosphate	-10.5
$\overset{\parallel}{-}\text{P}-\text{O}\sim$ with HO	$\overset{\text{O}}{\overset{\parallel}{\text{ROP}-\text{O}}}\sim \text{(P)}$ with HO	Pyrophosphate	Adenosine diphosphate	-7.6
$\overset{\parallel}{-}\text{C}\sim$	$\overset{\text{O}}{\overset{\parallel}{\text{RC}}}\sim \text{SR'}$	Acyl thioester	Acetyl CoA	-10.5

11.3.1 Biopolymer

Polymer usually indicates a high molecular weight of multiple sequences of repeating units. These repeating units are derived from the **monomer** from which the polymer is synthesized. Many monomers do have specially functional groups or unsaturation through which the polymer linkages can be formed either through condensation or addition. Polyester is a condensation polymer where the bonds are formed from difunctional alcohol (diol) and acids (dicarboxylic acid) as monomers. Polyvinyl chloride is an addition polymer that is made from the monomer vinyl chloride through the unsaturated double bonds of the vinyl functions (α-olefins).

Polymer chain entanglement in the solid state is the key characteristic to show that the mechanical properties of high molecular-weight large molecules are different from those of small molecules. The traditional use of polymers as flexible or high strength materials is based on the interactions of large molecules to assume certain configurations and conformations with backbone chain atoms as well as side groups. Other uses of polymers, such as electrical conductors or semiconductors (e.g., polyacetylenes), are still in the developing stages. The potential use for polymer-bond catalysts which combines the advantages of both homogeneous and heterogeneous catalysts at fixed sites are the better mediators now. Furthermore, the use of polymers which bear antigens, whole cells, and enzymes can separate reactive sites, prevent denaturation of proteins and allow sequenced reactions such as controlled releasing of chemotherapeutic agents. Of course the most important value of polymers of biological origin is their function as coding systems and templates of replications, as shown in Figure 11-9.

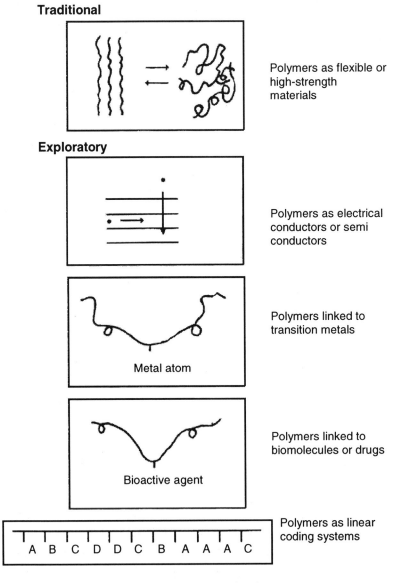

Figure 11-9. New uses of polymers.

Polymers can be classified as **homopolymers** with one type of repeating unit. If there is more than one type of repeating unit (two, for example), they are called **copolymers**; three, **terpolymers**; and so on until the repeating units reach n, for which the word is **multipolymer**. For giant bioactive molecules the biopolymers, such as proteins, are made from a combination of different amino acids. It is possible to prepare a polypeptide of the 22 different acids.

Linking of biomonomers to form low molecular weight polymers (**oligomers**) is also important in biology. Table 11-4 illustrates such oligomer formation; for example, **nucleoside** is formed by purine or pyrimidine and a simple sugar (scheme 1). The **nucleotide** is formed through nucleoside and phosphoric acid (scheme 2). Nucleotide can be further condensed into polynucleotide (scheme 3). A trinucleotide is generally referred to as nucleic acid (oligonucleotide), which can form with three different bases. Both RNA and DNA are tetranucleotides. The former is composed of ribose (sugar) and four bases, ACGT (A, adenine; C, cytosine; G, guanine; T, thymine), which is called **ribonucleic acid** (RNA). The latter consists of deoxyribose and four bases, ACGU (the ACG is same as above; U stands for uracil), and is called **deoxyribonucleic acid** (DNA). Both chemical structures of RNA and DNA are depicted in Figure 11-10. DNA exists in cells of all organisms; its sequences in every species can map the life phylogeny. DNA profiling can identify each individual human, and it has become a powerful tool in forensics. RNA is the messenger for replication and is essential for genetic information.

Let us proceed further on to the unique properties of polypeptides or proteins that were derived from amino acids, which have been discussed in the last section (section 11.3) on the properties of amino acids and their derivatives. Figure 11-11 shows two peptides in the completely extended β-conformation, and Figure 11-12 gives a conformational map for allowed values of the torsional angles. These conformation and conformation maps will help to understand why proteins possess the unique properties as listed in (4), (5), and (6). For an explanation of the torsional angle and bonding, refer to Chapter 2. Table 11-6 lists approximate torsion angles for some regular peptide structures. More unique properties are: (4) Hydrogen in the structure that can perform both intra- and inter-molecular hydrogen bonding. The hydrogen bonding also exists between side group bases of DNA. (5) Protein synthesis can be accomplished by measuring RNA (template synthesis) DNA $\rightarrow$ mRNA $\rightarrow$ Proteins. (6) A secondary structure is evolved through the hydrogen bonding of the base to form a double chain. (Watson–Crick double chain). (7) There are certain allosteric enzymes that help the protein fit the substrate, as shown in Figure 11-13. The combination of DNA and protein constitutes the basic design feature of all present terrestrial organisms. Replicable DNA messages are translated into functional protein sequences, as shown in Figure 11-14, where R, T, and C stand for different kinds of activity: replication, translation, and control. Figure 11-14 also summarizes all the unique properties of amino acid and polypeptides.

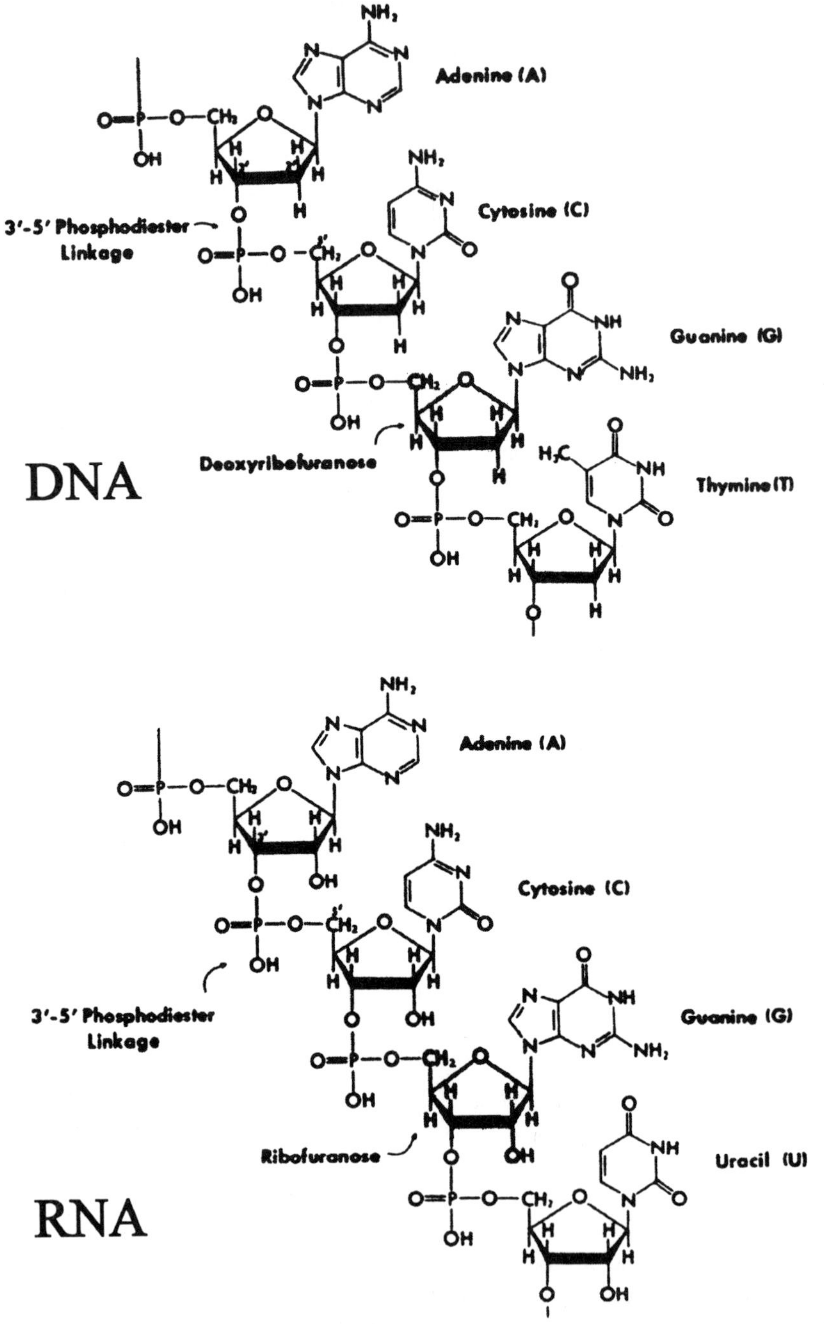

Figure 11-10. Structures of DNA and RNA.

Table 11-6. Approximate Torsion Angles for Some Regular Peptide Structures

Structure	ϕ	ψ
Hypothetical fully extended polyglycine chain	-180	+180
β-poly(L-alanine) in antiparallel-chain pleated sheet	-139	+135
Parallel-chain pleated sheet	-119	+113
Polyglycine II	-80	+150
Poly(L-proline) II	-78	+149
Collagen	-50, -76, -45	+153, +127, +148
Right-handed helix	-57	-47

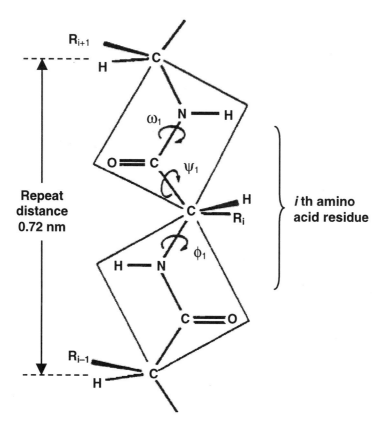

Figure 11-11. The peptide units in the completely extended β conformation. The torsion angles ϕ_1, ψ_1, and ω_1, are defined as 0° when the main chain atoms assume the cis or eclipsed conformation. The angles in the completely extended chain are all 180°.

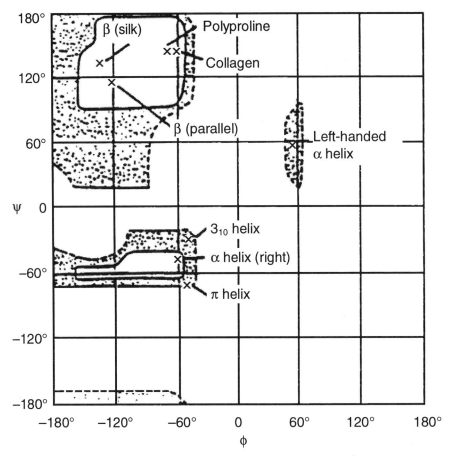

Figure 11-12. Calculated Ramachandran diagram or conformational map showing allowed values of ϕ and ψ. The two irregularly shaped boxes on the left side of the figure represent the regions in which no steric hindrance exists. Note that the β pleated sheet, collagen, and α helix structures fall within these regions. The stippled regions are those in which some hindrance exists, but for which real molecular structure are possible.

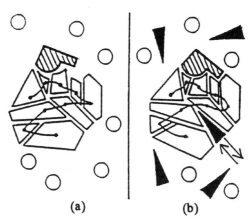

Figure 11-13. (a) Protein does not fit substrate. (b) In the presence of a third molecule, the protein may assume an alternative tertiary structure that does fit the substrate.

$$\text{R-CH} - \text{COOH} \rightleftharpoons \text{R-CH} - \text{COO}^{\ominus}$$

with NH_2 below the left and $NH_3^{\oplus}$ below the right.

(1) Amphoteric — be in any pH environment
    Zwiter ion

(2) hydrophobic — hydrophilic
    — adjust the unit of R

(3) polymeric — polyamino acids

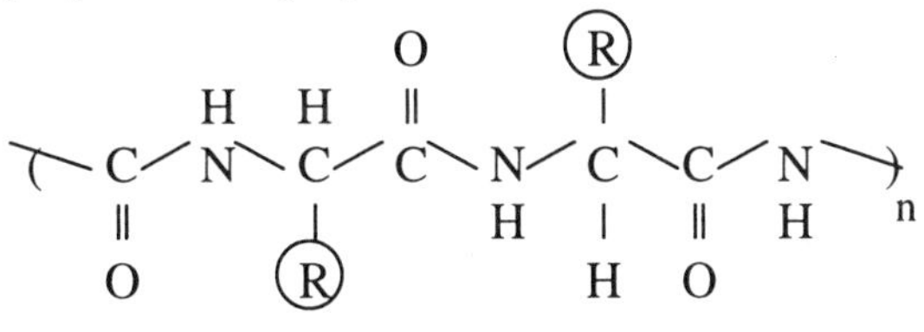

Polypeptides
α-helix, Φ, ψ torsional angles

(4) H-bonding

$$-\text{N}-\text{H}\cdots\text{O}=\text{C}$$

intramolecular vs intermolecular

Figure 11-14. Summary of the properties of amino acids.

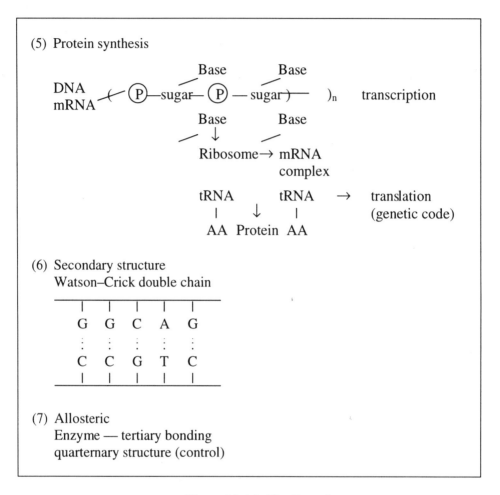

(5) Protein synthesis

(6) Secondary structure
Watson–Crick double chain

(7) Allosteric
Enzyme — tertiary bonding
quarternary structure (control)

Figure 11-14. (Continued)

11.3.2 Chemical Evolution

Other planets in the solar system also have a core, a mantle, a crust and an atmosphere. However, only on earth are there structures that can replicate themselves, change into different forms by mutation and genetic recombination, and transmit such changes to their descendants. The mineral analysis of the lunar samples from Apollo 11 and 12 showed that the soils of the moon are similar to those of the earth. However, no porphyrin, an indicator of living matter, was detected. In this section, we will discuss the formation of the biosphere.

Prior to the emergence of the first life-forms, the biomonomers and polymers necessary for building a living structure may have been synthesized through processes that were not mediated by any living system. That is to say, there may be a period of chemical evolution preceding the biological evolution in the development of the biosphere.

In 1957, Miller performed the first successful primitive earth simulation. The apparatus he used is shown in Figure 11-15. The chromatographic analysis indicates that many organic compounds essential to life were formed by the electrical charge of an inorganic mixture in a reducing primitive atmosphere, as shown in Table 11-7.

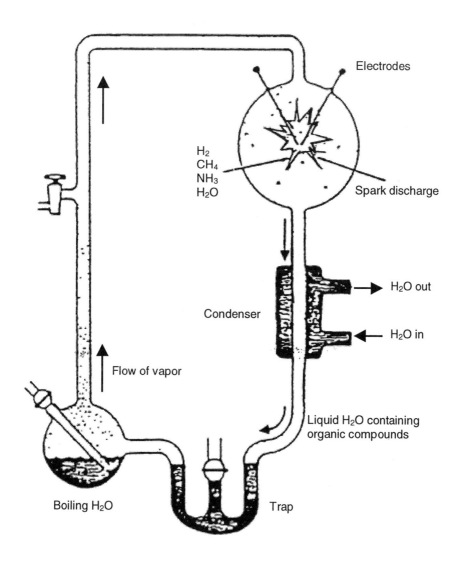

Figure 11-15. Apparatus used by Miller for simulation of primitive earth atmosphere. (After J.W. Moore and E.A. Morre, 1976)

Table 11-7. Structures and Yields of Some Compounds Formed by Electrical Discharge in a Reducing Primitive Atmosphere

Name	Yield(mol)	% Yield	Structure
Formic acid	233	3.9	HCOOH
Glycine	630	2.1	H_2NCH_2COOH
Glycolic acid	560	1.9	$HOCH_2COOH$
Lactic acid	390	1.8	$CH_3CH(OH)COOH$
Alanine	340	1.7	$CH_3CH(NH_2)COOH$
Propionic acid	126	0.6	CH_3CH_2COOH
Acetic acid	152	0.5	CH_3COOH
Glutamic acid	6	—	$HOOCCH_2CH_2CH(NH_2)COOH$
Aspartic acid	4	—	$HOOCCH_2CH(NH_2)COOH$

Based on initial carbon.

Source: S.L. Miller, J. Amer. Chem. Soc., 77, 2351 (1955); Biochim. Biophys. Acta, 23, 480 (1957).

Table 11-8. Representative Abiotic Synthesis Experiments

Compound Class	Reactants	Energy	Products
Amino Acids	CH_4, NH_3, H_2, H_2O	Electric discharge	Amino acids, hydroxy acids, HCN, urea
	Ammonium fumarate	Heat	Aspartic acid
	CO_2, NH_3, H_2, H_2O	Electric discharge	Amino acids
	CH_4, NH_3, H_2O, H_2, CO_2, N_2	x-ray	Amino acids
	Ammonium acetate	γ-ray	Glycine, aspartic acid, diaminosuccinic acid
	Ammonium carbonate	β-ray	Glycine
	CH_4, NH_3, H_2O	UV	Glycine, alanine
	NH_3, HCN, H_2	Heat (343 K)	Amino acids
	CH_4, NH_3, H_2O	Accelerated electron	Glycine, alanine
	CH_4, NH_3, H_2O	Heat (>1123 K)	Amino acids
	CO, H_2, NH_3 (Ni-Fe, Fe_3O_4, Al_2O_3, SiO_2 catalysts)	Heat (750-1000 K)	Amino acids
	HC-CN, HCN, NH_3, H_2O	Heat (373 K)	Aspartic acid

Table 11-8. (Continued)

Compound Class	Reactants	Energy	Products
Purine, pyrimidines	HCN, NH$_3$, H$_2$O	Heat (373 K)	Adenine
	Malic acid, urea, polyphosphoric acid	Heat (403 K)	Uracil
	CH$_3$, NH$_3$, H$_2$O	Accelerated electron	Adenine
	CO, H$_2$, NH$_3$, catalysts	Heat (700-1000 K)	Adenine, guanine
Sugars	HCHO, CH$_3$CHO; glyceraldehyde, acetaldehyde, Ca(OH)$_2$	Heat (323 K)	2-deoxyribose 2-deoxyxylose
	HCHO	UV	Ribose, deoxyribose
Nucleotides	Adenosine, polyphosphate ester	UV	AMP, ADP, ATP
	Nucleoside, phosphate	Heat (433 K)	Nucleotides
	Nucleosides, polyphosphoric acid	Heat (295 K)	Nucleotides
hydrocarbons	Methane	Electric discharge	Higher hydrocarbons
	Methane	Heat (1273 K; silica gel)	Higher hydrocarbons
	CO, H$_2$, catalysts	Heat (750-1000 K)	Linear alkanes
porphyrins	Pyrrole, benzaldehyde	γ-ray	Tetraphenylporphyrin
	Pyrrole, formaldehyde Ni^{2+}, Cu^{2+}		Porphyrin
	CH$_4$, NH$_3$, H$_2$O	Electric discharge	Porphyrin
	CO, H$_2$, NH$_3$, catalysts	Heat (750-1000 K)	Cyclic or linear pyrrole polymers

Source: S.W. Fox, K. Harada, G. Krampitz, and G. Mueller, Chemical origin of cells. *Chem. Eng. News*, June 22, p. 86 (1970)

Since then, many other investigators have also carried out primitive earth simulations. In addition to amino acids, purines, pyridines, sugars, nucleosides, nucleotides and porphyrins have been synthesized (Table 11-8).

Therefore, we may assume that all that is necessary for synthesis of the building blocks in biochemistry (one of the examples is the amino acids with different functional groups that provide specialty proteins as shown in Table 11-1) is the interaction of compounds containing the appropriate element with a source of energy.

Formaldehyde and hydrogen cyanide are the two most important reactive compounds in the initial synthesis of biomonomers. Carbohydrates may be regarded as polymers of formaldehyde. In organic compounds, the -C≡N group can undergo hydrolysis to produce carboxylic acids and consequently the amino acids. The highly unsaturated triple bonds allow many molecules to have additional reactions with HCN to produce high molecular weight compounds.

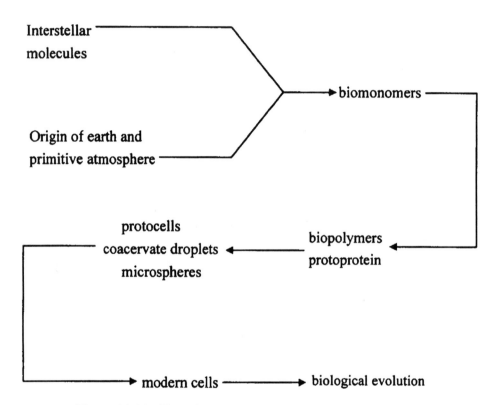

Figure 11-16. Flow sheet for one view of chemical evolution.

After the chemical evolution, the next step in the development of the biosphere is the initiation of the biological evolution, as shown in Figure 11-16. In addition, Figure 11-17 illustrates a diagrammatic representation of a possible scheme of primordial biogenesis. The first living organisms probably grew in shallow bodies of water under anaerobic condition. The stages in evolution of the atmosphere and hydrosphere with evolution of organisms are shown in Figure 11-18.

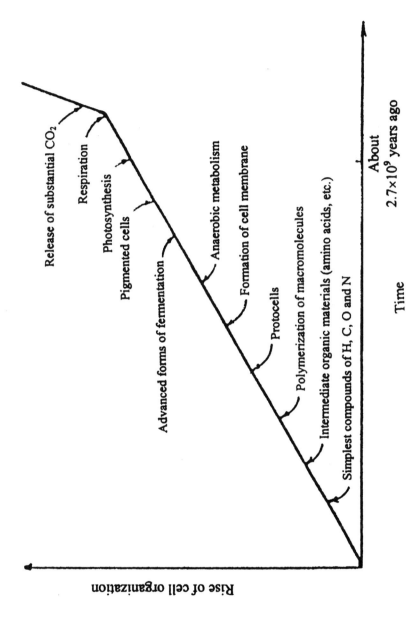

Figure 11-17.   Diagrammatic representation of a possible scheme of primordial biogenesis. This figure is highly over-simplified and is intended only as a convenient means of visualizing the probable sequence of phases in the overall process.

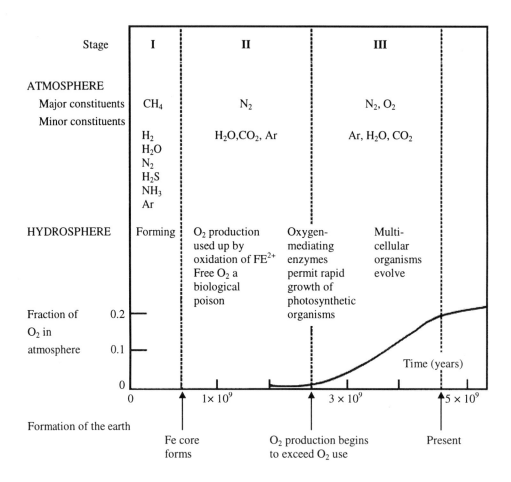

Figure 11-18. Stages in the evolution of the atmosphere and hydrosphere (After J.W. Moore and E.A. Moore, 1976).

11.3.3 Chemical Autopoiesis

The word **autopoiesis** is derived from the Greek "auto" (self) and "poiesis" (formation) in attempt to define the minimal life. An autopoietic unity is a unity that is self-generating and self-perpetuating as a consequence of its own activities within a boundary of its own making. Both self-replicating reverse micelles and self-replicating vesicles have been studied. It is possible through the studies of membrane-mimetic chemistry to mimic cellular systems under prebiotic conditions.

A hypothetical model of the origin of life is depicted in Figure 11-19. Using clay as an example in a stream, there are four particular distinct patterns that have printed themselves many times to create four regions in which the physical consistency of the clays is different. The so-called sloppy, sticky, lumpy, and tough characteristics will represent the disorder to be ordered gradually as the water flows. Recently biomolecules can be synthesized, which indicates definitely their self-replicating capabilities and mutant properties.

The pattern of molecular association and assemblage will yield specific functions in a number of stometric arrangements (tessellation), as seen in Figure 11-19.

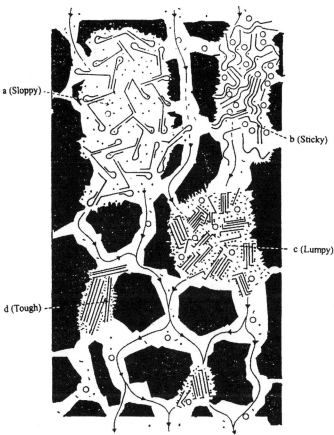

Figure 11-19. A hypothetical 'origin of life.' We imagine clay platelets forming replicatively from solutions flowing through porous rock and sand. Different substitution patterns give clays that have different folding habits and which stick to each other in different ways: They also adsorb and entrain different proportions of molecules in the environment. Thus, some patterns survive and replicate more effectively than others. (After A.G. Cairns-Smith, the life puzzle, University of Toronto press, 1971)

A few simple principles of bioorganic chemistry should be reviewed here as the molecular basis of autopoiesis: (1) **proximity effect**, (2) **molecular adaptation**, and (3) **molecular recognition** at a super molecular level. An example of the proximity effect is that 2,2'-tolane-carboxylic acid in ethanol is converted with ease to 3-(2-carboxy-benzilidene)phthalate, as shown in Figure 11-20.

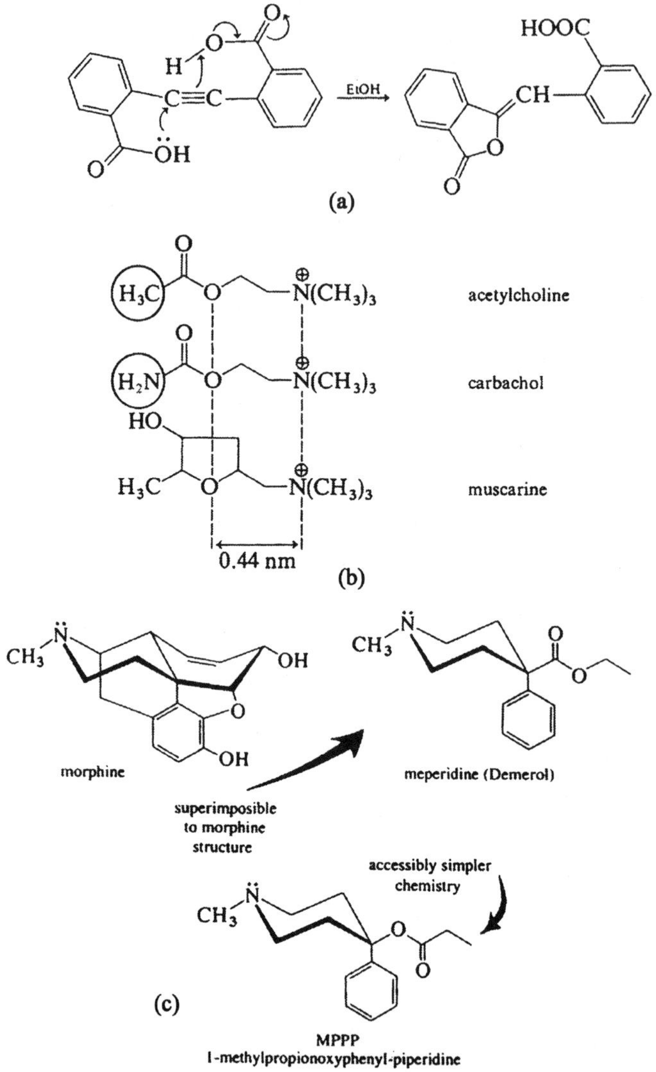

(a)

acetylcholine

carbachol

muscarine

0.44 nm

(b)

morphine

superimposible
to morphine
structure

meperidine (Demerol)

accessibly simpler
chemistry

(c)

MPPP
1-methylpropionoxyphenyl-piperidine

Figure 11-20. (a) example of proximity effect; (b) difference between two neurotransmitters and muscarine; and (c) MPPP.

In contrast the corresponding reaction of 2-tolanecarboxylic or 2,4'-tolanecarboxylic is 10^4 slower. Therefore, the 2,2'-position must participate in the transition state. This type of complementary bifunctional catalysis is a simple model for enzyme activity. Molecular adaptation can be illustrated with pharmacological activities. For example, the difference between two neurotransmitters and the mushroom poison, muscarine (in *Amanita muscaria*), is such that the biostatic equivalence is clear (unblocked) or blocked by competition, as shown in Figure 11-20. Another example of molecular adaptation is the analgesic properties of morphine, demerol, and MPPP (Fig. 9-27). Actually, MPPP can be hydrolyzed to MPTP (similar to herbicide; see Fig. 11-20), which can become MPTP, a neurotoxic pyridine metabolite that can cause Parkinson's disease. Molecular recognition in the sensory response can be illustrated by the optical isomerism. For example, the L-L isomer, aspartame, is 200x sweeter than sugar, and the corresponding L-D isomer is bitter. Another example is P-4000, in that a slight change from NH_2- to NO_2- groups will switch the quality from extremely sweet to extremely bitter, as shown in Figure 11-21. To summarize, no matter whether the formations involved are recognition (receptor), or release and transport (transporter), or molecular catalysis (catalyst), these functional approaches to the formation of supermolecular devices that can aid the organized assemblies are most helpful, as shown in Figure 11-22.

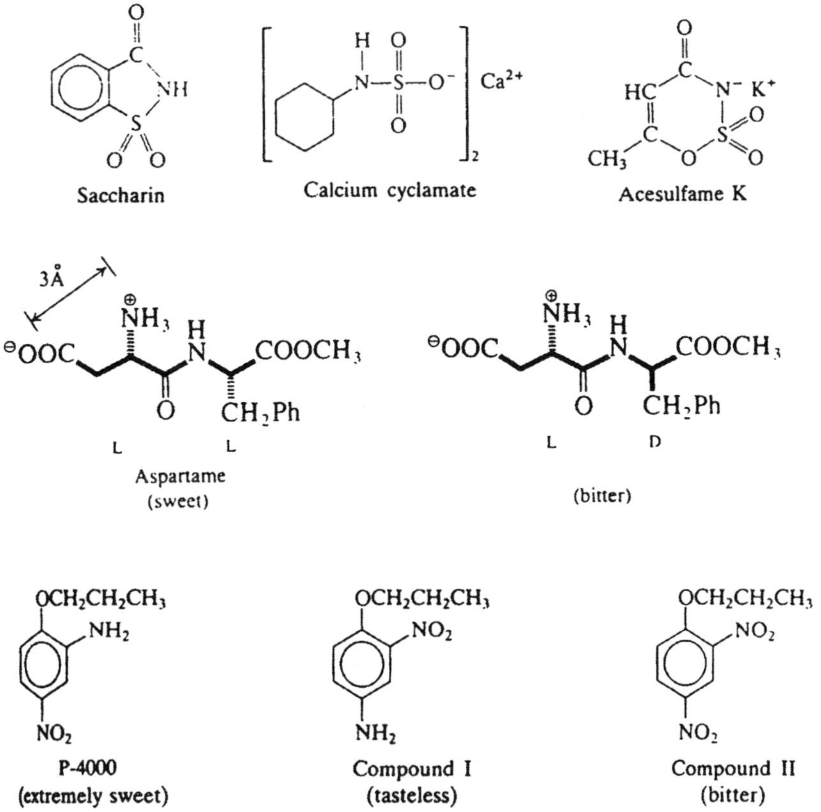

artificial sweeteners.

Figure 11-21. Artificial sweeteners.

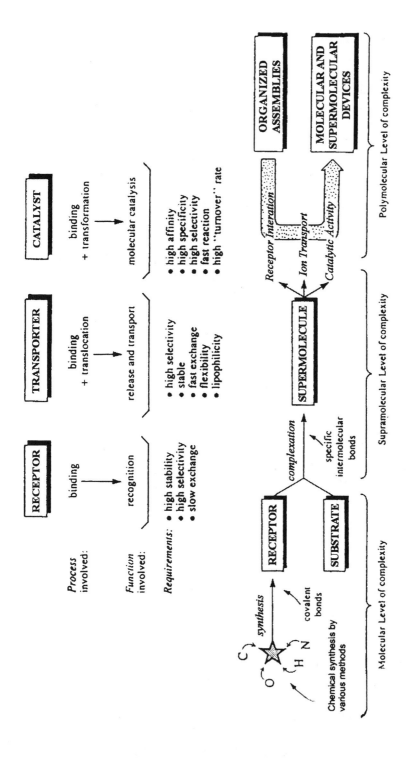

Figure 11-22.   From atoms to molecules to supramolecules. (Adapted from Dugas.)

11.4 ENVIRONMENTAL GEOCHEMISTRY

As mentioned, the sun and planets in the solar system started with the same initial chemical composition. However, the current composition of the earth's crust varies considerably from that estimated for the universe as a whole. The deficiency factor, the logarithm of the ratio of cosmic abundance/terrestrial abundance, commonly serves as an indicator. The values of the deficiency factor for H and He are 6.6 and 14.2, respectively, meaning that the universe has a much higher concentration of these two elements than the earth. Similarly, a lot of volatile, inert elements such as N_2 and the noble gases have escaped from the earth. From the kinetic theory of gases, the average molecular speed at a given temperature T is $(3RT/M)^{1/2}$, where M = molecular weight. To escape the gravitational field of the earth, a particle's vertical velocity must exceed

$$V_{escape} = (2GM_b/r)^{1/2}$$

where

$G (= 6.672 \times 10^{-11} \text{ m}^3\text{s}^{-2}\text{kg}^{-1})$ is the gravitational constant

M_b = mass of the earth $(6.0 \times 10^{24} \text{ kg})$

r = radius of the earth $(6.4 \times 10^6 \text{ m})$.

It was found from detailed calculations that only those gases with an average molecular speed less than one fifth of the escape velocity are expected to remain on the earth over the billions of years.

During the formation of the earth, most of the oxygen was present as oxides, while sulfur existed in the form of sulfides. Of the dozen or so compounds, metals including iron were present in an excess over oxygen and sulfur, and they remained in a molten, metallic phase. Because of its greater density, iron associated with other metals migrated toward the center of the earth, while the less denser silicates, oxides, and sulfides floated to the surface. As a result, the three main structural components of the earth: core, mantle, and crust, were formed.

The formation of the earth and the loss of gaseous elements are essential and a brief outline is listed here:

- Formation of Earth

  α-β-γ theory, this theory is named after the 3 physicists: Alpher, Bethe, and Gamow. Actually, the theory represents α-recombination and β-emission accompanied by γ-emission (isomerization). Thus, successively through neutron capture to build up heavy elements, planets, asteroids, and meteorites originated from an interstellar cloud.

- Loss of Gaseous Elements

Deficiency factor of elements

$\quad = \log[(\text{cosmic abundance})/(\text{territorial abundance})]$

e.g.,   H = 6.6

$\quad\quad$ He = 14.2

Average Molecular Speed $= (3RT/M)^{\frac{1}{2}}$

$$V_{escape} = (2GM_b/r)^{\frac{1}{2}}$$

G = gravitational constant = 6.67×10^{-11} m^3/s^2/kg

M_b = mass of body where gravitational field is to escape = 6.0×10^{24} kg

$r = 6.4 \times 10^6$ m (surface to center of body)

$$\text{A.M.S.} < 1/5 \; V_{escape}$$

- Remaining Elements

Ne, He — most deficient

H, N — next

C in CH_4, CO, CO_2 — intermediate

Mg, Al, Si, Fe close to cosmos

Goldschmidt classified the metal elements according to their preference to (1) Siderophile — dissolve in the core of the iron, (2) Charcophile — be in the crustal phases with sulfide as major anionic constituents, (3) Lithophile — be in the oxide and silicate lattices of the mantle and crust, and (4) Atmophile — gaseous component, shown in Table 11-9. The relationship between Goldschmidt's geochemical classification and periodic table is shown in Figure 11-23. In general, the metallic lithophiles are more readily oxidized; that is, they have more negative reduction potentials than iron, as shown in Table 11-10. Their presence in the oxide phase results from the inability of metallic iron to replace them from their compounds.

The lithophiles tend to be bound to oxide ions and form three-dimensional arrays. Four basic rules govern the formation of stable ionic lattices:

- The principle of hard and soft acids and bases applies; that is, hard acids prefer to bond to hard bases, and soft acids prefer soft bases. Hard acids and bases are small and not polarizable, while soft acids and bases have large radii and are easily polarized. Because O^- is a hard base, hard cations (such as Ca^{2+} and Rb^+) will be preferred and soft cations (such as Cu^{2+} and Ag^+) will scarcely be found.

- The smaller ions will form stronger bonds.
- If the two cations have nearly the same radius, the one with the greater charge will form a more stable lattice.
- When substitutions of ions having different charges (isomorphous substitution) are made, electroneutrality must be maintained.

After the earth's core, mantle, and crust differentiated themselves, a variety of processes acted to transform the igneous rock. In some cases, these processes have served to concentrate compounds to favorable deposits; resulting in the ores. The processes of metamorphosis, weathering, hydrothermal transport, and sedimentation have all contributed to ore formation. **Metamorphosis** refers to changes induced in crustal rock adjacent to a molten magma. For example, iron (II) oxide can be oxidized in the presence of carbonate minerals to form magnetite, an important iron ore. Most ores of chalcophiles (such as Zn, Cu, Pb, Hg, and Ag) are formed through the hydrothermal transport process. Using ZnS as an example,

$$ZnS + 2\ HCl = ZnCl_2 + H_2S$$

Table 11-9. Goldschmidt's Geochemical Classification of the Elements

Term	Characteristic Property	Elements Included [*]
Siderophile	Association with metallic iron	Fe, Co, Ni, Ru, Rh, Pd, Os, Ir, Pt, Mo, W, Re, Au, Ge, Sn, C, P, (Pb, As, S)
Chalcophile	Tendency to form sulfide minerals	Cu, Ag, Zn, Cd, Hg, Ga, In, Tl, (Ge, Sn,), Pb, As, Sb, Bi, S, Se, Te, (Fe, Mo, Cr)
Lithophile	Tendency to be bound to oxide ions	Li, Na, K, Rb, Cs, Be, Mg, Ca, Sr, Ba, B, Al, Sc, Y, Rare Earths, (C), Si, Ti, Zr, Hf, Th, (P), V, Nb, Ta, Cr, (W), U, F, Cl, Br, I, Mn, (H, Tl, Ga, Ge, Fe)
Atmophile	Tendency to occur as a gaseous component of the atmosphere	N, He, Ne, Ar, Kr, Xe

[*] Elements in parentheses show the indicated characteristic to a lessen extent.

Data from B. Mason. "Principles of Geochemistry," 3rd Ed. Wiley, New York, 1966; L.H. Ahrens, "Distribution of the Elements in Our Planet." McGraw-Hill, New York, 1965

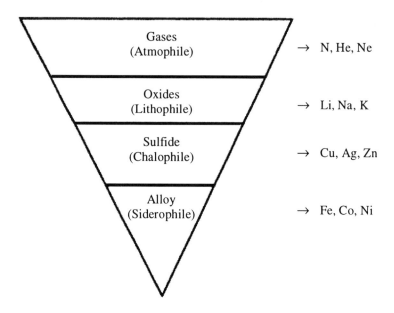

Figure 11-23. Goldschmidt geochemical classification of the elements.

Because $ZnCl_2$ is volatile and water soluble, the hydrothermal transport process allows it to be transported for some distance before the gas is cooled or the acid is neutralized by some relatively basic rock, where ZnS is precipitated. Weathering can wash off soluble minerals and leave insoluble ones. Bauxite, the main source of aluminum, is commonly found in tropical regions where high rainfall prevails. In the oceans, sedimentation of low solubility minerals is also a concentration process. The following outline summarizes this section.

- Primary Differentiation of Elements

3-layer cake   crust - mantle – core
Fe has a greater density – sinks
less dense metal Ni, Au, Pt - along the way
less dense - float on surface
metal carbides, oxides, sulfides

- Secondary Differentiation of Elements

Isomorphous substitutes - close pack lattice
Silicates and oxides
1) hard acid or base (small radius) as stable
2) if same, greater charge is more stable

- Concentration of Elements

1) segregation of molten magma
2) metamorphosis $CO_2 + 3FeO \rightarrow Fe_3O_4(ore) + CO$
3) weathering
4) hydrothermal transport

$$ZnS + 2HCl = ZnCl_2 + H_2S \text{ (more soluble in water)}$$

5) sedimentation

$$PO_4^{-3} + Ca^{2+} + H_2O \rightarrow Ca_5(PO_4)_3OH\downarrow$$

Table 11-10. Electrode Potential and Solubilities of Selected Elements Related to Geochemical Classification

Element	Reduction Potential	Solubility of Sulfide in H_2O (M)
		Lithophiles
Li/Li$^+$	-3.045	Soluble
Ca/Ca^{2+}	-2.866	—
Al/Al^{3+}	-1.662	Hydroxide precipitates more readily than sulfide
Cr/Cr^{3+}	-0.774	Hydroxide precipitates more readily than sulfide
		Chalcophiles
Zn/Zn^{2+}	-0.7628	1.58×10^{-11}
Ga/Ga^{3+} (borderline lithophile)	-0.529	—
Cu/Cu2+	+0.337	8.94×10^{-19}
Ag/Ag+	+0.7991	1.76×10^{-17}
		Siderophiles
Fe/Fe2+	-0.429	6.32×10^{-10}
Ni/Ni2+	-0.250	5.48×10^{-11}
Sn/Sn2+	-0.136	1.00×10^{-13}
Rh/Rh3+	+0.80	—
Pd/Pd2+	+0.987	—
Au/Au+	+1.691	—

11.4.1   Geochemical Biomarkers

There are molecules originating from the biosphere that are relatively stable toward biodegradation (refractoric) and can survive in geological time spans. These molecules can behave as tracers or indicators of a given geological age. Another term is **molecular fossils** because they can be relatively isolated or determined in a complex environment.

We will discuss two types of biomarkers: the **porphyrins**, and the **terpenoids**. The first type is the pigment used widely by the organisms for energy transport; for example, chlorophyll a or heme (cytochrome). The second type includes isoprenoids (C_5), monoterpenes (C_{10}), sesquiterpenes (C_{15}), diterpenes (C_{20}), sesterterpenes (C_{25}), triterpenes (C_{30}), and tetraterpenes (C_{40}). For the first type, the metalloporphyrins are extremely stable; for example, vanadyl porphyrin can be found in the Nonesuch formation of the PreCambrian era. The geochemical transformation from chlorophyll to such geoporphyrins is illustrated by Figure 11-24. For the second type, the isoprene units are originated from energy storage; for example, the isopentenyl pyrophosphate and dimethylallyl pyrophosphate can be easily compiled or aggregated through either head-to-head or head-to-tail orientation in straight chains or in a cyclic fashion.

A system for treating terpene systems has been developed. Using the symbol I (J) for the number of isoprene units (e.g., if I = 2, monoterpenes), I is A, B, C, indicating the sequence of isoprene units and (J) in which J = 1, 2, ...5 (1 and 5 are equivalent) shows the positions at which the isoprene units are coupled. A detailed method is illustrated in Figure 11-25. In practice, the diterpenes and the triterpenes are frequently employed. In many cases, the nor- series (meaning one carbon less) or derivatives of -OH and -COOH are involved.

Geochemical biomarkers have been used in characterization of natural waters, e.g., black trona water in the northern Green River Basin. Both pentacyclic triterpenoids and steranes have been isolated, as shown in Figures 11-26 and 11-27. In this case, the tri- and tetracyclic terpanes have also been identified, as shown in Figure 11-28.

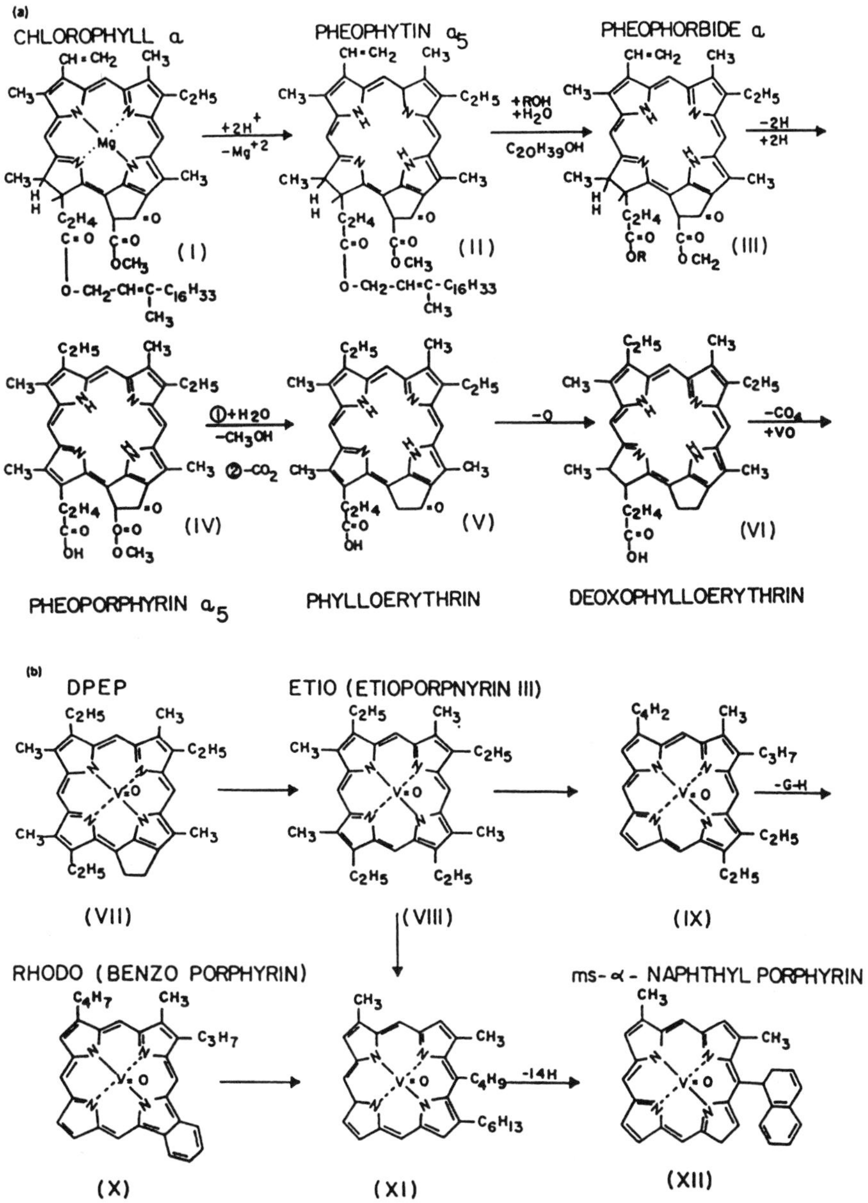

Figure 11-24. Proposed geochemical transformation of chlorophyll to vanadyl DPEP and other stable vanadyl chelates. (a) Chlorophyll to dexophylloerythin. (b) DPEP to ms-α-naphthyl porphyrin. (Source: T. F. Yen and G.V. Chilingarian, 1994.)

Terpenes, I_i (j), $I_i = A, B \ldots$ $(i = 2, AB; i = 3, ABC; i = 8; A \ldots H)$

$j = 1, 2 \ldots 5$ (1 and 5 are equivalent)

$i = 1$, isoprene

$$\overset{5}{\underset{1\ 2\ 3\ 4}{C-C-C-C}}$$

$i = 2$, monoterpene

(i) $A(1-4)B$ myrcene

(ii) $A\binom{4-3}{1-4}B$ menthol

(iii) $A\binom{2-4}{4-3}B$ thujone
       $\binom{}{1-4}$ 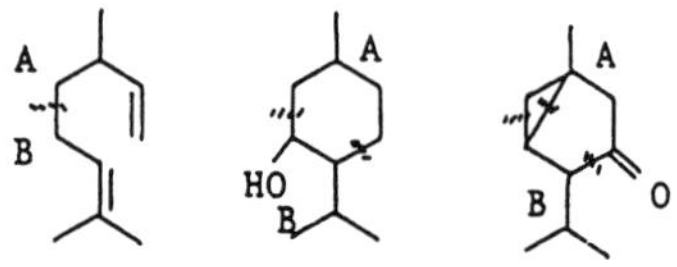

$i = 3$, sesquiterpene

$A\binom{3-4}{4-1}B\binom{3-4}{4-1}C$ Bulgarene

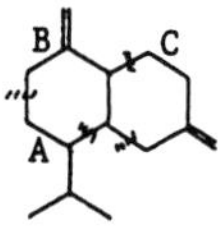

$i = 4$, diterpenes

(i) $A(4-1)B(4-1)C(4-1)D$ phytane

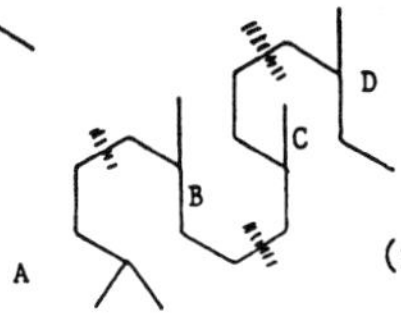

(ii) $A\binom{2-3}{4-1}B\binom{2-3}{4-1}C\binom{5-3}{4-4}D$ Abietane

(iii) $A\binom{1-3}{4-1}B\binom{2-3}{4-1}C\binom{2-1}{4-4}D$ Kawane
       $\binom{}{5-3}$

$i = 5$, sesterterpene

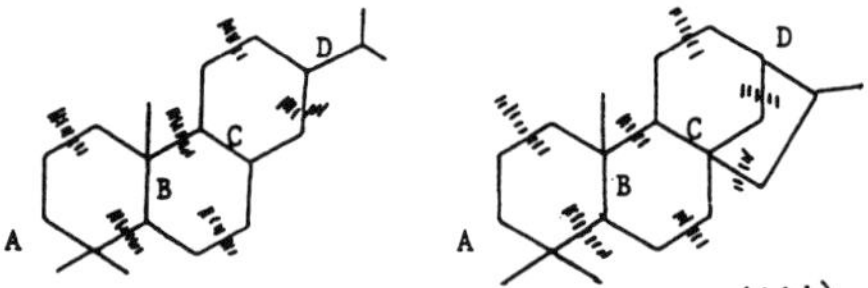

$A(4-1)B\binom{4-4}{3-2}C\binom{3-3}{1-1}D(4-1)E(4-3)C$

Gascardic acid

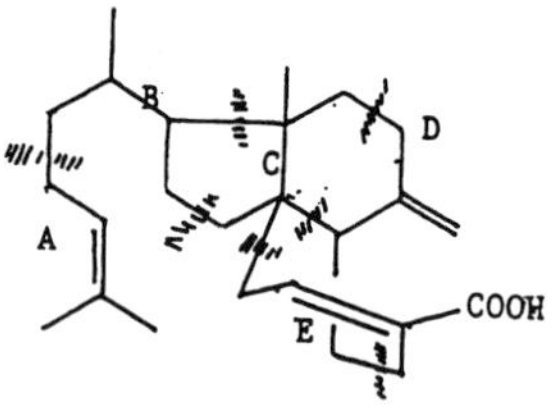

Figure 11-25. Generalized terpenoid structure (After T. F. Yen, Y. Li and M. C. Liu, Geochem. Div., ACS, 1990).

i = 6, triterpenes

A(4-1)B(4-1)C(4-4)D(1-4)E(1-4)F̄ squalene

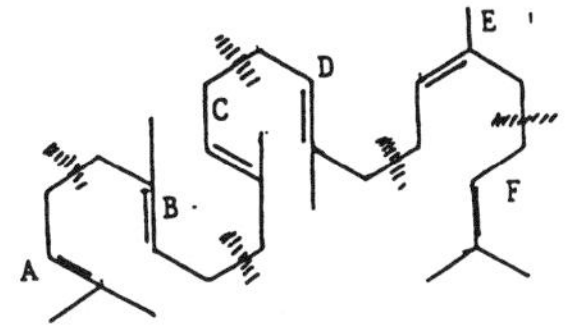

$A\binom{4-1}{2-3}B\binom{4-1}{2-3}C\binom{4-1}{2-3}D\binom{4-4}{2-3}E(1-4)F - A(1) - A(5) - C(5)$

cholesterol

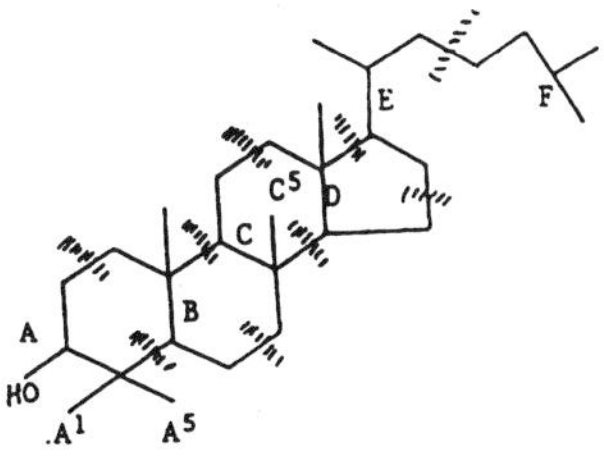

$A\binom{4-1}{2-3}B\binom{4-1}{2-3}C\binom{4-4}{2-2}D\binom{3-2}{1-4}E\binom{3-3}{1-4}F$ hopane

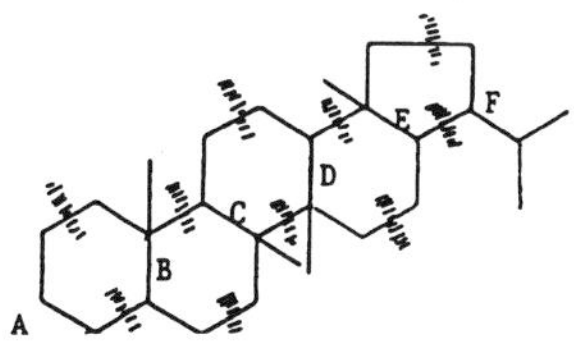

i = 8, tetraterpene

A(4-1)B(4-1)C(4-1)D(4-4)E(1-4)F(1-4)G(1-4)H

lycopene

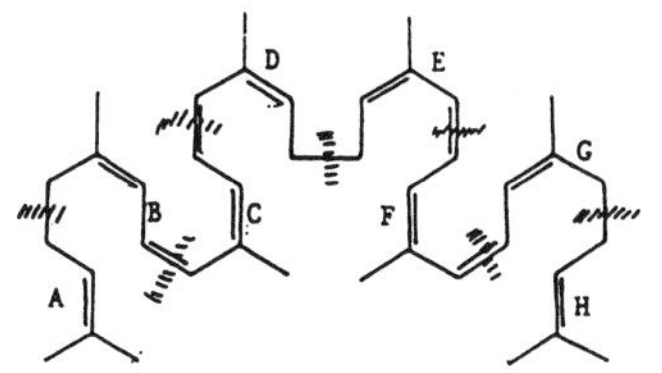

$A\binom{4-1}{2-3}B(4-1)C(4-1)D(4-4)E(1-4)F(1-4)G\binom{3-2}{1-4}H$

β-carotene

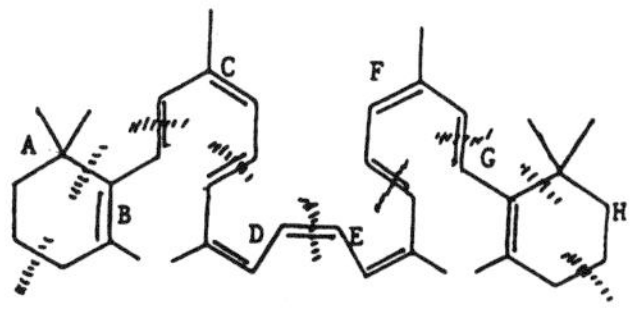

Figure 11-25. (Continued)

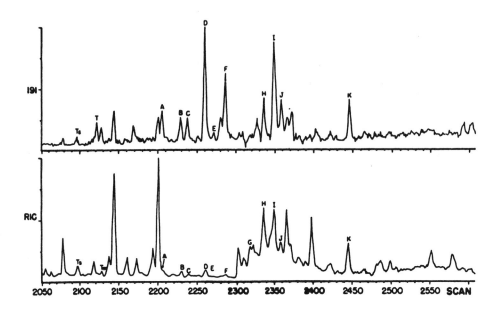

Compound	Name	M.W.	Identification
$C_{27}H_{46}$	18α(H)-22,29,30-trisnorhopane	370	Ts
$C_{27}H_{46}$	17α(H)-22,29,30-trisnormeohopane	370	Tm
$C_{29}H_{50}$	17α(H),21β(H)-30-norhopane	398	A
$C_{30}H_{50}$	hop-17(21)-ene	410	B
$C_{29}H_{50}$	17β(H),21α(H)-normoretane	398	C
$C_{30}H_{52}$	17α(H),21β(H)-hopane	412	D
$C_{30}H_{52}$	unknown C_{30} triterpane	412	E
$C_{30}H_{52}$	17β(H),21α(H)-moretane	412	F
$C_{31}H_{54}$	22S-17α(H),21β(H)-30 homohopane	426	G
$C_{31}H_{54}$	22R -17α(H),21β(H)-30 homohopane	426	H
$C_{30}H_{52}$	gammacerane	412	I
$C_{30}H_{52}$	17β(H),21β(H)-hopane	412	J
$C_{30}H_{54}$	17β(H),21α(H)-homomoretane	426	J
$C_{31}H_{54}$	17β(H),21β(H)-30 homohopane	426	K

Figure 11-26. Spectral records of the triterpanes identified in black trona water (Source: T.F. Yen and J.M. Moldowan. Geochemical Biomarkers, Harwood Academic Pub., 1988, p. 442).

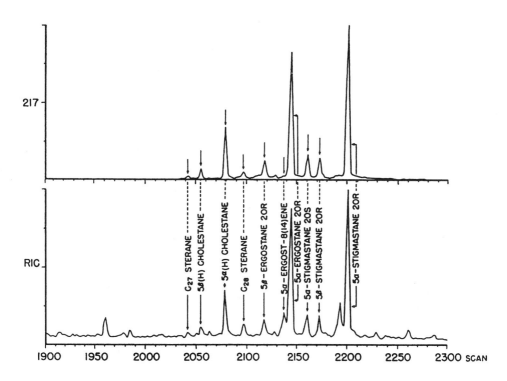

Figure 11-27. Distribution pattern of steranes (m/z at 217) with the compounds identified (Source: T. F. Yen and J.M. Moldowan, 1988).

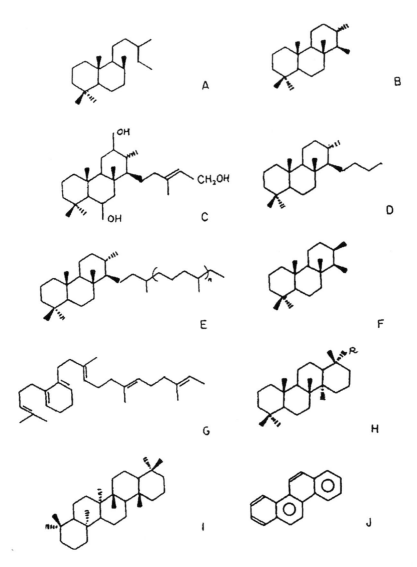

Figure 11-28. Tri- and Tetracyclic terpanes and related compounds. a, labden C_{20}; b, C_{20} Tricyclic diterpane, b can be found through a; c, cheilanthatriol; d, cheilanthane found as 18, 19-bisnor 13 β (H). 14α(H) cheilanthane; e, C_{30} tricyclic triterpane (n=1), C_{40} tricyclic tetraterpane (n=3), when n=4 a C_{45} tricyclic terpane is obtained; f, isocopa-lane; g, an hexaisoprenoid; h, tetracyclic terpanes, R = H, C_{24} 17, 21-secohopane; i, gamacerane; j, chrysene. (Source: Wang, Y., Wang, L.S. and Yen T. F., 1988).

11.4.2 Long-Range Geochemical Cycles

The dynamic nature of the common geochemical processes is universally important in dealing with global kinetics. The long-range carbon cycle based on Holland, as shown in Figure 11-29, is illustrated here. There are a total of nine reservoirs whose contents and fluxes are in units of PgC and PgC/yr, respectively.

$$A_1 \overset{k_{12}}{\underset{k_{21}}{\rightleftarrows}} A_2$$

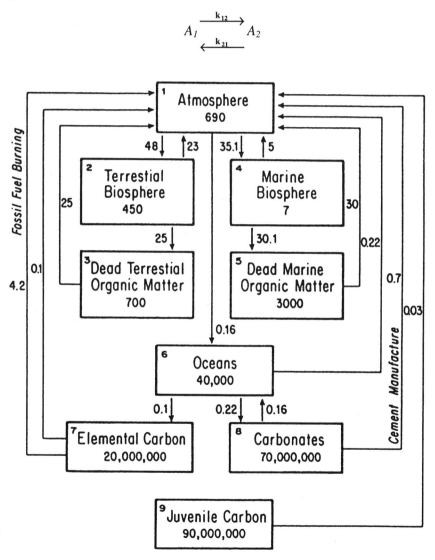

Figure 11-29. The long-term carbon cycle (Modified from Holland, 1978). The reservoir contents of PgC and the fluxes in units of PgC/yr.

For simplicity of the kinetic treatment of multiple reservoirs, we will limit our calculation to two, A_1 and A_2, as shown in the scheme that follows the treatment formulated by Lasaga, where A_i is reservoir contents (a particular element in reservoir i) and k_{ij} is the rate constants of first order. Flow can be expressed as $K_{ij}A_j$. We can write the simple rate equations:

$$\frac{dA_1}{dt} = -k_{12}\,A_1 + k_{21}\,A_2 \qquad\qquad [11\text{-}8]$$

$$\frac{dA_2}{dt} = k_{12}\,A_1 - k_{21}\,A_2 \qquad\qquad [11\text{-}9]$$

or in matrix form,

$$\frac{d}{dt}\begin{pmatrix} A_1 \\ A_2 \end{pmatrix} = \begin{vmatrix} -k_{12} & k_{21} \\ k_{12} & -k_{21} \end{vmatrix}\begin{pmatrix} A_1 \\ A_2 \end{pmatrix} \qquad\qquad [11\text{-}10]$$

The solution of equation [11-10] takes the form

$$A(t) = a_1\,e^{E_1 t}\,\Psi_1 + a_2\,e^{E_2 t}\,\Psi_2 \qquad\qquad [11\text{-}11]$$

where E_1 and E_2 are the eigenvalues and $\Psi_1\ \Psi_2$ are the eigenvectors of the matrix K defined by

$$K = \begin{vmatrix} -k_{12} & k_{21} \\ k_{12} & -k_{21} \end{vmatrix} \qquad\qquad [11\text{-}12]$$

To obtain the eigenvalues of K, we solve the determinantal equation,

$$\begin{vmatrix} -k_{12}-E & k_{21} \\ k_{12} & -k_{21}-E \end{vmatrix} = 0 \ \text{ or } \ (k_{12}+E)(k_{21}+E) - k_{12}k_{21} = 0 \qquad [11\text{-}12a]$$

The solution to this quadratic in E is

$$E_1 = 0 \qquad\qquad E_2 = -(k_{12}+k_{21}) \qquad\qquad [11\text{-}13]$$

To obtain the eigenvectors, we must solve the equation,

$$K\,\psi_i = E_i\psi_1 \qquad\qquad [11\text{-}13a]$$

or

$$\begin{vmatrix} -k_{12} & k_{21} \\ k_{12} & -k_{21} \end{vmatrix}\begin{pmatrix} a \\ b \end{pmatrix} = E\begin{pmatrix} a \\ b \end{pmatrix} \qquad\qquad [11\text{-}13b]$$

If E = 0, then

$$\begin{vmatrix} -k_{12} & k_{21} \\ k_{12} & -k_{21} \end{vmatrix} \begin{pmatrix} a \\ b \end{pmatrix} = 0$$ [11-13c]

or

$$-k_{12}a + k_{21}b = 0; \ k_{12}a - k_{21}b = 0$$ [11-13d]

Either equation yields

$$b = \frac{k_{12}}{k_{21}} a$$ [11-13e]

Because a can be arbitrary, we can set $a = 1$, as follows:

$$\Psi_1 = \begin{pmatrix} 1 \\ \dfrac{k_{12}}{k_{21}} \end{pmatrix}$$ [11-13f]

Likewise, to find ψ_2, we must solve for $K \psi_2 = E_2 \psi_2$

$$\begin{vmatrix} -k_{12} & k_{21} \\ k_{12} & -k_{21} \end{vmatrix} \begin{pmatrix} a \\ b \end{pmatrix} = -(k_{12} + k_{21}) \begin{pmatrix} a \\ b \end{pmatrix}$$ [11-13g]

or

$$-k_{12}a + k_{21}b = -(k_{12} + k_{21})a; \ k_{12}a - k_{21}b = -(k_{12} + k_{21})b$$ [11-13h]

These equations reduce to $a = -b$. Hence, setting $a = 1$ once more

$$\Psi_2 = \begin{pmatrix} 1 \\ -1 \end{pmatrix}$$ [11-13i]

Having obtained the eigenvalues and eigenvectors of K, the general solution, Equation [11-11] becomes

$$A(t) = \begin{pmatrix} A_1(t) \\ A_2(t) \end{pmatrix} = a_1 \exp(0 \cdot t) \begin{pmatrix} 1 \\ \dfrac{k_{12}}{k_{21}} \end{pmatrix} + a_2 \, e^{-(k_{12} + k_{21})t} \begin{pmatrix} 1 \\ -1 \end{pmatrix}$$ [11-14]

To obtain the coefficients a_1 and a_2, we need the initial condition of our cycle, that is, $A_1{}^0$ and $A_2{}^0$. From Equation [11-14] it follows that, setting $t = 0$,

$$A_1^0 = a_1 + a_2 \ ; \ A_2^0 = \frac{k_{12}}{k_{21}} a_1 - a_2$$ [11-15a]

Equation [11-14] also can be expressed as

$$A^0 = \Psi a \qquad [11\text{-}15b]$$

where the Ψ matrix is comprised of the eigenvectors Ψ_1 and Ψ_2

$$\Psi = \begin{vmatrix} 1 & 1 \\ \dfrac{k_{12}}{k_{21}} & -1 \end{vmatrix} \qquad [11\text{-}15c]$$

The solution to Equation [11-15a] is

$$a_1 = \frac{k_{21}\left(A_1^0 + A_2^0\right)}{k_{12} + k_{21}} \;\; ; \;\; a_2 = \frac{k_{12}\,A_1^0 - k_{21}\,A_2^0}{k_{12} + k_{21}} \qquad [11\text{-}16a]$$

or

$$a = \Psi^{-1} A^0 \qquad [11\text{-}16b]$$

where the inverse of the matrix in Equation [11-15c] is

$$\Psi^{-1} = \frac{1}{k_{12} + k_{21}} \begin{vmatrix} k_{21} & k_{21} \\ k_{12} & -k_{21} \end{vmatrix} \qquad [11\text{-}16c]$$

Combining equations [11-14] and [11-16a], we have the final equations

$$A_1(t) = \frac{k_{21}\left(A_1^0 + A_2^0\right)}{k_{12} + k_{21}} + \frac{k_{12}\left(A_1^0 - k_{21}\,A_2^0\right)}{k_{12} + k_{21}}\exp\!\left[-\left(k_{12} + k_{21}\right)t\right] \qquad [11\text{-}17a]$$

$$A_2(t) = \frac{k_{12}\left(A_1^0 + A_2^0\right)}{k_{12} + k_{21}} - \frac{k_{12}\left(A_1^0 - k_{21}\,A_2^0\right)}{k_{12} + k_{21}}\exp\!\left[-\left(k_{12} + k_{21}\right)t\right] \qquad [11\text{-}17b]$$

Notice there is only one zero eigenvalue signifying that the cycles will return to a unique steady state. Also, the none-zero eigenvalue of the 2 reservoir cycle is $-(k_{12} + k_{21})$, a negative eigenvalue is required for the cycle to be stable. Furthermore, the response time of the cycle can be obtained from the exponential term of Equation [11-17].

$$\tau_{\text{response}} = \frac{1}{k_{12} + k_{21}} \qquad [11\text{-}18]$$

The residence time of our element (at steady state) in Reservoirs one (1) and two (2) are given by

$$\tau_{\text{residence}}^{(1)} = \frac{A_1^{\text{steady}}}{(dA/dt)_{\text{input}}} = \frac{A_1^{\text{steady}}}{(dA/dt)_{\text{output}}} = \frac{A_1^{\text{steady}}}{k_{12}\,A_1^{\text{steady}}} \qquad [11\text{-}19]$$

Thus,

$$\tau^{(1)}_{residence} = \frac{1}{k_{12}} \qquad\qquad [11\text{-}20a]$$

$$\tau^{(1)}_{residence} = \frac{1}{k_{21}} \qquad\qquad [11\text{-}20b]$$

Now we will turn our attention to the 9-reservoir Holland's model in Figure 11-29. Reservoirs 1-5 represent the interactions between the atmosphere and the biosphere; because carbon cycles are much faster in these reservoirs, they comprise the short-term carbon cycle. The long-term cycle involves important additional processes. These processes include the burial of a small fraction (<0.01%) of the dead marine organic matter, which escapes oxidation to CO_2 in the water column and is incorporated into sediments. It is this small fraction that is responsible for the net supply of oxygen to the atmosphere. The other major geochemical process in the long-term cycle is the weathering of carbonates and silicates in ancient sediments and the deposition of authigenic carbonates and silicates in new sediments. CO_2 is the weathering agent in the crust. For example, carbonates weather according to the reaction

$$CaCO_3(s) + CO_2(g) + H_2O \leftrightarrow Ca^{2+} + 2HCO_3^-$$

(calcite)

For each mole of calcite dissolved, one mole of CO_2 is consumed from the atmosphere. The reverse holds during the deposition of carbonates. Likewise, the dissolution and deposition of Mg-, Na-, and K-silicates follow reactions similar to

$$MgSiO_3 + H_2O + 2CO_2 \leftrightarrow Mg^{2+} + H_4SiO_4 + 2HCO_3^-$$

The carbonate reactions are more important than the silicate reactions in the long-term carbon cycle.

Figure 11-29 also includes the oxidation of elemental carbon (7→1 flux) in ancient sediments. It also gives the anthropogenic contributions from fossil fuel combustion and cement manufacturing. The juvenile carbon reservoirs are planned for balancing the fluxes only.

Using linear approximation and ignoring the anthropogenic sources, the rate constants, as listed in Table 11-11 for the carbon cycle in Figure 11-29, are obtained. Next, we diagonalize the resulting k matrix as Equation 11-12a for the first eight reservoirs (the 9→1 flux is ignored), and the resulting eigenvalues of E are listed in Table 11-12.

Table 11-11. Long-Term Carbon Cycle. Nonzero Rate Constants (in yr^{-1})

$k_{12} = 6.9565 \times 10^{-2}$	$k_{14} = 5.0870 \times 10^{-2}$	$k_{16} = 2.3188 \times 10^{-4}$
$k_{21} = 5.1111 \times 10^{-2}$	$k_{23} = 5.5556 \times 10^{-2}$	$k_{31} = 3.5714 \times 10^{-2}$
$k_{41} = 7.1429 \times 10^{-1}$	$k_{45} = 4.300$	
$k_{51} = 1.0 \times 10^{-2}$	$k_{56} = 3.3333 \times 10^{-5}$	
$k_{61} = 5.50 \times 10^{-6}$	$k_{67} = 2.500 \times 10^{-6}$	$k_{68} = 5.50 \times 10^{-6}$
$k_{71} = 5.000 \times 10^{-9}$	$k_{86} = 2.2857 \times 10^{-9}$	$k_{91} = 3.333 \times 10^{-10}$

Source: Lasaga, 1981, with the permission from Mineralogical Society of America.

The resulting eigenvectors are also listed in Table 11-12. Assuming A^0 vector are the initial reservoir contents, then a vector can be obtained by Equation [11-16b] or

$$A(t) = -1.5780 \times 10^{-4} \Psi_1 e^{E_1 t} - 0.3848 \Psi_2 e^{E_2 t} + 0.6637 \Psi_3 e^{E_3 t} + 1.2437 \Psi_4 e^{E_4 t}$$
$$+ 1452.04 \Psi_5 e^{E_5 t} - 8181.80 \Psi_6 e^{E_6 t} + 6.3893 \times 10^6 \Psi_7 e^{E_7 t} - 7.61157 \times 10^7 \Psi_0 e^{E_0 t} \quad [11\text{-}21]$$

To find the steady state of the carbon cycle, we only need to allow $t \to \infty$ in Equation [11-21], thus,

$$A_{steady} = -7.61157 \times 10^7 \Psi_0 \quad [11\text{-}22]$$

By multiplying the Ψ_0 eigenvector as listed in Table 11-12 by -7.61157×10^7, we obtain the steady state of the reservoir content as follows:

A_i^{steady}	in Pg C
A_1	657.5
A_2	428.8
A_3	667.1
A_4	6.67
A_5	2858.9
A_6	30971
A_7	1.5485×10^7
A_8	7.4524×10^7

This type of calculation can be used to assess each reservoir. Examples include the atmospheric carbon computed by Lasaga for the connection of anthropogenic influence. Assuming

$$k_{71} = 2.15 \times 10^{-7} \text{ yr}^{-1}$$

$$k_{81} = 1.0 \times 10^{-8} \text{ yr}^{-1} \quad [11\text{-}23]$$

Table 11-12. Long-Term Carbon Cycle (After Lasaga, 1981)

Eigenvalues (yr^{-1})

$E_1 = -5.022$	$E_2 = -1.612 \times 10^{-1}$	$E_3 = -8.488 \times 10^{-2}$
$E_4 = -1.963 \times 10^{-2}$	$E_5 = -5.989 \times 10^{-5}$	$E_6 = -7.157 \times 10^{-6}$
$E_7 = -4.152 \times 10^{-9}$	$E_0 = 0$	

Eigenvectors

$$\Psi_1 = \begin{bmatrix} 0.10865 \\ -0.15378 \times 10^{-2} \\ 0.17135 \times 10^{-4} \\ -0.75444 \\ 0.64732 \\ -0.93140 \times 10^{-5} \\ 0.46370 \times 10^{-11} \\ 0.10201 \times 10^{-10} \end{bmatrix} \quad \Psi_2 = \begin{bmatrix} 0.57395 \\ -0.73223 \\ 0.32419 \\ 0.60161 \times 10^{-2} \\ -0.17114 \\ -0.79032 \times 10^{-3} \\ 0.12257 \times 10^{-7} \\ 0.26966 \times 10^{-7} \end{bmatrix} \quad \Psi_3 = \begin{bmatrix} -0.20176 \\ -0.64424 \\ 0.72796 \\ -0.20821 \times 10^{-2} \\ 0.11962 \\ 0.50429 \times 10^{-3} \\ -0.14853 \times 10^{-7} \\ -0.32676 \times 10^{-7} \end{bmatrix}$$

$$\Psi_4 = \begin{bmatrix} -0.18228 \\ -0.14568 \\ -0.50307 \\ -0.18565 \times 10^{-2} \\ 0.83215 \\ 0.74080 \times 10^{-3} \\ -0.94364 \times 10^{-7} \\ -0.20760 \times 10^{-6} \end{bmatrix} \quad \Psi_5 = \begin{bmatrix} -0.10627 \\ -0.69349 \times 10^{-1} \\ -0.10806 \\ -0.10782 \times 10^{-2} \\ -0.46485 \\ 0.86517 \\ -0.36116 \times 10^{-1} \\ -0.79451 \times 10^{-1} \end{bmatrix} \quad \Psi_6 = \begin{bmatrix} -0.1839 \times 10^{-1} \\ -0.83736 \times 10^{-2} \\ -0.13028 \times 10^{-1} \\ -0.13025 \times 10^{-3} \\ -0.55860 \times 10^{-1} \\ -0.76266 \\ 0.26660 \\ 0.58630 \end{bmatrix}$$

$$\Psi_7 = \begin{bmatrix} 0.12884 \times 10^{-4} \\ 0.84024 \times 10^{-5} \\ 0.13070 \times 10^{-4} \\ 0.13070 \times 10^{-6} \\ 0.56016 \times 10^{-4} \\ 0.23993 \times 10^{-3} \\ 0.70694 \\ -0.70727 \end{bmatrix} \quad \Psi_0 = \begin{bmatrix} -0.86386 \times 10^{-5} \\ -0.56339 \times 10^{-5} \\ -0.87638 \times 10^{-5} \\ -0.87638 \times 10^{-7} \\ -0.37559 \times 10^{-4} \\ -0.40689 \times 10^{-3} \\ -0.20344 \\ -0.97909 \end{bmatrix}$$

The new k matrix, E_1 to E_4 remain the same, other eigenvalues are changed $E_5 = -5.988 \times 10^{-5}$ /yr, $E_6 = -7.241 \times 10^{-6}$ /yr, and $E_7 = -1.503 \times 10^{-7}$ /yr. Also, assuming that fossil burning is around 500 years, then evolution of CO_2 gas is as follows:

$$A_1(t) = -0.0015e^{-5.02t} - 18.253e^{-0.0612t}$$
$$- 11.074e^{-9.08488t} - 18.748e^{-0.01963t}$$
$$- 10377.96e^{-5.988 \times 10^{-5}t} - 11508.78e^{-7.241 \times 10^{-6}t}$$
$$+ 16150.79e^{-1.503 \times 10^{-7}t} + 6474.029 \qquad [11\text{-}24]$$

The preceding Equation [11-24] is already corrected with fossil fuel combustion and cement manufacturing.

A plot for atmospheric CO_2 with long-term global cycle is indicated in the upper graph of Figure 11-30 with years expressed in log scale using Equation [11-20]. Notice in this instance that the near term of 100 years does not show any increase of temperature. If the fossil fuel and other chemicals are added, the near-term picture will be changed dramatically. For example there is a constant increase if anthropogenic influence is modified. In the lower graph of Figure 11-30, the Equation [11-24] is used for calculation.

As global warming is a problem for the future, can we, as human ingenuity extends, find another pathway for reduction of the atmospheric carbon? Many versions of carbon cycles illustrate the distribution of carbon in the biosphere with similar features as the Holland version, as shown in Figures 11-31 and 11-32 from Bolin. Because the dominant carbon reservoir is in the sediment, a tremendous amount of which is locked in and is not being circulated, can we explore this resource as a food to relieve the world's hunger problem?

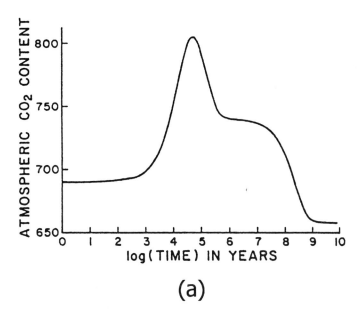

(a)

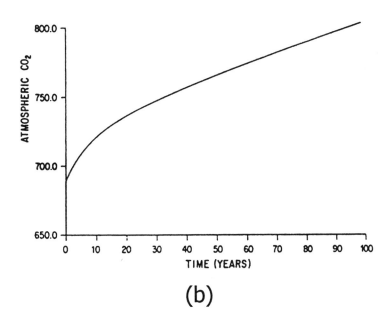

(b)

Figure 11-30. (a) Long range vs. (b) short range atmospheric CO_2 concentration (After Lasaga).

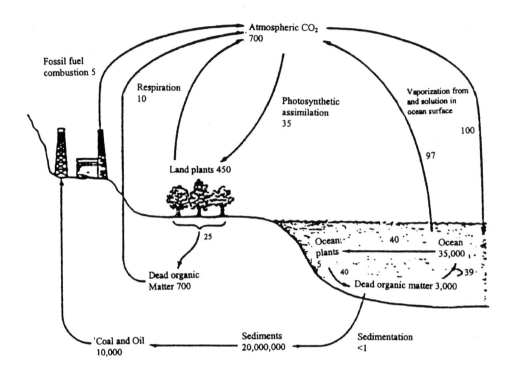

Figure 11-31. The carbon cycle. Numbers represent quantities of C in various reservoirs or annual flows from one reservoir to another. All are in units of 10 tonne. [Data from B. Bolin, *Sci. Amer.* 223(3), 125–132 (1970)].

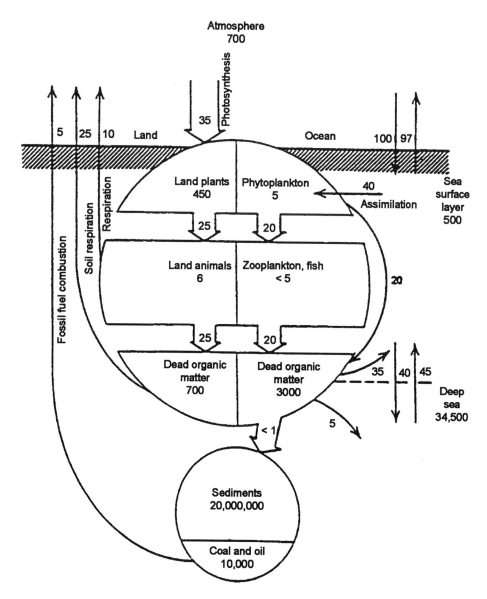

Figure 11-32. Distribution of carbon. Reservoir numbers are in billions of tons (metric) and flux numbers are in billions of tons per year (values, except for land animals, are from Bolin, 1970).

REFERENCES

11-1. A. G. Cairns-Smith, *The Life Puzzle*, University of Toronto Press, Toronto, 1971.

11-2. T. F. Yen, "Terrestrial and Extraterrestrial Stable Organic Molecules," in *Chemistry in Space Research*, (R. F. Landel & A. Rembaum, eds.), American Elsevier, New York, 1972.

11-3. I. Thornton (ed.), *Applied Environmental Geochemistry*, Academic Press, New York, 1983.

11-4. R. Balian, J. Aadouze, and D. N. Schramm (eds.), *Physical Cosmology*, North-Holland, New York, 1980.

11-5. D. H. Kenyon and G. Steinman, *Biochemical Predestination*, McGraw-Hill, New York, 1969.

11-6. R. S. Kandel, *Earth and Cosmos*, Pergamon, New York, 1980.

11-7. A. H. Brownlow, *Geochemistry*, Prentice-Hall, Englewood Cliffs, New Jersey, 1979.

11-8. A. C. Lasaga and R. J. Kirpatrick, *Kinetics of Geochemical Processes*, Mineralogical Society of America, Washington, DC, 1981.

11-9. T. F. Yen and J. M. Moldowan, *Geochemical Biomarkers*, Harwood Academic, Chur, Switzerland, 1988.

11-10. R.A. Horne, *The Chemistry of Our Environment*, Wiley-Interscience, New York, 1978.

11-11. J.J. W. Rogers, A History of the Earth, Cambridge University Press, Cambridge, England, 1994.

11-12. A. Braibanti, *Bioenergetics and Thermodynamics: Model Systems*, Reidel, Dordrecht, Holland, 1980.

11-13. B. Mason, *Principle of Geochemistry*, Wiley, New York, 1960.

11-14. D. W. Waples, *Geochemistry in Petroleum Exploration*, International Human Resource Development Corp., Boston, Massachusetts, 1985.

11-15. S. S. Bucher, R. J. Charlson, G. H. Orians, and G. V. Wolfe, *Global Biogeochemical Cycles*, Academic Press, London, 1992.

11-16. R. M. Garrels, F. T. Mackenzie, and C. Hunt, *Chemical Cycles and Global Environment, Assessing Human Influences*, William Kaufmann, California, 1973.

11-17. J. L. Coman and F. K. Thielemann, R-Process Nucleosynthesis in Supernovas, Physics Today 57(10), 47-53(2004).

PROBLEM SET

1. Find the average molecular speed of N_2 at 25°C. Calculate velocity required for any gas/particle to escape from the earth. Can we expect N_2 to remain on the earth for the next billion years?

2. Write the following chemical equations:

   a. β – decay of $_{28}Ni^{56}$
   b. β – decay of $_{14}Si^{28}$
   c. $_{13}Al^{27}$ (α,n) $_{15}P^{30}$

3. What will be the stage of evolution of the atmosphere and hydrosphere, if the fraction of O_2 in the atmosphere is 0.1? Also, predict approximately how many years ago this was.

TOXICOLOGY AND RISKS

12.1 THE QUALITY OF LIFE

As the world becomes more industrialized in an attempt to maximize the **quality of life** (QOL) for society, many adverse environmental problems (which tend to worsen the quality of life) arise as well. There is actually a trade-off between a new public policy and a new invention, which is designed to bring up the living standard, and it's associated with environmental problems. Coping with the total environment requires individual and group activity. In addition, various technologies and ideologies for coping with the material and spiritual environments have been developed. Taking these into consideration, we can see the quality of life as an operational concept as shown here.

The **Quality of Life** (QOL) can be expressed as material and spiritual aspects:

	Individual	Societal
Material	P, Provision of materials	E, Environment Quality
Spiritual	F, Fulfillment in psychic realm	J, Justice in righting wrongs

$$R = E + P \text{ if F, J are fixed (R = resource)}$$

If S is a satisfactory index, then

$$QOL = f(S_P, S_E, S_F, S_J) \qquad\qquad [12\text{-}1]$$

Resources, R, are often limited at a given moment, and when R decreases, at least one other goal variable (P, E, F, or J) must decrease. On the other hand, when R increases, it is possible to increase one or more of the goal variables without an accompanying decrease in any of them. Figure 12-1 illustrates an example of this analysis. For this special case, F and J are fixed, while P is plotted as the y-axis and E as the x-axis. At any given time, P and E cannot be chosen independently, but are constrained by a fixed amount of resources, for example, R". If, for any reason, E increases, P will have to decrease. For instance, money spent for air pollution control devices will not be available for food or clothing. Only new technology that increases R would allow both P and E to increase. It is often the case that resources are decreasing, say from R" to R', and sacrifices must be made in E or P or both. A decision has to be made to determine the trade-off point.

If factors affect societal decisions when there occurs a change in resources, a first step might be to suggest alternatives. For example, the following is a list of the consequences of possible strategies to meet a decrease in the availability of gasoline.

	ΔP	ΔE	ΔF	ΔJ
Price rationing	(+)	(-)	+	-
Coupon rationing	(-)	(+)	-	+
Travel restriction	0	+	-	-
Relax environmental constraints	+	-	+	-

However, there is no basis for decision; the preceding table does not show that a decrease in E, for example, is to be weighted against an increase in P. It is instructive to find a function of P, E, F, and J that could sense the signal of the quality of life. We assume that the quality of life can be positively correlated with a satisfaction index, S: The greater the quality of life is, the greater S will be. As a first approximation, the overall satisfaction, S, can be taken as the sum of the four partial satisfaction indices, S_P, S_E, S_F, and S_J, each index corresponding to the satisfaction with respect to the degree of attainment of the respective goal variable when the other three goal variables are held constant. Figure 12-2, as an example, shows the general feature of the function S_P. In the figure, as P increases, S_P increases, but at a decreasing rate. At a very low P value, satisfaction tends to reach negative infinity. The overall satisfaction index should be maximized by choosing a balance among P, E, F, and J, making each partial satisfaction function not too small and above all positive, so that the sum of the partial satisfaction indices is as favorable as practicable.

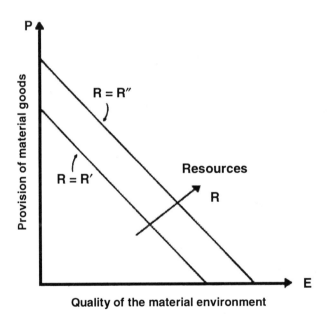

Figure 12-1. Resource in material aspects of quality of life.

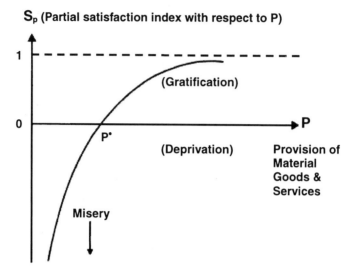

Figure 12-2. Properties of satisfactory parameters.

Usually the **parameter impringed unit** (PIU), which has to do with human interest, aesthetics, biology, and environmental pollution, is an important factor to evaluate the environmental impact — so-called the **environmental impact unit** (EIU). It can be shown as follows:

$$\text{Environmental Impact Unit (EIU)} = \text{(EQI)(PIU)} \qquad [12\text{-}2]$$

$$\text{EQI (Environmental Quality Index)} = \text{(expected quality)/(ideal quality)} \quad [12\text{-}3]$$

PIU (Parameter Impringed Unit) consists of:

Ecology (species, habitats, ecosystems)

Aesthetics (biota, wetlands, forestry)

Environmental Pollution (water, air, soil)

Human Interest (education, culture, life patterns)

12.1.1   Optimization for the Infrastructure

As population increases, many of the estimated services projects (such as wastewater treatment plants) should have constructed to handle additional demand (such as interceptor sewers). In theory, the capacity expansion should stay ahead of the growth of wastewater flow or demand in MGD. The question is, what is the optimal expansion size, and what should the timing be.

Considering the facility cost, there is strong argument for building a large size. Counteracting this approach is that expeditious are committed to the future, the present worth cost is decreased. In general, both wastewater plants and interceptor sewers are controlled by economics as follows:

$$C(x) = ax^b \qquad [12\text{-}4]$$

where $0 < b < 1$

a = constant

b = economy of scale factor

x = hydraulic design capacity (e.g., MGD)

$C(x)$ = present worth cost of capital plus operation and maintenance ($M)

The trade-off is between economic of scale (building large now) and discounting (building small now). A method approach to this problem is using unconstrained optimization by nonlinear programming. To optimize any unconstrained function; for example,

$$z = f(x_1, x_2, ... x_n) \qquad [12\text{-}5]$$

The condition is for

$$\nabla f = \left(\frac{\partial f}{\partial x_1}, \frac{\partial f}{\partial x_2} ... \frac{\partial f}{\partial x_n} \right) = 0 \qquad [12\text{-}6]$$

Thus, the function has a critical point, crystallizing point, at x_0 is $\nabla f = 0$. If $f^{n+1}(x_0) \neq 0$, then $f(x)$ has a maximum of $f^{n+1}(x) < 0$ (n is odd), and $f(x)$ has a minimum of $f^{n+1}(x) > 0$ (n is odd).

Solving the trade-off problem, if we designate:

D = rate of linear demand increase (MGD per year)

t = time in years

r = rate of discount selected by Congress for use in water resource project (in %).

If the point of capacity expansion is t_0, t_1, t_3,...n with the corresponding expansion capacity identical, then

$$x_1^* = x_2^* = x_3^* ... = x_n^* \qquad [12\text{-}7]$$

and

$$(t_1 - t_0) = (t_2 - t_1) = (t_3 - t_2) = t' \qquad [12\text{-}8]$$

Therefore,

$$x_i^* = Dt' \qquad [12\text{-}9]$$

and $\exp(-rt')$ is part of the present worth cost.

Thus, for all expansions over n periods of the length of t' years

$$C(x) = ax^b e^{-n_o} + a x^b e^{-r't} + ax^b e^{-2r't} + \ldots + ax^b e^{-n'rt}$$

If $t_0 = t$,

$$C(x) = ax^b (1 + e^{-rt'} + e^{-2rt'} + \ldots + e^{-nrt'}) + ax^b [1 + e^{-rt'} + (e^{-rt'})2 + \ldots + (e^{-rt'})n] \quad [12\text{-}10]$$

and

$$0 < \frac{1}{e^{rt'}} < 1 \text{ for } r > 0, \ t' > 0 \qquad [12\text{-}11]$$

As $n \to \infty$, the geometric series converges to $1/(1-e^{-rt'})$.

Thus, equation [12-4] becomes

$$C(x) = \frac{ax^b}{\left(1 - e^{-rt'}\right)} \qquad [12\text{-}12]$$

Substitute $x = Dt'$ and express equation [12-12] in log scale

$$\ln C(x) = \ln a + b \ln (D't) - \ln \left(1 - e^{-r't}\right) \qquad [12\text{-}13]$$

or after differentiation,

$$\frac{\partial}{\partial t'} [\ln C(x)] = \frac{b}{t'_0} - \frac{re^{-rt'_0}}{1 - e^{-rt'_0}} = 0$$

Employing this equation, t_0' can be solved for fixed values of b, economy of scale factor and the discount rate. Table 12-1 lists the results for such a calculation. Considering the federal discount rate, low interest rates (2-4%) prevailing in the 60s, peaked in the late 80s at 8.875% and in fiscal 1996, rates reached 7.625%. In comparison, funds for general obligation bonds and revenue bonds are lighter. The traditional design periods for wastewater treatment plants and interceptor sewers have been 20 and 50 years respectively.

Table 12-1. Optimal Expansion Time for (a) Treatment Plants (years; USEPA, 1975); (b) Interceptor Sewers (years; USPEA, 1975)

Treatment Plants			
Discount	Economy of Scale Factor		
Rate (%)	0.6	0.7	0.8
5	19	13	9
7	13	10	6
10	9	7	4
12	8	6	4
Interceptor Sewers			
Discount	Economy of Scale Factor		
Rate (%)	0.3	0.4	0.5
5	40	31	24
7	28	22	17
10	20	16	12
12	18	14	10

12.2 CHEMICAL TOXICOLOGY

This section deals largely on the chemical toxicity originated from **internal pollution**. This term defines the harmful materials that are used by man, including through ingestion, inhalation, and application as cosmetics to the body.

For example, esophageal cancer formed in Casping region is correleable to the high salinity of the lithoral soils. Food additives as well as the nutrition supplements are involuntary intakes for necessary food and medicine. At this time, we should include a large number of toxic substances which are subjected to voluntary intake. Both involuntary and voluntary introduction to the body are the causes of internal pollution.

This list will include the socially acceptable use of alcohol, tobacco, and even many hallucinogenic drugs from psychotomimetic plants, such as marijuana. A great number of hallucinogens and neurohumors are known to be used in religious ceremonies, such as peyote rites of the American Indians. The active component is mescaline derived from peyote cactus. Chemical structures of narcotics, such as morphine, can also be comparable to many of the hallucinating drugs such as psilocybin, harmine, muscimole, PCP, etc. that contain two nitrogens with four carbons apart. Activity is enhanced by placing an aromatic ring α to the nitrogen to form an indole ringlike system. An exception to this rule is morphine, the N-containing ring is a six carbon caged structure, yet the dibenzofuran-like base is similar to that of Δ'-tetrahedrocannabinol, the active component of marijuana. A potent synthetic substance called LSD conforms with the rule that four carbons between two nitrogens with one nitrogen in an indole ring are required to give hallucinogenic properties. This physiological property may be due to the interference or competition of serotonin in human brains.

The chemical structures of most internal pollutants so far discussed can be found in Figure 12-3.

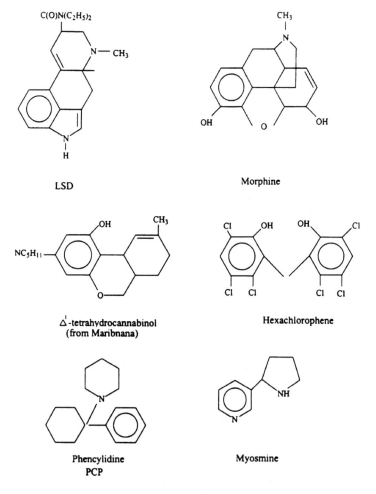

Figure 12-3. Structure of DES and other pollutants.

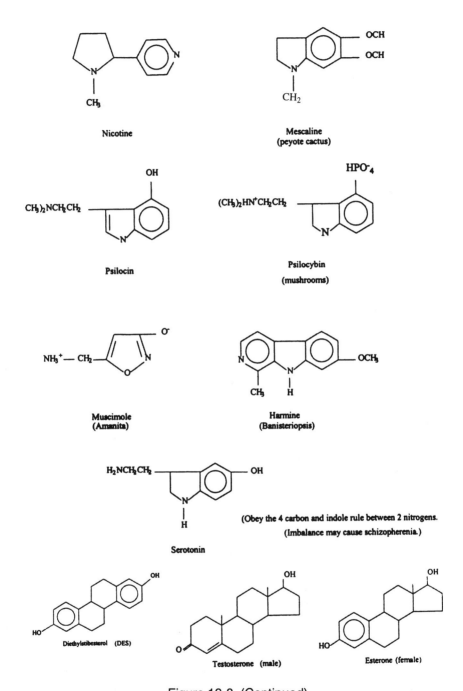

Nicotine

Mescaline
(peyote cactus)

Psilocin

Psilocybin
(mushrooms)

Muscimole
(Amanita)

Harmine
(Banisteriopsis)

Serotonin

(Obey the 4 carbon and indole rule between 2 nitrogens.
(Imbalance may cause schizopherenia.)

Diethylstibesterol (DES)

Testosterone (male)

Esterone (female)

Figure 12-3. (Continued)

Food additives and the above mentioned drugs cause internal pollution via oral route. One should not forget the intakes from the lung and skin. For example, hexachlorophene have been legitimately used as hygienic spray or vaginal deodorant for years. Even hospitals washed human babies with a dilute solution of hexachlorophene to ensure infections against diaper rash. As developed in this section, most of the internal pollutants cause genetic changes which become mutagenic and may eventually become carcinogenic.

12.2.1   Food Safety

In this section we will briefly review some environmental aspects of food. The public is confused and concerned about the safety of the food it consumes because it receives mixed views from presumably reputable people on the safety of food. Some experts claim that one element or another of our diet is unsafe; others say this is untrue. The following are some collected facts for your own judgment:

Table 12-2. Confirmed Foodborne Diseases, 1979

Microbial Agents	Outbreaks	Cases
Clostricium botulinum	7	9
Clostridium perfringens	20	1110
Salmonella sp.	44	2794
Shegella sp.	7	356
Staphylococcus aureus	34	2391
Vibrio parahaemolyticus	2	14
Other bacteria	5	133
Viruses	6	229
Trichinella spiralis	11	93
Chemical Agents		
Naturally occurring seafood toxins	30	217
Toxic mushrooms	1	2
Heavy metal	1	18
Other chemicals	4	13
Total	172	7379

Table 12-2 gives the data for foodborne diseases in 1979. There are 172 incidents involving almost 7,400 cases in which the cause of illness could be established. From time to time we discover microbial hazards that we did not know of before. Potentially, one of the most serious biological agents of all are the fungal poisons called **mycotoxins**. The best known and probably the most important of these is **aflatoxin**, which was discovered over 25 years ago in moldy peanut meal. It is a potent liver carcinogen in a broad range of animal species. Epidemiological studies in several tropical areas support the assumption that **aflatoxin** can cause liver cancer in humans.

Food additives perform a variety of functions in foods. The principal benefits of the use of food additives are cost reduction, user convenience, nutrition, food quality and attractiveness, prevention of microbial contamination, and an increase in the variety of foods available to the consumers. A sampling of additives according to function includes anticaking agents, antioxidants, colors, curing and pickling agents, emulsifiers, enzymes, flavoring agents, leavening agents, nutrient supplements, thickeners, and texturizers. Table 12-3 lists all the chemical additives in a loaf of bread. An important factor in evaluating the possible hazard of a food ingredient is the quantity ingested.

Table 12-3. Additives in a Loaf of Bread

Preservatives	Calcium propionate, sodium diacetate, sodium propionate, acetic acid, lactic acid
Leavening Agents	Potassium acid tartrate, monocalcium phosphate, sodium acid pyrophosphate
Bleaching Agents	Benzoyl peroxide, chlorine dioxide, nitrosyl chloride, oxides of nitrogen
Bread "Improvers"	Potassium bromate, potassium iodate, calcium peroxide
Antioxidants	Butylated hydroxytoluene, butylated hydroxyanisole, propyl gallate, nordihydroquaiaretic acid
Emulsifiers	Lecithin, monohlycerides, diglycerides, sorbitan, and polyoxethylene fatty acids

In addition,

The Wheat is grown contaminated with fertilizers, herbicides, and pesticides

The Grain on storage is treated with rodenticides, fungicides (some of them have mercury compounds), and insecticides

The Water added to the flour may have been purified with alum, chlorine, copper sulfate, and serta ash, and fluoride compounds may have been added (recently epidemics of acute hemolytic anemia in uremic patients have been traced to chlorination of urban drinking water by chloramine bacteriacides

The Sugar (or dextrose) added to the flour may have been refined with lime, sulfur dioxide, posphates, and charcoal

The Salt may contain added iodide and carbonates of calcium and magnesium

The Yeast has been fed on ammonia salts

The Shortening may contain trans fats, is refined, bleached, and decolorized and has been exposed to traces of nickel; anti-oxidants; citric, ascorbic, and phosphoric acids; and has been glycerinated

Source: R.M. Linton, *Terricide*, Little Brown Co., Boston, 1970

Table 12-4 lists the 15 **GRAS** (generally recognized as safe) food ingredients used in the food industry annually. Use of sucrose appears to have been under-reported. Including use in beverages, the total use was 70 lb. per capita in 1971 and probably about the same in the 1990s. Table 12-5 lists the natural flavoring substances used in the largest amounts in the six years following 1976. Table 12-6 lists the 15 synthetic substances used in the greatest amounts in flavoring by the processed food industry. To emphasize the modern-day diet, the beverages and the foods we take contain many chemical additives, as well as DDT and its derivatives (referring back to Section 9.3). The latter insecticides have been banned years ago, but the residue survives in the soil environment for a long duration.

Table 12-4. Annual GRAS Food Ingredients Used by Food Industry

	Annual Use	
Substance	Million lb	lb/Capita
Sucrose	5000	23
Corn syrup	1530	7.2
Sodium chloride	1420	6.7
Dextrose	266	1.3
Mono- and diglycerides	86	0.40
Hydrochloric acid	82	0.38
Caramel	74	0.35
Sodium bicarbonate	60	0.28
Yeasts	57	0.27
Citric acid	57	0.27
Calcium phosphate, monbasic	48	0.23
Monosodium glutamate	28	0.13
Carbon dioxide	27	0.13
Hydrolyzed vegetable proteins	23	0.11
Sodium aluminum phosphate	15	0.07

Meat eaters do have to worry about the artificial **estrogens**, such as stibesterol and diethystibesterol (DES), because capsules or pellets of DES are implanted in chicken's necks or in the ears of cattle to produce tender meat. These artificial estrogens are similar in structure to human sex hormones. The chemical structure of DES and sex hormones are also shown in Figure 12-3. They have the potential to alter secondary sex characteristics and are also strongly carcinogenic. Many natural toxicants are found in plant foods. **Coniine** is present in berries. The chemical structures of a number of nitrogen bases that are toxic are presented in Figure 12-4. Mycotoxins, such as the aflatoxins in groundnut meal, which are the secondary metabolic products of certain fungi that are toxic to man and domestic animals, are the cause of the acute haemolytic symptoms of some individuals after eating foods containing fababeans which may contain **vicine** and **convicine**. Many naturally occurring antioxidants, such as the active components in rosemary leaves, or **sinigrin** (for chemical structures please refer to Figure 12-4), and in mustard, radish, and brassica vegetables, are flavoring agents acquainted by man.

The movement of agriculture toward the use of lower levels of agrochemicals and the exploration of integrated systems of pest management will be accompanied by the search for new crop varieties possessing enhanced natural resistance that may be termed as natural pesticides. These natural pesticides will include tannins, lectins, alkaloids, antioxidants, and glycosides. The current fashion of increasing one's intake of green vegetables and dietary fiber will also result in an increased intake of biologically active food constituents whose long-term effects on human health has not yet been determined. It is certain that the intake of such compounds by vegetarians may be 10–100 fold greater than that for an equivalent omnivore population.

Mutagens do associate with food. The relationship between cancer and food will be discussed in the following section.

Table 12-5. Natural Flavoring Substances Reported Used by Food Industry in Largest Amounts in Six Years Period after 1970

	Usage	
Substance	1000 lb/year	mg/(capita day)
Mustard, yellow	31,000	170
Pepper, black	19,051	108
Malt extract	7,050	40
Pepper, red	2,334	14
Caffeine	2,000	11
Lemon oil	1,547	9
Cassia	1,147	7
Oregano	920	5
Peppermint oil	870	5
Nutmeg	770	4
Caraway seed	634	4
Cocoa extract	536	3
Cloves	519	3
Allspice	495	3

NAS/NRC Surveys

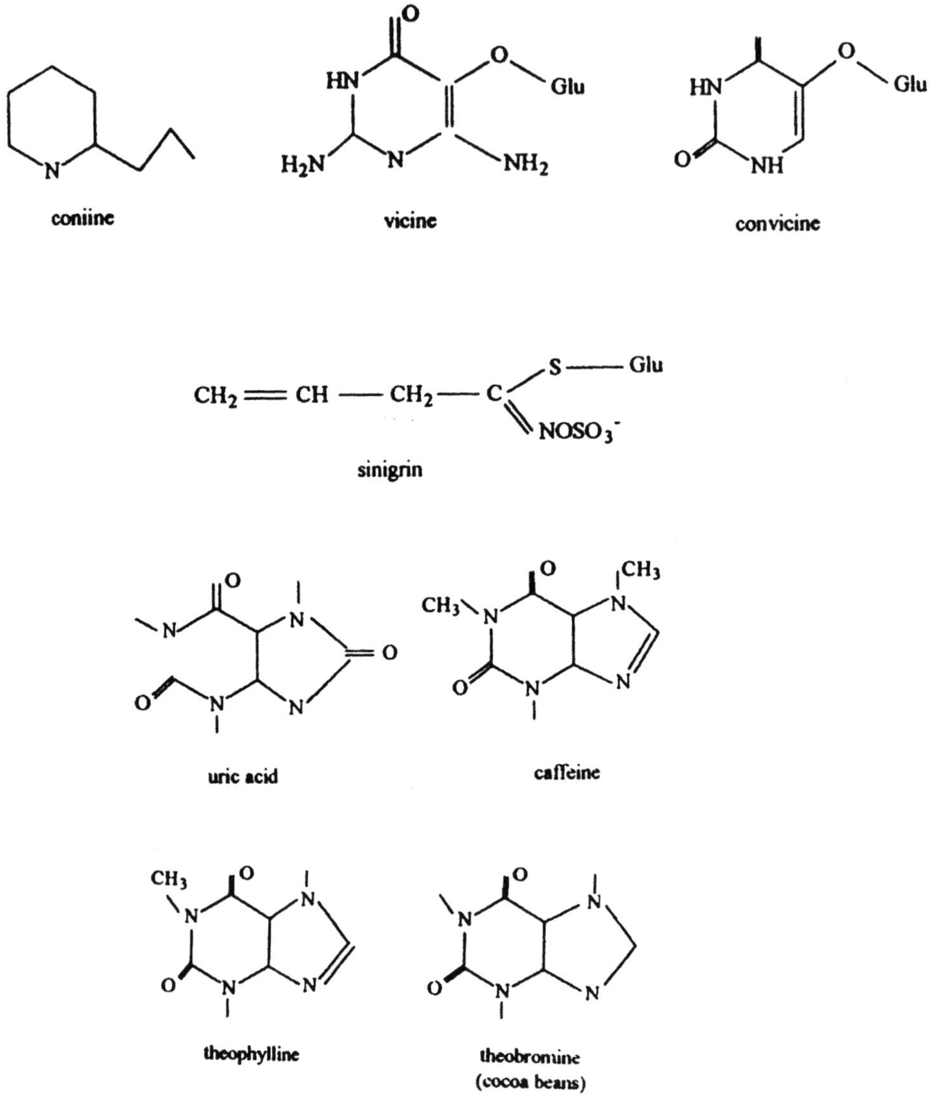

Figure 12-4. Structure of toxic nitrogen bases including caffeine and related compounds.

Table 12-6. Synthetic Flavoring Substances, Adjuncts and Adjuvents, Reported Used in Largest Amounts

	Usage	
Substance	1000 lb/year	mg/(capita day)
Monosodium glutamate	18,000	104
Isopropyl alcohol	4,000	
Malic acid	3,000	17
Acetone	570	3.6
Ethyl acetate	560	3.2
Methyl salicylate	280	1.6
Ethyl acetoacetate	68	0.36
4-(Methylthio)-2-butanone	59	0.34
Thiamine hydrochloride	57	0.33
Isobutyl acetate	44	0.25
Ethyl maltol	38	0.22
Isoamyl butyrate	36	0.21
Butyric acid	27	0.15
Triacetin	20	0.16
Acetaldehyde	19	0.11

1977 NRC survey

12.2.2   Carcinogenesis and Mutagenesis

Although there are numerous toxic effects, either acute or long-term, associated with chemicals, one of the most dreaded is cancer. Cancer was responsible for almost 20% of all deaths in the United States in the 1980s. Above all, there is a strong likelihood that most cancers have environmental causes. One of the most striking pieces of evidence for this comes from the pattern of the incidences of cancer in different parts of the body in different countries. Japanese people have much higher death rates from stomach and liver cancer and lower rates from colon and prostate cancer than do white residents of California. However, these differences are much less pronounced in Japanese immigrants to California. In addition, there is a marked correlation between the increase in cigarette smoking among men early in the century and the incidence of lung cancer some 20 years later. Many chemicals are capable of causing problems. Table 12-7 gives a list of chemical carcinogens along with the types of cancer they produce.

Table 12-7. Chemicals Recognized as Human Carcinogens

Chemical Mixtures	Site of Cancers
Soots, tars, oils	skin, lungs
Cigarette smoke	lungs
Industrial chemical	
2-Naphthylamine	urinary bladder
Benzidine	urinary bladder
4-Aminobiphenyl	urinary bladder
Chloromethyl methyl ether	lungs
Nickel compounds	lungs, nasal sinuses
Chromium compounds	lungs
Asbestos	lungs
Arsenic compounds	skin, lungs
Vinyl chloride	liver
Drugs	
N,N'-bis(2-chloroethyl)-2-naphthylamine	urinary bladder
Bis(2-chloroethyl)sulfide(mustard gas)	lungs
Diethylstilbestrol	vagina
Phenacetin	renal pelvis
Naturally occurring compounds	
Betel nuts	buccal mucosa
Aflatoxins	liver
Potent carcinogens in animals to which human populations are exposed	
Cyclamates	bladder
Sterigmatocystis	liver
Cycasin	liver
Safrole	liver
Pyrrolizidine alkaloids	liver
Nitroso compounds	esophagus, liver, kidney, stomach

Source: C. Heidelberge, "Chemical Carcinogenesis," Annu. Rev. Biochem., 44 (1975)

We do not know exactly how chemical carcinogens produce cancer, but it is likely that they attack the DNA molecules. The following summarizes some types of chemical carcinogenesis that chemically alter the base pair duplications in DNA. Some of the alteration mechanisms are listed here:

Chemical Carcinogenesis

- change of base by nitrous acid — conversion of NH_2- to OH-

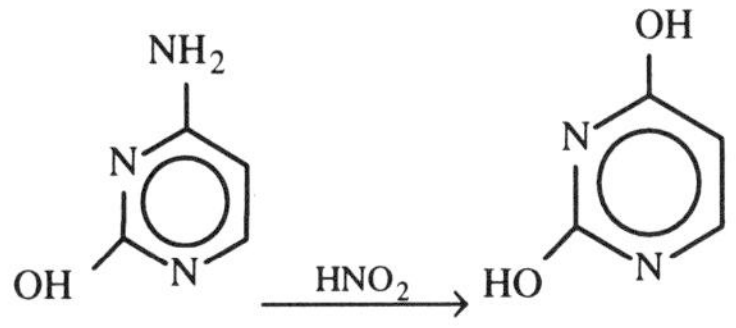

Cytosine   Unacil

- alkylation by alkylating agent to the base — e.g., by N-nitrosoamine or ethylene oxide

$$RNH - H = 0 \rightarrow RN = N - OH \rightarrow R - N+ \equiv N$$

N-nitrosoamine

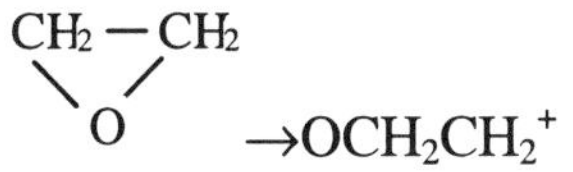

$$\rightarrow OCH_2CH_2^+$$

ethylene oxide

- intercalation by agents such as heterocyclics or PAH; e.g., acridine

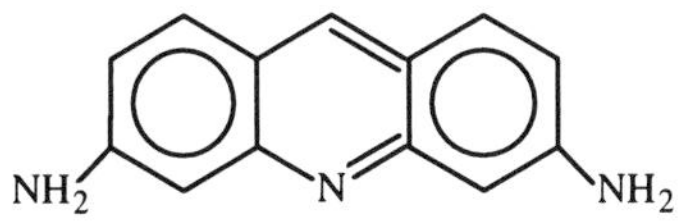

Any damage to DNA can result in mistakes in the genetic code. Often this results in the death of the cell, but if the cell survives, the mistake can be passed on to the daughter cells when the DNA is copied. This is called **mutation** and might cause an alteration in some function of the organism. Cancer is a condition in which a cell grows and divides uncontrollably, invading normal body tissues. Most cancer-producing chemicals have to be activated in the organisms to express their carcinogenic properties.

There are about 500 new chemicals that are marketed each year; therefore, singling out the potentially carcinogenic ones is a primary concern in cancer control. Studying the effects of chemicals on rats has proven to be the most valuable screening method to date, but it is time-consuming and the results are often inconclusive. The most widely known of the new tests is one developed by Ames at University of California, Berkeley. The **Ames test** is based on the fact that, with a few exceptions, carcinogens are also mutagens. The test is inexpensive, takes only a few days to complete, and only needs microgram amounts of the chemical. It is performed in a petri dish on various strains of *Salmonella typhimurium*, bacteria that are developed for detecting and classifying mutagens. Each tester strain has a specific mutation in one of the genes of the operon for histidine synthesis, which makes it require added histidine for growth. A chemical with mutagenic activity can cause a back-mutation in the operon, which restores the capacity for histidine synthesis. The indication of a positive result is the growth of the revertant bacteria around the spot where the chemical was applied.

The detection of carcinogens or mutagens in foods was not studied extensively until the development of the Ames test. Using the Ames test, mutagens were found in coffee, tea, alcoholic spirits, wines, and spices, as shown in Table 12-8. Mutagens are found to be formed during cooking, because pyrolysis of protein has been shown to cause the formation of mutagens (Figure 12-5 shows the increase of mutagens with the cooking time).

Table 12-8. Mutagenic Activities Induced by One Cup of Coffee or One Glass of Whisky or Brandy

	Number of Revertants [a]
Coffee [b]	135,000
Whisky [c]	3,600
Brandy [c]	8,000

[a] *C. typhimurium* TA 100 without 69 mix was used.

[b] one cup contained 150 ml coffee

[c] one glass contained 30 ml whisky or brandy

For a large number of toxic compounds, the correlation between the mutagenic potential and carcinogenic potential is not linear, as shown in Figure 12-6. For example, some heterocyclic amines are mutagenic, but their carcinogenic potentials are not very high. It is clear from Figure 12-6 that these mutagens deviate greatly from those obtained for some typical carcinogens.

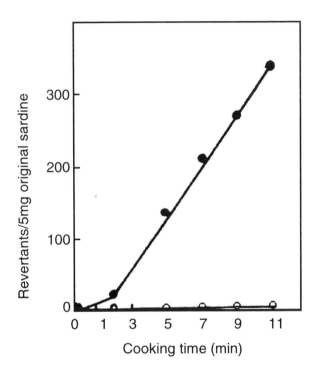

Figure 12-5. Mutagenicity of sun-dried sardines broiled at various times. *S. typhimurium* TA98 with (•) and without (o) S9 mix was used for the assay.

It is also important to note here that **teratogens** are different from either mutagens or carcinogens. Table 12-9 depicts the complex sequence of events in teratogenesis. Although some parts in the table may be held in common with carcinogenesis (e.g., certain causes) and mutagenesis (e.g., somatic mutations and chromosomal aberrations), many aspects are unique to teratogenesis, thus further emphasizing the basic differences existing among these processes. Table 12-10 shows conditions of exposures that determine the response in carcinogenesis, mutagenesis, and teratogenesis.

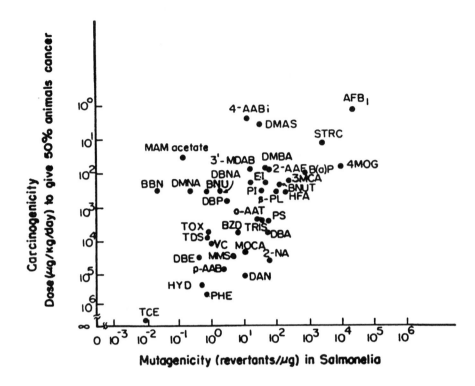

Figure 12-6. Relationship between mutagenicity in *S. typhimurium* and carcinogenicity. Abbreviations used: p-AAB, p-aminoazobenzene; 4-AABi, 4-acetylaminobiphenyl; 2-AAf, 2-acetylaminofluorene; o-AAT, o-aminoazotoluene; AFB1, aflatoxin B1; BBN, N-butyl-N-butanolnitrosamine; BNU, N-n-butyl-N-nitrosourea; BNUT, N-n-butyl-N-nitrosourethane; B(a)P, beno[a]pyrene; BZD, benzidine, DAN, 2,4-diaminoanisole; DBA, dibenz[a,h]anthracene; DBE, 1,2-dibromoethane; DBNA, N,N'-di-n-butylnitrosamine; DBP, 1,2-dibromochloropropane; DMAS, 4-dimethylaminostrilbene; DMBA, 7,12-dimethylbenz[a]anthracene; DMNA, N, N-dimethylnitrosamine; EI, ethyleneimine; HFA, N-hydroxy-2-acetylaminofluorene; HYD, hydralarine; MAM acetate, methylazoxymethanol acetate; 3MCA, 3-methylcholanthrene; 3'-MDAB, 3-methyl-4 dimethylaminoazbenzene; MMS, methyl methanesulfonate; MOCA 4, 4'methylene-bis (2-chloroaniline); 2NA, 2-naphthylamine; PHE, phenacetin; PI, prophyleneimine; β-PL, β-propholactone; PS, propane sultone; STRC, sterigmatocystin; TCE, trichloroethylene; TDS, toluenediamino sulfate; TOX, toxaphene; TRIS, tris (2,3-dibromopropylphosphate); VC, vinyl chloride.

Table 12-9. Summary of Teratogenesis

Causes	→ Mechanisms	→ Manifestations
Action of an agent from the environment on the embryo or the germ cells, e.g.,		
Radiations	→ Reaction within the embryo or germ cells, such as one or more of the following:	
Chemicals		
Dietary deficiency	Mutation	→ Pathogenesis, initiated by one or more of the following:
Infection	Chromosomal nondusjunction	
Hypoxia, etc.	Mitotic interference	Cell death
Temperature	Altered nucleic acid integrity or function	Mitotic rate change
Endocrine imbalance		Reduced biosynthesis
Physical trauma	Lack of precursors, aubstrats, etc.	Altered differentiation schedules
Placental failure	Altered energy sources	
	Changed membrane characteristics	Impeded morphogenetic movement
	Water-electrolyte imbalance	
	Enzyme inhibition	Etc.
		And leading to abnormal tissue and organ development which determine the nature and incidence of final defect

Reprinted with permission from Wilson, 1971

Table 12-10. Conditions of Exposure That Determine the Response in Carcinogenesis, Mutagenesis, and Teratogenesis

	Susceptible Tissues	Optimal time of Exposure	Duration and Level of Dosage
Carcinogen	Proliferating tissues	Uncertain, probably all states capable of mitosis	Usually chronic, possibly all doses
Mutagen	Germinal tissues	All stages of remetogenesis	Either acute or chronic, possibly all doses
Teratogen	Possibly all immature tissues	Highest during early differentiation	Acute only, above usual no effect level

12.2.3 Radon

As discussed previously in Section 3.1.5 (on nuclear power), naturally occurring radioactivity may originate from the decay series of uranium (4n + 2), thorium (4n), and actinium (4n + 3) as shown in Table 12-11. Because mass lost in the decay series is due almost exclusively to α-emission (mass of 4), the mass numbers of the members of a given series may be represented by 4n, of which n is an integer. The sources of radon and other isotopes are also listed in Table 12-11 (e.g., ^{222}Rn is radon, ^{220}Rn is thoron and ^{219}Rn is actinon). Each of the series begins with a very long half-life time (~4.5 billion years) equivalent to the age of the earth. The 4n + 1 series is not found in nature in any significant amount, but laboratory tests indicate that ^{237}N$_p$ (also long-lived) can also produce it. Whatever amount was present would have decayed by now. All these led to the end product of lead. Radon also can give daughters, for example, ^{218}Po (α-emitter), ^{214}Pb (β- and γ-emitter), etc.

Table 12-11. Some Characteristics of the Natural Radioactive Decay Series

Series	Uranium	Thorium	Actinium
Mass number code	4n+1	4n	4n+3
Long-lived parent and half-life	^{238}U 4.51×10^9 y	^{232}Th 1.39×10^{10} y	^{235}U 7.13×10^8 y
Radium parent and half-life	^{226}Ra 1,600 y	^{223}Ra 11.4 d	^{224}Ra 3.66 d
Radon isotope and half-life	^{222}Rn (radon) 3.82 d	^{220}Rn (thoron) 55.6 s	^{219}Rn (actinon) 4.0 s
Potential alpha energy in short-lived radon decay chain * (OECD-1983)	19.2 Mev per atom	20.9 Mev per atom	20.8 Mev per atom
Stable end product	^{206}Pb	^{208}Pb	^{207}Pb

*The potential alpha energy is the total alpha energy emitted during the decay of an atom along its decay chain.
Source: M. Wilkening

Radon and its decay products can be delivered to sensitive tissue in the lung respiratory system. Usually an adult can be expected to breathe at a rate of 0.75 m^3/min. Typical aerosol concentrations are approximately 10^{10}/m^3 with radii ca. 0.5 μm. Assuming that the average amount of indoor air contaminants is 50 Bq/m^3 of radon, 40 Bq of radon are taken in lung per minute. This activity will reach the bronchial tube and damage the basal cells underneath. In many locations, especially underground tunnels and mines, the radioactivity can exceed up to 10,000 times this amount (Table 12-12). Even the radon level in a popular spa, Badgastein, in the Austrian Alps is 1.1×10^5 Bq/m^3. Lung cancer is a common illness in underground mining because many of the mines have radon levels up to 100,000 Bq/m^3, or 2900 pC$_i$/L.

Table 12-12. Radon Released from Mines and Mills

Metal	Type of Mine/Mill	Number of Sources	Radon Release Per Source (Bq y^{-1}×10^{10})
Uranium	Underground	305	25,000
Uranium	Open pit	63	7,250
Uranium	Mill	20	10,400
Iron	Underground	11	560
	Open pit	57	70
Copper	Underground	15	20
	Open pit	46	1,500
Zinc	Underground	36	850

Adapted from R.H. Johnson Jr., et al., in Natural Radiation Environment (Bombay), Wiley Eastern Limited (1982) pp. 182–183.

Radon concentration in an underground cavity can be evaluated so that ventilation processes can be utilized. As shown in Figure 12-7, radon-rich air within the cavity is replaced by outside air from the outside having a low radon content. For a cavity of interior surface S and volume V,

$$SE = \lambda VC + Q(C - C_{od}) \qquad [12\text{-}14]$$

where E is the net inward flux of atoms/sec/m^2, λ is the decay constant for ^{222}Rn, C and C_{od} are the radon concentrations within the cavity and outdoors respectively in atoms of ^{222}Rn/m^3, and Q is airflow m^3/sec. If one brings the outdoor air (C_{od} ~ 0.2 pC$_i$/L) relative to that within the cave (C ~ 40 p C$_i$/L), and then allows C_{od} to be negligible,

$$C = \frac{SE}{\lambda V + Q} \qquad [12\text{-}15]$$

In this manner, for the Carlsbad Caverns in New Mexico, the radon-rich air in the cave is displaced by cooler air from the surface through natural ventilation, which is caused by the temperature differences between the cave and outside air. Some methods used for measuring **radon flux** are explained by Figure 12-8 which shows, the accumulator, flow, and adsorption methods. The flux density or exhalation rate can be expressed in the transfer rate as Bq/m^2/sec. To reduce the hazard of radon in domestic home is to use either the drain-tile suction or sub-slab suction, gas-proof liners should be installed.

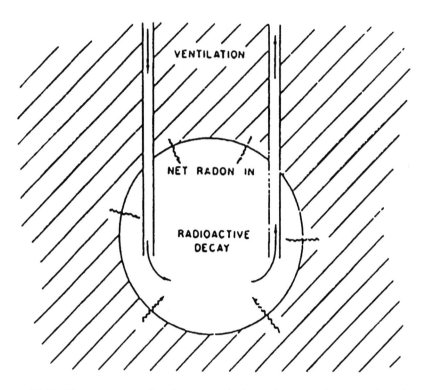

Figure 12-7. Factors governing the radon balance in an underground cavity. (After M.H. Wilkening and D.E. Watking, Health Phys. 31, 139 (1976))

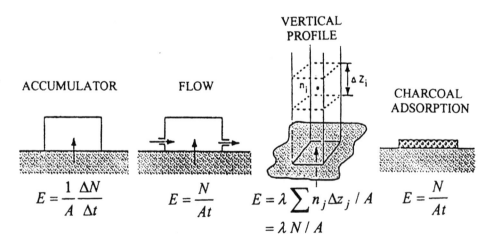

Figure 12-8. Methods used in measuring radon flux. E is the flux, A is the area covered by the device or system, t is the accumulation time, N is the quantity of radon, and λ is the decay constant for radon. (M. Wilkening, Ref.1)

12.3 RISK ASSESSMENT AND OCCUPATIONAL TOXICOLOGY

The risk can be simply expressed in **standard mortality ratio** (SMR)

$$SMR = \frac{\text{observed death}}{\text{expected death}} \qquad [12\text{-}16]$$

For example, the ratio of lung cancer death in a population of smokers to the lung cancer death in a nonsmoking population of the same size is 11. Consequently the SMR is 11, meaning that the risk of death from lung cancer is 11 times as high for a heavy smoker as for a nonsmoker.

Actually the lifetime risk can be obtained from the dose-response curve shown in Figure 12-9.

Lifetime risk is a probability function of dose, d, such that (from multistage model)

$$Risk = P(d) = 1 - \exp[-(q_0 + q_1 d + q_2 d_2 + ...q_n d_n)] \qquad [12\text{-}17]$$

$q_0, q_1, q_2,...$ are parameters to fit the data.

The background rate for cancer incidence is when d=0, or

$$P(0) = 1 - e^{-q_0} \approx \left\{1 - \left[1 + (-q_0)\right]\right\} = q_0 \qquad [12\text{-}18]$$

because

$$e^x = 1 + x + \frac{x^2}{2!} + ... + \frac{x^n}{n!} \approx 1 + x$$

if only two terms in the exponential function are used.

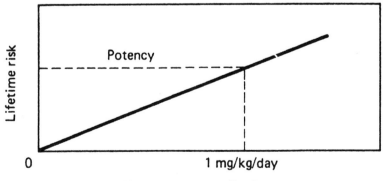

* The potency factory is the slope of the dose-response curve at low doses.

Figure 12-9. Dose-response curve (G.M. Masters, 1991, Ref.2).

Therefore,

$$P(d) = 1 - [1 - (q_0 + q_1 d)] = q_0 + q_1 d = P(0) + q_1 d \qquad [12\text{-}19]$$

or

$$\text{additional risk} = P(d) - P(0) = q_1 d \qquad [12\text{-}20]$$

Therefore, lifetime probability is linearly related to dose.

The slope of risk versus dose curve at low dosage is called the **potential factor** (PF),

and

$$PF = -q_1 \qquad \text{in } (mg/kg/day)^{-1} \qquad [12\text{-}21]$$

where d is in mg/kg/day

Lifetime risk= (PF)(chemical daily intake) = (PF)(CDI) for 70 yrs.

Some PF values are as follows:

Chemical	PF (oral route) (mg/kg/day)-1
chloroform	$6.1 \times 10\text{-}3$
benzene	$2.9 \times 10\text{-}2$
Dioxin	$1.56 \times 10\text{-}5$
PCB	7.7
TCE	$1.1 \times 10\text{-}2$

[**Example 12-1**]   THM in drinking water is 70 µg/L for an adult of 70 kg who drinks 2 L/day. What is the risk? (After G. Masters)

$$CDI = \frac{(70 \times 10^{-6} \, g/L)(10^3 \, mg/g)(2 \, L/day)}{70 \, kg}$$

$$= 0.002 \, mg/kg/day$$

Using the RF values of chloroform listed in preceding tables for the RF values of THM, we obtain

$$\text{Risk} = (\text{PF})(\text{CDI}) = (6.1 \times 10^{-3})(2 \times 10^{-3}) = 12.2 \times 10^{-6}$$

Probability is 12.2 per million people for 70 yr
For population of 0.5 million

$$\text{Cancer/yr} = (500,000)(12.2/10^6)(1/70)$$

$$= 0.09 \approx 0.1 \text{ cancer/yr}$$

The Occupational Safety and Health Act (OSHA) has set the **threshold limit values** (TLV) of many inorganic and organic contaminants that have been determined for the occupational worker being exposed to a certain chemical as illustrated in Table 12-13. The lifetime exposure of a worker is considered as 40 hours/week for 50 years. The **time-weighted average** exposure (T.W.A.) can be obtained as follows:

Calculation of Time-Weighted Average Exposure (T.W.A.) is as follows:

$$\frac{(\text{Exposure Time})(\text{Conc. "A"}) + (\text{Exposure Time})(\text{Conc. "B"}) + ...}{(\text{Total Work Time per shift})} = \text{T.W.A.} \quad [12\text{-}22]$$

Or

$$\frac{\sum_{i=1}^{n} (T_i) \cdot (C_i)}{T_{total} \text{ (work time)}}$$

= Time - weighted average concentration (T.W.A.) for this particular material

where

T_i = Duration of exposure period
C_i = Level of a specific contaminant during time period T_i,
Ttotal = Total work time per shift (7- to 8-hour workday)

A number of internal pollution, for example, asbestos and the lead paints on walls, the lead remains on highways have been reduced.

Table 12-13. Threshold Limit Values (TLV) of Fossil Fuel Contaminants

Substance	ppm	mg/m^3
Coal dust (bituminous)	—	2
Coal tar pitch volatiles (benzene soluble fraction)	—	0.2
Anthracene	NC [a]	
Benzo(a)pyrene	C [b]	
Phenanthrene	NC [a]	
Acridine	NC [a]	
Chrysene	C [b]	
Pyrene	NC [a]	
Naphtha (coal tar)	100	400
Asphalt fumes	—	5
Phenol – skin	5	19
Cresol – skin	5	22
Naphthalene	10	50
Oil mist, particulate	—	5
Organic solvent		
Acetone	1,000	2,400
Allyl alcohol – skin	2	3
Butyl alcohol	100	300
Benzene – skin	25	80
Carbon tetrachloride – skin	10	65
Carbon disulfide – skin	20	60
Chloroform	25	120
Cyclohexane	300	1,050
Dichloromethane	100	210
Heptane	500	2,000
Hexane	500	1,800
Methyl alcohol	200	260
Petane	500	1,500
Pyridine	5	15
Toluene	100	375
Xylene	100	435

Table 12-13. (Continued)

Trace metals, gas, and inorganics		
Antimony (Sb)	—	0.5
Arsenic (As)	—	0.5
Cadmium (Cd)	—	0.2
Lead (Pb)	—	0.15
Chromium (Cr)	—	0.5
Mercury (Hg)	—	0.05
Tin (Sn)	—	0.1
Vanadium (V_2O_5) – dust	—	0.5
Ammonia	25	18
Carbon dioxide	5,000	9,000
Carbon monoxide	50	55
Chlorine	1	3
Hydrogen chloride	5	7
Hydrogen sulfide	0.05	0.2
Nitrogen dioxide	5	9
Sulfur dioxide	5	13

[a] NC: negative carcinogenicity [b] C: positive carcinogenicity

12.4 SOUND, MICROWAVE AND OTHER ELECTROMAGNETIC EXPOSURE

Humans live surrounded by low-level electric and magnetic fields from electric powerlines and appliances and electronic devices. The frequencies exposed range from 1 through 1022 Hz as indicated by Figure 12-10. The biological effects from low-level nonionizing fields are well established; for example, there are reports of cancer clusters in electromagnetic fields (EMFs) which affect houses near a 500 kV line, testicular cancer caused by police radar guns, spontaneous abortions by computer video display terminals (VDTs), alleged brain cancers caused by cellular phones, ill effects due to electric blankets of 30-300 mG, and so on. However, the health effects of electromagnetic fields still remain unresolved.

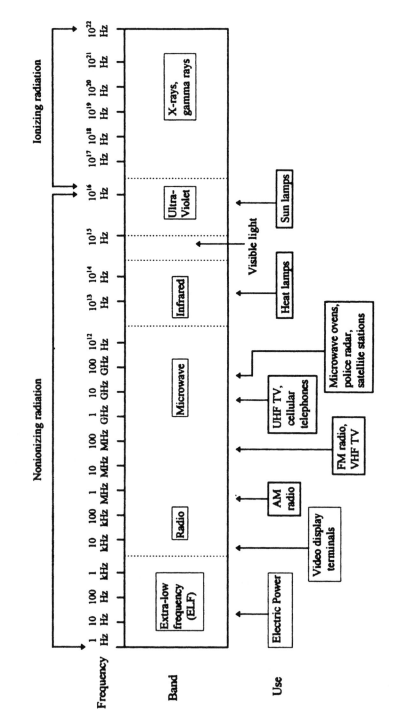

Figure 12-10.   Broad range of electromagnetic frequencies for human exposure.

12.4.1 Noise Pollution

Noise is commonly defined as unwanted sound. Energy in the form of sound waves may be produced in the form of energy residuals from manufacturing processes such as waste heat; fortunately, it is short-lived and relatively nonextensive. The vibration of solid objects will cause the surrounding air to undergo alternating compression and rarefaction in such a way as to effect changes in density and pressure. The pressure wave is obtained as **root mean square** (rms) sound pressure.

$$P_{rms} = \left(\overline{p^2} \right)^{\frac{1}{2}} = \left[\frac{1}{T} \int_0^T p^2(t)\,dt \right]^{\frac{1}{2}} \qquad [12\text{-}23]$$

where p is the amplitude of the wave and T is the time period. The sound intensity, I, is defined as the time-weight average sound power per unit area normal to the direction of sound propagation and

$$I = \frac{(P_{rms})^2}{\rho c} \qquad [12\text{-}24]$$

where

I = intensity in W/m^2

P_{rms} = root mean square sound pressure in Pa

ρ = density of medium in kg/m^3

c = speed of sound in medium in m/s

usually c = 20.05 T$^{1/2}$

where T is absolute temperature in K.

The scale of sound pressure ranging from the faintest sound to a loud sound of a rocket takeoff can be best expressed on the logarithm of the ratios of the measured quantities. Measurements on this scale are termed levels, and the unit is **bel** (named after Alexander Graham Bell).

$$L' = \log \frac{Q}{Q_0} \qquad [12\text{-}25]$$

where

L' is level in bels

Q is measured in quantity

Q_0 is reference quantity

Normally, for convenience of usage, a **bel** is divided into 10 subunits called **decibels** (dB) as

$$L = 10 \log \frac{Q}{Q_0} \qquad [12\text{-}26]$$

Now, if $Q_0 = 10^{-12} \text{W}$
$L = L_W$ (sound power level)
if $Q_0 = 10^{-12} \text{W/m}^2$
$L = L_I$ (sound intensity level)
if $Q_0 = (P_{rms})_0^2$
$L = L_P$ (sound pressure level)

Usually,

$$L_P = 10 \log \frac{(P_{rms})^2}{(P_{rms})_0^2} = 20 \log \frac{P_{rms}}{(P_{rms})_0} \qquad [12\text{-}27]$$

In general, the reference pressure is 20 micropascals, 20μPa. The scale established is in Figure 12-11. The average value of a collection of sound pressure level measurements is

$$\overline{L_P} = 20 \log \frac{1}{N} \sum_{j=1}^{N} 10^{\frac{L_j}{20}} \qquad [12\text{-}28]$$

where

$\overline{L_P}$ = average sound pressure level in dB re: 20μPa

N = number of measurements

L_j = the jth sound pressure level in dB re: 20μPa where j = 1,2,...,N

The mathematical addition of decibels is quite complicated. A simplified procedure is the use of a graph as shown in Figure 12-12. As a rule of thumb, the equal sounds are added in terms of the L_p by 3 dB, and if one sound is more than 10 dB louder than a second sound, the contribution of the latter is negligible.

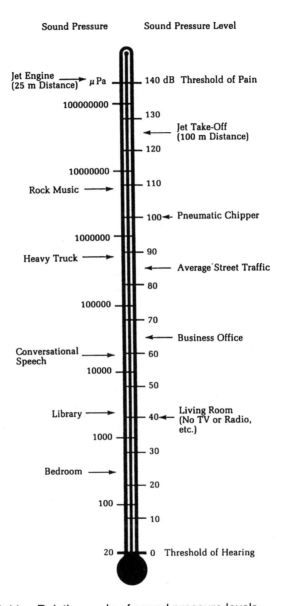

Figure 12-11. Relative scale of sound pressure levels.

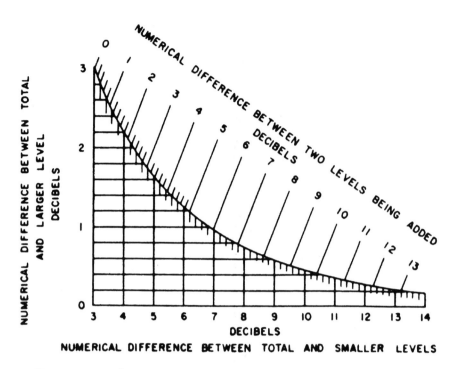

Figure 12-12. Chart for combining different sound pressure levels. For example: Combine 80 and 75 dB. The difference is 5 dB. The 5-dB line intersects the curved line at 1.2 dB. Thus the total value is 81.2 dB. (Courtesy of General Radio.)

[**Example 12-2**]   A ground crew standing 50 ft. away from a 4-engine jet when the first engine is turned on will hear a sound intensity level of 80 dB. What L_p does the crew hear when the second, the third, and the fourth engines are turned on? Also, after the fourth engine is turned on, an arrival plane hovers over the crews head given 90 dB. What does the crew hear?

Assume each engine has a sound intensity level of 80dB. From the chart of Figure 12-12, the difference between the two engine intensity levels is

$$80 - 80 = 0$$

A numerical difference of 0 is equivalent to adding 3 dB, thus

$$80 + 3 = 83 \text{ dB}$$

When the 3rd engine is on

$$83 - 80 = 3$$

and simultaneously

$$83 + 1.8 = 84.8 \text{ dB}$$

Thus for all four engines that are turned on

$$84.8 - 80 = 4.8$$

$$84.8 + 1.2 = 86 \text{ dB}$$

If at this time the hovering plane noise is added, then

$$90 (+86 - 86) = 90 \text{ dB}$$

American National Standard Institute (ANSI) prescribes three basic weighing networks (A, B, and C) when a sound level meter is used, the resulting level being dBA, dBB, etc. For a noise rating system, a statistical analysis indicates how frequently a particular sound level is exceeded; for example, if $L_{40} = 72 \text{dBA}$, then we know that 72dBA is exceeded for 40% of the measured time. Thus, we can write L_N, and a cumulative distribution curve can be constructed when L_N is plotted against N and N = 1%, 2%, etc. (see Fig. 12-13). A probability distribution curve also can be made (Fig. 12-14). The equivalent continuous equal energy level, L_{eq}, can be applied to any fluctuating noise level. It assumes a constant noise level that, over a given time, expends the same amount of energy as fluctuating levels over the same time period.

$$L_{eq} = 10 \log \frac{1}{t} \int_0^t 10^{\frac{L(t)}{10}} \, dt \qquad [12\text{-}29]$$

where

t = time over which L_{eq} is determined

$L(t)$ = the time varying noise level in dBA

$$L_{eq} = 10 \log \sum_{i=1}^N 10^{\frac{L_i}{10}} t_i \qquad [12\text{-}30]$$

where

n = total number of samples taken

L_i = noise level in dBA of the i th sample

t_i = fruition of total sample time

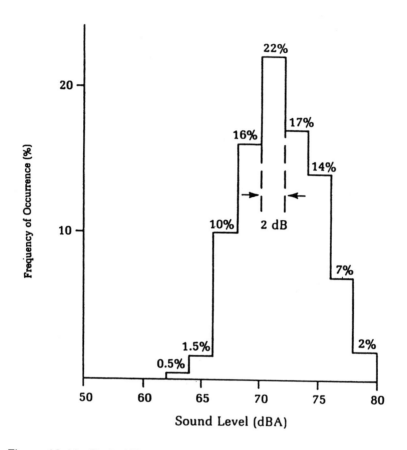

Figure 12-13. Probability distribution plot. (Courtesy of B & K Instruments, Inc., Cleveland.)

If sound is transmitted from a source outward, an inverse square is used:

$$I = \frac{W}{4\pi \, r^2}$$

[12-31]

where

I = sound intensity in W/m^2

W = sound power of source in W

r = distance between source and receiver in m

This measures L_W rather than L_P. For L_P one can derive

$$L_P \approx L_W - 20 \log r - 11$$

[12-32]

where

L_P = sound pressure level in dB re: $20\mu\text{Pa}$
L_W = sound power level in dB re: 10^{-12}W
$20 \log r$ = decibel transform = $10 \log r^2$
11 = decibel transform $\approx 10 \log (4\pi)$ = 10.99

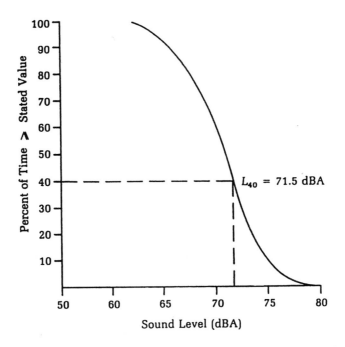

Figure 12-14. Cumulative distribution curve. (Courtesy of B & K Instruments, Inc., Cleveland.)

The directivity factor is a measure of the directivity of a spherical sound source by polar coordinates, or

$$DI_0 = L_{P\theta} - L_{Ps} \qquad [12\text{-}33]$$

where $L_{P\theta}$ is sound pressure level measured at distance r' and angle θ from a directive source radiating power W into anechoic space in dB, and L_{Ps} is sound pressure. The level is measured at distance r' from a nondirective source radiating power W into anchoric space in dB. For a source located on or nearby a hard and flat surface, the directivity index is

$$DI_\theta = L_{P\theta} - L_{PH} + 3 \qquad [12\text{-}34]$$

where L_{PH} is unique to the hard surface and the 3 dB addition is that measurement made over a hemisphere instead of a sphere. The intensity at a radius r is twice as large if a source radiates into a hemisphere rather than a sphere. Assuming the directivity pattern does not change its shape regardless of distance, then the directive index may be added to the inverse square law.

$$L_{P\theta} \approx L_N + DI_\theta - 20\log r - 11 - A_e \qquad [12\text{-}35]$$

The last term A_e is the excess attenuation beyond wave divergence. For example, the attenuation may be by absorption by the air or rain, sleet, snow, fog, or by barriers, grass, shrubbery, trees, or by atmospheric turbulence, characteristics of the ground, and so on.

For **noise control** there are three different approaches for abatement. The first is the control modification and reduction of the source or the noise output. There are various ways for reduction of impact forces, the reduction of speeds and processes and frictional resistance, or the noise breakages. The isolation and dampening of the vibrating elements as in the provision of mufflers and silencers would involve material science and the proper mechanical design. The next way is to alter or to control the transmission path and environment. Here the nullification of the amplitudes by another source signal out of phase in the noise path or by the use of barriers, screens, or deflectors becomes important. The sound-absorbing materials (such as acoustic tile, carpets, drapes, and asphalt or other polymeric foams) can be very useful. The **Sabin absorption coefficients**, α_{SAB}, at 125, 250, 500, 1000, 2000 and 4000 Hz, are in general use. Finally, the protection of the receiver by earplugs and muffs can actually reduce the noise by 15 to 25 dB as shown in Figure 12-15.

The readers are asked to look at the generalized electromagnetic series. As the frequency increases from 1 to 10^{22} Hz, the energy of radiation or effect to human exposures also increases. In the audible range, sound and noise directly impact on the hearing. In this instance, discordant sound resulting from nonperiodic vibrations in air is defined as **noise**. Temporary or permanent damages leading to hearing loss results from noise exposure of sufficient energy. The secondary effect is that the **acoustic pollution** (or noise pollution) draws emotional responses on conscious and subconscious levels and annoys, awakens, angers, distracts, frustrates, and creates stresses that result in psychological problems.

Even working with ultrasound region, for example, in the 25 to 40 KHz range, the overtones associated with the sonication still fall in the audible region of hearing. In working places, control and protection measures are made, such as sound isolation, absorption and damping. Biological and physiological effects in radio frequency and microwave regions are not clear. With the exception of selective frequencies in the x-band region,, the microwave will cause the thermal vibration of water molecules present in cells with the consequence of fatal destruction. Although the health effects of the low-level of the nonionizing field remains an issue, damages due to high dosages in a prolonged exposure remain a risk.

Molecular vibration and rotation resulting to bond scission of weak linkages will cause cell disfunctions. The bombardment of outer electrons will harm the eye and skin, such as the well-known ionizing radiation from ultraviolet beam. The incursion of the biosphere system with electron, neutron, helium, gamma and other particles through x-ray, γ-ray and cosmic radiation causes irreversible injury as often referring to the **ionizing radiation**. A good example is that the wrinkles in the face of farmers are caused by the crosslinking of the keratin under UV radiation from the sun.

Humans as a part of the entire biosphere are adapted to the natural environment; however, when anthropogenic factors are imposed on, imbalance will occur with the consequence of mutagenicity as well as carcinogenicity. The internal pollution from the chapter will include radon, noise and unwanted nutritional supplements. At the end of the present volume, the reader should be able to use chemical principles discussed in various chapters for answering many of the present-day issues.

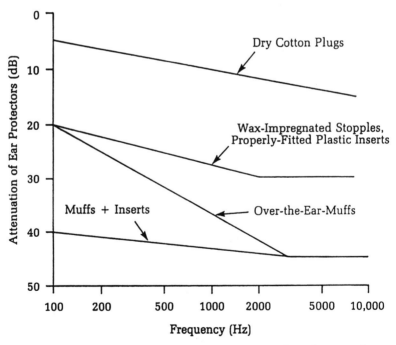

Figure 12-15. Attenuation of ear protectors at various frequencies. (Source: National Bureau of Standards Handbook 119, 1976.)

For example, to answer the question why active or passive tobacco smoking is harmful, the reader can at least state:

- Smoke contains millions of Aitken particles. These particles have the right sizes to wedge into the pulmonary region to impair the exchange of oxygen from inhalation. (See Section 5.1)

- Tar consists of a number of PHA that will be in the smoke. Many of the PHA are carcinogenous such as benz (3,4) pyrene.

- Tobacco contains nicotine, which contains two nitrogens, one in pyrrole, the other in pyridine, having four carbons apart. The original intention is for the tobacco plant to fight against aphids. Instead, humans who inhale it will have the equivalence of hallucinatory properties of other psychrotomimetic plants. (See Section 12.2)

- Tobacco contains large amounts of CO, HCN acetaldehyde, and acrolein. The toxicity of CO is well known. (See Section 5.3.1). Both hydrogen cyanide and acrolein have metal-binding powers.

- Tobacco leaves have the concentration power to enrich the daughter neuclide of radon in such a way that the aerosol in the smoke may be radioactive. (See Sections 3.1.5, 3.4.2 and 12.2.3)

- Cigarette paper contains cadmium as sizing material. Cadmium will cause hypertension if diffused into the blood stream through smoking. (See Section 10.3.3)

REFERENCES

12-1.   M. Wilkening, Radon in the Environment, Elsevier, Amsterdam, 1990.

12-2.   G. M. Masters, Introduction to Environmental Engineering and Science, Prentice-Hall, Englewood Cliffs, New Jersey, 1991.

12-3.   C.S. ReVelle, E.E. Whitlatch, Jr., and J.R., Wrigeet Civil and Environmental Systems Engineering, Prentice-Hall, Upper Saddle River, New Jersey, 1997.

12-4.   D.J. Montgomery, "Making the Quality of Life an Operational Concept," p. 46, Phi Kappa Phi Journal, Winter (1975).

12-5.   R. Fenwick and M. Morgan, "Natural Toxicants in Plant Food," Chem. Gt. Brit. 27, 1027-1029 (1991).

12-6.   B. Hideman, "Health Effects of Electromagnetic Fields Remain Unresolved," C & ZN, Nov. 8, 15-29 (1993).

12-7.   D.T. Allen and K.S. Rosselot, *Pollution Prevention for Chemical Processes*, Wiley, New York, 1997.

12-8.   National Academy of Sciences, *Engineering with Ecological Constraints*, National Academy of Engineering, Washington D.C., 1996.

12-9.   L. L. Beranek, Noise and Vibration Control, McGraw-Hill, New York, 1971.

12-10. P. N. Cheremisinoff and F. Ellerbusch, *Guide for Industrial Noise Control*, Ann Arbor Science, Ann Arbor, Michigan, 1982.

12-11. E. Miles, R. Pealy, and R. Stokes, Natural *Resources Economics and Policy Applications*, University of Washington Press, Seattle, 1986.

12-12. S. E. Manahan, *Environmental Chemistry*, Lewis Publishers, Chelsea, Michigan, 5th ed., 1991.

12-13. J. W. Moore and E. A. Moore, *Environmental Chemistry*, Academic Press, New York, 1976.

PROBLEM SET

1.   A 70-Kg man is exposed to 5 x 10^{-5} mg/L of TCE in the air at his workplace. If he works for 8 hours per day, 5 days per week, 40 weeks per year in a 10-year period and inhales 2 m^3/hr of air, what would be his lifetime risk (assuming the lifetime is 70 years)? What would be the risk to a 30-Kg child similarly exposed?

2.   What is the concentration of VOC in drinking water that would result in a 10^{-5} risk for a 50-Kg person who drinks 2L/day throughout his lifetime? Assuming the slope of a carcinogenic VOC with the dose-risk curve is 0.01 $(mg/Kg/day)^{-1}$.

3.   A researcher is operating two sets of ultrasounds with different frequencies. He hears a sound intensity level of 70dB when the first ultrasound with lower frequency is turned on. The L_p that he hears when the second ultrasound is turned on is 72dB. What is the intensity of second ultrasound?

APPENDIX

Decimal Multipliers that Serve as SI Unit Prefixes

Prefix	Origin	Symbol	Multiplying Factor
yotta	Greek or Latin *octo*, "eight"	Y	10^{24}
zetta	Latin *septem*, "seven"	Z	10^{21}
exa	Greek *hex*, "six"	E	10^{18}
peta	Greek *pente*, "five"	P	10^{15}
tera	Greek *teras*, "monster"	T	10^{12}
giga	Greek *gigas*, "giant"	G	10^{9}
mega	Greek *megas*, "large"	M	10^{6}
kilo	Greek *chilioi*, "thousand"	k	10^{3}
hecto	Greek *hekaton*, "hundred"	h	10^{2}
deka	Greek *deka*, "ten"	da	10^{1}
deci	Latin *decimus*, "tenth"	d	10^{-1}
centi	Latin *centum*, "hundred"	c	10^{-2}
milli	Latin *mille*, "thousand"	m	10^{-3}
micro	Latin *micro* (Greek *mikros*), "small"	μ	10^{-6}
nano	Latin *nanus* (Greek *nanos*), "dwarf"	n	10^{-9}
pico	Spanish *pico*, "a bit," Italian *piccolo*, "small"	p	10^{-12}
femto	Danish-Norwegian *femten*, "fifteen"	f	10^{-15}
atto	Danish-Norwegian *atten*, "eighteen"	a	10^{-18}
zepto	Latin *septem*, "seven"	z	10^{-21}
yocto	Greek or Latin *octo*, "eight"	y	10^{-24}

Common Unit Conversions

Gas Constant	Volume	Density
0.082 atm^{-1}/g-mole K	1 ft^3 = 28.316 liter	1g/cm^3 = 1000 kg/m^3
62.36 mmHg^{-1}/g-mole K	= 7.481 gal	= 62.428 lb/ft^3
8.314 Joule/g-mole K	1 in^3 = 16.39 cc	= 8.345 lb/gal
1.314 atm-ft^3/lb-mole K	= 5.787 × 10^{-4} ft^3	= 0.03613 lb/in^3
1.987 cal/g-mole K	1 gal = 3.785 liter	
1.987 Btu/lb-mole °R	= 8.34 lb H$_2$O	
0.73 atm-ft^3/lb-mole °R	1 m^3 = 35.32 ft^3	
10.73 psi-ft^3/lb-mole °R	= 264.2 gal	
1545 ft-lb$_f$/lb-mole °R		

Length	Viscosity	Conversion Factor
1 mile = 1609 m = 5280 ft	1 poise	1 cal/g-mole = 1.8Btu/lb-mole
1 ft = 30.48 cm = 12 in	= 6.7197 × 10^{-2} lb$_m$/ft-sec	1 amu = 1.66063 × 10^{-24}g
1 in = 2.54 cm	= 2.0886 × 10^{-3} lb$_f$-sec/ft^2	1 eV = 1.6022 × 10^{-12}erg
1 m = 3.2808 ft	= 2.4191 × 10^2 lb$_m$/ft-hr	1 radian = 57.3°
= 39.37 in	= 1 g/cm-sec	1 cm/sec = 1.9685 ft/min
1 nm = 10^{-9}m = 10 A		1 rpm = 0.10472 radian/sec

Pressure	Constant	Mass
1 atm = 101325 N/m^2	h = 6.6262 × 10^{-27}erg-sec	1 kg = 2.2046 lb
= 14.696 psi	k = 1.38062 × 10^{-16}erg/K	1 lb = 453.59 g
= 760 mmHg	N$_0$ = 6.022169 × 10^{23}	1 ton = 2000 lb
= 29.921inHg	C = 2.997925 × 10^{10}cm/sec	= 907.2 kg
(32 °F)	F = 96487 coul/eq	1 B ton = 2240 lb
= 33.91 ftH$_2$O	e = 1.60219 × 10^{-19}coul	= 1016 kg
(39.1 °F)	g = 980.665 cm/sec^2	1 tonne = 2205 lb
= 2116.2 lb$_f$/ft^2	=32.174 ft/sec^2	= 1000 kg
= 1.0133 bar		1 slug = 32.2 lb
= 1033.3 g$_f$/cm^2		= 14.6 kg

Area	Power	Force
1 m^2 = 10.76 ft^2 = 1550 in^2	1 HP = 550 ft-lb$_f$/sec	1N = 1 kg-m/sec^2
1 ft^2 = 929.0 cm^2	= 745.48 watt	= 10^5 dyne
	1 Btu/hr = 0.293 watt	= 0.22481 lb$_f$
		= 7.233 lb$_m$-ft/sec^2

Common Unit Conversions (continued)

Transfer Coefficient	Energy & Work	Stress
1 Btu/hr-ft^2 °F	1 cal $\quad= 4.184$ Joule	1 MPa $= 145$ psi
$= 5.6784$ Joule/sec-m^2 K	1 Btu $\quad= 1055.1$ Joule	1 MPa $= 0.102$ kg/mm^2
$= 4.8825$ Kcal/hr-m^2 K	$\qquad\quad= 252.16$ cal	1 Pa $= 10$ dynes/cm^2
$= 0.45362$ Kcal/hr-ft^2 K	1 HP-hr $= 2684500$ Joule	1 kg/mm^2 $= 1422$ psi
$= 1.3564\times10^{-4}$cal/sec-cm^2 K	$\qquad\quad= 641620$ cal	1 psi $= 6.90\times10^{-3}$ MPa
1 lb/hr-ft^2	$\qquad\quad= 2544.5$ Btu	1 kg/mm^2 $= 9.806$ MPa
$= 1.3562\times10^3$kg/sec-m^2	1 KW-hr $= 3.6\times10^6$ Joule	1 dyne/cm^2 $= 0.10$ Pa
$= 4.8823$ kg/hr-m^2	$\qquad\quad= 860565$ cal	1 psi $= 7.03\times10^{-4}$ kg/mm^2
$= 0.45358$ kg/hr-ft^2	$\qquad\quad= 3412.75$ Btu	1 psi in$^{1/2}$ $= 1.099\times10^{-3}$ MPa m$^{1/2}$
1 cal/g °C $= 1$ Btu/lb$_m$ °F	1 l-atm $\quad= 24.218$ cal	1 MPa m$^{1/2}$ $= 910$ psi in$^{1/2}$
$\qquad\quad= 1$ Pcu/lb$_m$ °C	1 ft-lb$_f$ $\quad= 0.3241$ cal	
1 Btu/hr-ft °F	1 Pcu $\quad= 453.59$ cal	
$= 1.731$ W/m K	1 kg-m $\quad= 2.3438$ cal	
$= 1.4882$ kcal/hr-m K		

Some Useful Constants

Atomic mass	$m_u \approx 1.6605402 \times 10^{-27}$
Avogadro's number	$N \approx 6.0221367 \times 10^{23}$ mol^{-1}
Boltzmann's constant	$k \approx 1.380658 \times 10^{-23}$ J· K^{-1}
Elementary charge	$e \approx 1.60217733 \times 10^{-19}$ C
Faraday's constant	$F \approx 9.6485309 \times 10^4$ C · mol^{-1}
Gas (molar) constant	$R = k \cdot N \sim 8.314510$ J· mol^{-1}· K^{-1}
	≈ 0.08205783 L· atm· mol^{-1}· K^{-1}
Gravitational acceleration	$g = 9.80665$ m· s^{-2}
Molar volume of an ideal gas at 1 atm and 25°C	$\overline{V}_{ideal\ gas} \approx 24.465$ L· mol^{-1}
Permittivity of vacuum	$\varepsilon_0 = 8.854187 \times 10^{-12}$ C· V^{-1}· m^{-1}
Planck's constant	$h \approx 6.6260755 \times 10^{-34}$ J·s
Zero of the Celsius scale	$0°C = 273.15$ K

Source: IUPAC, 1988.

Some Useful Constants for Lithosphere, Hydrosphere, Atmosphere and Biosphere

Quantity	Unit of Measurement	Symbol	Numerical Value
Area of continents	km^2	S_C	149×10^4
Area of world oceans	km^2	S_O	361×10^4
Mean height of continents above sea level	m	h_C	875
Mean depth of world oceans	m	h_O	3794
Mean position of earth's surface with respect to sea level	m	h_m	2430
Mean thickness of lithosphere within the limits of the continents	m	$h_{c.l.}$	35
Mean thickness of lithosphere within the limits of the ocean	km	$h_{o.l.}$	4.7
Mean rate of thickening of continental lithosphere	km	$\dfrac{\Delta h}{\Delta t}$	$10 - 40$
Mean rate of horizontal extension of continental lithosphere	m/ 10^6hr	$\dfrac{\Delta l}{\Delta t}$	$0.75 - 20$
Mass of lithosphere	km/ 10^6yr	m_t	2.267×10^{25}
Mass of water released from mantle and core in course of geological time	gr		3.400×10^{24}
Total reserve of water in the mantle	gr		2×10^{26}
Present day content of free and bound water in the earth's lithosphere	gr		$2.2 - 2.6 \times 10^{24}$ $1.8 - 2.7 \times 10^{24}$
Mass of hydrosphere	gr	m_h	1.664×10^{24}
Amount of oxygen bound in the earth's crust	gr		1.300×10^{24}
Amount of free oxygen	gr		1.5×10^{21}
Mass of atmosphere	gr	m_a	5.136×10^{21}
Mass of biosphere	gr	m_b	1.148×10^{19}
Mass of living matter in the biosphere	gr		3.6×10^{17}
Density of living matter on dry land	gr/ cm^3		0.1
Density of living matter in ocean	gr/ cm^3		15×10^{-8}

Some Components of Cigarette Smoke

Component	Emission (mg/ cigarette)	
	Mainstream Smoke	SidestreamSmoke
Tar	10.2 – 20.8*	34.5 – 44.1
Nicotine	0.46 – 0.92*	1.27 – 1.69
Carbon monoxide	18.3	86.3
Ammonia	0.16	7.4
Hydrogen cyanide	0.24	0.16
Acetone	0.58	1.45
Phenols	0.23	0.60
Formaldehyde	—	1.44
Toluene	0.11	0.60
Acrolein	0.084	0.825
NO_x	0.014	0.051
Opolonium–210 (pCi)	0.07	0.13
Fluoranthenes	7.7×10^{-4}	1.6×10^{-3}
Benzofluorenes	2.5×10^{-4}	1.0×10^{-3}
Pyrenes	2.7×10^{-4}	1.5×10^{-3}
Chrysene	1.9×10^{-4}	1.2×10^{-3}
Cadmium	1.3×10^{-4}	4.5×10^{-4}
Perylenes	4.8×10^{-5}	1.4×10^{-4}
Dibenzanthracenes	4.2×10^{-5}	1.4×10^{-4}
Anthanthrene	2.2×10^{-5}	3.9×10^{-4}

*Range is for filtered to unfiltered cigarettes.

Source: Data from S.A. Glantz, "Health Effects of Ambient Tobacco Smoke," in Indoor Air Quality, ed. P.J. Walsh, C.S. Dudney, and E.D. Copenhaver (Boca Raton, Fla.: CRC Press, 1984), Table 1, p.160.

Periodic Table of the Elements. (From IUPAC)

1	2	3	4	5	6	7	8	9	10	11	12	13	14	15	16	17	18
1 H 1.00794																	2 He 4.002602
3 Li 6.941	4 Be 9.012182											5 B 10.811	6 C 12.0107	7 N 14.00674	8 O 15.9994	9 F 18.9984032	10 Ne 20.1797
11 Na 22.989770	12 Mg 24.3050											13 Al 26.981538	14 Si 28.0855	15 P 30.973761	16 S 32.066	17 Cl 35.4527	18 Ar 39.948
19 K 39.0983	20 Ca 40.078	21 Sc 44.955910	22 Ti 47.867	23 V 50.9415	24 Cr 51.9961	25 Mn 54.938049	26 Fe 55.845	27 Co 58.933200	28 Ni 58.6934	29 Cu 63.546	30 Zn 65.39	31 Ga 69.723	32 Ge 72.61	33 As 74.92160	34 Se 78.96	35 Br 79.904	36 Kr 83.80
37 Rb 85.4678	38 Sr 87.62	39 Y 88.90585	40 Zr 91.224	41 Nb 92.90638	42 Mo 95.94	43 Tc (98)	44 Ru 101.07	45 Rh 102.90550	46 Pd 106.42	47 Ag 107.8682	48 Cd 112.411	49 In 114.818	50 Sn 118.710	51 Sb 121.760	52 Te 127.60	53 I 126.90447	54 Xe 131.29
55 Cs 132.90545	56 Ba 137.327	57 La 138.9055	72 Hf 178.49	73 Ta 180.9479	74 W 183.84	75 Re 186.207	76 Os 190.23	77 Ir 192.217	78 Pt 195.078	79 Au 196.96655	80 Hg 200.59	81 Tl 204.3833	82 Pb 207.2	83 Bi 208.98038	84 Po (209)	85 At (210)	86 Rn (222)
87 Fr (223)	88 Ra (226)	89 Ac (227)	104 Rf (261)	105 Db (262)	106 Sg (263)	107 Bh (262)	108 Hs (265)	109 Mt (266)	110 Ds (269)	111 Rg (272)							

58 Ce 140.116	59 Pr 140.90765	60 Nd 144.24	61 Pm (145)	62 Sm 150.36	63 Eu 151.964	64 Gd 157.25	65 Tb 158.92534	66 Dy 162.50	67 Ho 164.93032	68 Er 167.26	69 Tm 168.93421	70 Yb 173.04	71 Lu 174.967
90 Th 232.0381	91 Pa 231.03588	92 U 238.0289	93 Np (237)	94 Pu (244)	95 Am (243)	96 Cm (247)	97 Bk (247)	98 Cf (251)	99 Es (252)	100 Fm (257)	101 Md (258)	102 No (259)	103 Lr (262)

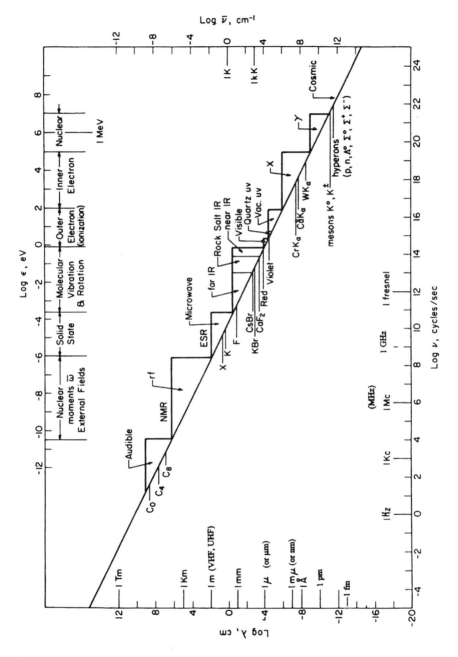

Generalized electromagnetic series

I N D E X

One of the greatest French poets, Paul Valéry, indicated in his poem "The Footsteps" that your footsteps are actually my heart beats. We are anticipating the arrival of *Our Savior* who can release us from the evil bondages of the deterioration and degradation of our environment.

Les Pas

Tes pas, enfants de mon silence,
Saintement, lentement placés
Vers le lit de ma vigilance
Procèdent muets et glacés.

Personne pure, ombre divine,
Qu'ils sont doux, tes pas retenus!
Dieux! … tous les dons que je devine
Viennent à moi sur ces pieds nus!

Si, de tes lèvres avancées,
Tu prépares pour l'apaiser,
A l'habitant de mes pensées
La nourriture d'un baiser,

Ne hâte pas cet acte tendre,
Douceur d'être et de n'être pas,
Car j'ai vécu de vous attendre,
Et mon coeur n'était que vos pas.

The Footsteps

Your footsteps, issue of my silence,
slowly make their hallowed way
chill and soundless to the bed
where I've kept constant vigil.

Oh, Pure One, Shadow Divine,
how soft Your tread in memory!
Lord! Every blessing I've surmised
has come to me on these bare feet.

If with lips inclined,
You mean to satisfy
the tenant of my mind
by feeding it a kiss,

don't rush this gentle grace imbued,
sweet state of being and not being.
For I have lived awaiting all of you,
my heart but imprint of your many feet.